ROOSEVELT UNIVERSITY LIBRARY
W9-DGB-396
3 3311 00126 526 5

BIOLOGICAL CONTROL OF PLANT PATHOGENS

The first demonstration that the total antagonistic microflora of a pathogen-suppressive soil could be successfully transferred to a conducive soil. Representative potato tubers grown in soils infested with the common scab pathogen. *Streptomyces scabies*. A. Suppressive soil alone. B. Autoclaved suppressive soil. C. Conducive soil plus 1% alfalfa meal plus 10% suppressive soil. D. Conducive soil plus 1% alfalfa meal. (From Menzies, 1959.)

BIOLOGICAL CONTROL OF PLANT PATHOGENS

Kenneth F. Baker
UNIVERSITY OF CALIFORNIA, BERKELEY

R. James Cook
U. S. DEPARTMENT OF AGRICULTURE
and
WASHINGTON STATE UNIVERSITY

with a Foreword by
S. D. Garrett
UNIVERSITY OF CAMBRIDGE

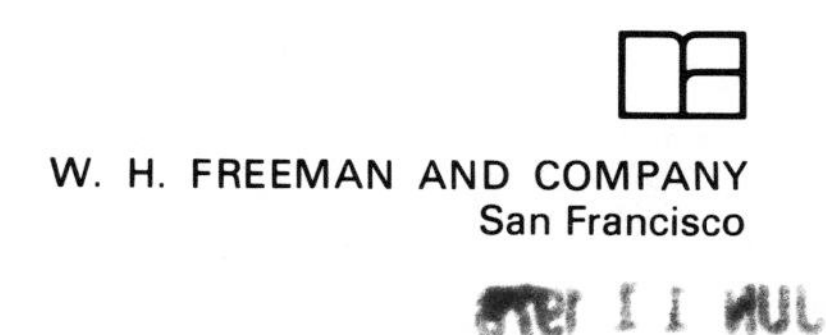

W. H. FREEMAN AND COMPANY
San Francisco

A SERIES OF BOOKS IN THE BIOLOGY OF PLANT PATHOGENS

Cover: Enlargement of Figure 8.2. Photo courtesy of R. C. Foster, CSIRO, Melbourne, Australia.

Library of Congress Cataloging in Publication Data

Baker, Kenneth Frank, 1908–
Biological control of plant pathogens.

Bibliography: p. 32
1. Pest control—Biological control. 2. Plant diseases. I. Cook, R. James, 1937– joint author. II. Title. [DNLM: 1. Pest control, Biological. 2. Plant diseases—Prevention and control. SB975 B167b 1974]
SB975.B34 632'.96 73-18420
ISBN 0-7167-0589-3

Printed in the United States of America

1 2 3 4 5 6 7 8 9

Trade names are used in this publication solely to provide specific information. Mention of a trade name does not constitute a warranty of the product by the University of California, the U.S. Department of Agriculture, or Washington State University, nor does it constitute an endorsement by them over other products not mentioned.

This book reports the current status of research involving use of certain chemicals that require registration under the Federal Environmental Pesticide Control Act (FEPCA). It does not contain recommendations for the use of such chemicals, nor does it imply that the uses discussed have been registered. All uses of these chemicals must be registered by the appropriate State and Federal agencies before they can be recommended.

CONTENTS

FOREWORD

This is the first book to be devoted wholly to the biological control of plant pathogens; more significantly, perhaps, it is the first publication of sufficient length to do full justice to this large and expanding field of research. To remark to plant pathologists today that most plant diseases are affected by microorganisms other than the pathogen, and sometimes by macroorganisms other than the host plant, is to voice a platitude. Less than half a century ago it was not so, and I have lived long enough to see one of yesterday's most important discoveries in plant pathology become one of today's platitudes. This profoundly important realization, which came in the late nineteen-twenties, has naturally had the greatest impact on pathologists studying soilborne diseases, simply because the soil is the most commodious and satisfactory habitat for terrestrial microorganisms. Nevertheless, as the authors have shown in this book, no pathologist studying either a seedborne or an airborne disease can afford to exclude from consideration microorganisms other than the pathogen.

Platitudes rarely come singly, and I am going on to say that all branches of science have both a fundamental and an applied aspect, though the degree to which these two aspects of the subject have been developed has varied widely from one science to another. Plant pathology has always been a practical science, based on the study of diseases as they occur in the field. This is not to say that the subject lacks its fundamental side; some plant pathologists of great distinction in the academic sense of the word, have spent their lives with no more contact with the field than has been gratuitously afforded by the view from their laboratory window. But to write a book about *control* of plant diseases, more than a knowledge of plant pathology is needed; it is equally essential to have a thorough understanding of crop ecology and crop husbandry, based on a long and wide experience in the field. I have known both Ken Baker and Jim Cook for a long time, and I know of no two better qualified in this second respect, as well as in the first, for the writing of this particular book. Indeed, their not infrequent asides, based on personal observations in the field, have been not the least of the pleasures for me in the reading of this book.

The study of microbial ecology, in the soil and elsewhere, is a thoroughly fundamental branch of science, and the authors have not been unduly pessimistic in emphasizing the complexity of microbial interrelationships, nor even in expressing some doubt as to whether complete understanding ever will be achieved. But they do point out, and I agree with them, that shortcuts in the application of science to practice are not merely permissible, but even essential. The history of science abounds with illustrations of the situation in which a research worker following a logically planned route to a certain goal is overtaken by another who has had the capacity to exploit incidental and quite unplanned observations.

The final point that I wish to make about this book can really be inferred from what I have already said about the widespread influence of microorganisms other than the pathogen on the incidence and development of plant diseases. This theme runs not only right through this book, but also through plant pathology as we know it today. Thus it happens that these two authors, while writing a book about biological control, have also written a very stimulating book about plant pathology in general. It is stimulating because the authors have thought more deeply than most about a very complex and difficult subject. Plant pathologists at the outset of their careers will find in this book a great deal of well-organized information. Yet I fancy that readers of my age who know, or ought to know, most of the information therein reviewed will gain even more from this book. In reading it I found myself forced, again and again, to reexamine my own interpretations of the various phenomena discussed by the authors. From this experience I emerged an insignificantly older man but, I believe, a significantly wiser one. And what more can one ask of any book?

July 1973 *S. D. Garrett*

PREFACE

This volume is presented to those interested in biological control of plant pathogens as a research area or as a fascinating frontier of biological knowledge, and to those who wish to apply it to agricultural practice. There has been no attempt at completeness, since the book is intended to be read. The volumes of the First (1963), Second (1968), and Third (1973) International Symposia on Factors Determining the Behavior of Plant Pathogens in Soil cover the literature of the field.

Space limitations have restricted the number of examples used to illustrate the operation of biological control. It was considered better to explain a number of such examples in detail and to refer to these familiar examples repeatedly than to introduce new ones each time. The examples selected are among the best of the large number available. A fairly complete story on each of these diseases may be obtained through the Index.

The observant reader will note that sometimes we have drawn from published data conclusions that differ from those of other reviewers. Although this is expected in a new subject, which encourages flexing of predilections, we have in most such cases conferred with the original investigator concerning our interpretation. In all instances the views expressed represent our considered judgment.

Although the aspects presented are preponderantly subterranean, a chapter on Biological Control of Pathogens of Aerial Parts is included to encourage diffusion of ideas across this soil-air interface. The space devoted to each fairly represents the published work, but does not, in our judgment, indicate actual potentials.

We sincerely think that biological control is working, and that it can supply a rational answer to many present problems. We make no apology for these convictions. In the words of E. Minkowsky, we "present here a subjective work, a work, however, which strives with all its might toward objectivity."

We do not pretend to have developed a unique viewpoint, although to some it will appear "far out." We feel, with G. Evelyn Hutchinson, that "The fertility, rather than the originality, of an idea is what is important." If this volume stimulates someone again to try biological control, suggests some new approach to another, causes yet another to come out of the laboratory to the field where the clues and the action are, or even irritates someone into proving us wrong, we will be content. If none of these effects is forthcoming, it will still have been a worthwhile and pleasant exercise in mind-stretching along an important biological frontier.

August 1973 *Kenneth F. Baker*

R. James Cook

ACKNOWLEDGMENTS

The authors are aware that their concepts, philosophies, and interpretations in matters of biological control have been influenced over the years by many discussions with their peers. This has been particularly true at the nineteen annual Pacific Coast meetings of the Conference on the Control of Soil Fungi and the fifteen meetings of Western Regional Project W-38 on the Nature of the Influence of Crop Residues on Fungus-Induced Root Diseases, as well as the Berkeley, London, and Minneapolis international symposia on Factors Determining the Behavior of Plant Pathogens in Soil. To all those persons who have, directly and indirectly, contributed to our understanding of the subject, we are deeply grateful.

It is a pleasure to acknowledge more specifically the many helpful suggestions of R. E. Allan, P. Broadbent, G. W. Bruehl, L. E. Caltagirone, R. A. Fox, R. G. Grogan, A. Kelman, C. Leben, F. J. Newhook, R. I. Papendick, L. P. Reitz, L. C. Schisler, C. L. Schoulties, L. Sequeira, R. W. Smiley, and A. M. Smith, who read part or all the manuscript. The authors are, however, solely responsible for the interpretations and views expressed.

We appreciate the thoughtful editing of the manuscript by Doris Dickman of the United States Department of Agriculture.

The many hours of checking, correcting, typing, and proofing by Katharine C. Baker during preparation of this volume are gratefully acknowledged.

BIOLOGICAL CONTROL OF PLANT PATHOGENS

1

BIOLOGICAL BALANCE

Our approach to nature is to beat it into submission.
We would stand a better chance of survival if we accommodated
ourselves to this planet and viewed it appreciatively
instead of skeptically and dictatorially.
—E. B. WHITE

As planning for this volume was begun, man had successfully completed a trip to the moon—a classical demonstration of his ability to solve thousands of small problems and to blend the answers into one magnificent achievement. Many thousands of people contributed to this unified effort, demonstrating beyond challenge that man can comprehend and control systems with an infinite number of variables if he wills to do so. This accomplishment has also shown that such complex research programs abound in surprises and unexpected factors, each requiring further study.

With this successful exploration of outer space, it is ironic that the soil which man has trod and which has nourished him on the long path of evolution to this moment, largely remains unknown. True, it is studied by many scientists in diverse fields, and notable progress is being made, but the essential mysteries of the ecology of the myriad living things in the soil still defy complete analysis. Man does not know the total roster of microorganisms in a single tiny lump of soil, let alone their complex interactions with each other and with their physical and chemical environment. Man's utter dependence on the produce of the land understandably has focused his attention on the *effects* of this subterranean world on his crop plants rather than on the interactions themselves. Furthermore, living things present an incredibly more complex situation than do the physical and chemical laws that control them. Because a living organism is far greater than the sum of its parts, complete understanding of it is still in the distant future, and understanding its interactions with other organisms more distant still.

In the course of growing millions of acres of crops in the ten thousand years of agriculture, man has, however, found many examples of the effects of these interactions. In the four to five thousand years of written records, many such observations have been made, and some of them recorded. Most have been taken for granted and therefore not studied. Insufficient attention has been paid to situations in which (a) a disease is absent or unimportant in one area but serious in another similar locality, (b) a pathogen has been extensively introduced into an area without becoming established, (c) a pathogen is present in a soil without producing disease, or (d) a disease has steadily declined with continuous monoculture. Similarly, comparative study of the microorganisms in soils known to have different rates of disease intensification would be instructive. In short, man seems to have been preoccupied with the quick kill and less interested in more complex studies on biological control.

Man has greatly decreased his options by polluting his environment—by accumulating, in the soil and water, chemical residues harmful to himself, to his crops and animals, and to soil microorganisms. He has finally begun to realize that lasting success cannot be achieved by poisoning his environment, and is increasingly turning to "natural control" by restoring a biological balance favorable to his crops. Although perhaps difficult to achieve, the goal fully justifies the effort. The urgent need for increased crop production to feed the world's multitudes will increase the pressure for quick results—perhaps away from biological control. Man thus faces a dilemma. If diseases and pests are controlled by the usual quick methods, the resulting environmental pollution may well eventually reduce, rather than increase, production of food. If biological control is employed, its effect may be too slow to achieve the necessary results. Is this to be another demonstration that man cannot avoid catastrophe, but must devote his energies instead to recovering from the last one?

Recent promising nonpolluting cultivation methods of increasing the quantity and quality of food have been used which are compatible with biological control and may provide the time necessary for its adoption. The semidwarf wheats recently developed make possible for the first time the economic use of fertilizers and irrigation on this crop, with tremendously increased yield. The greater return from this makes possible methods of disease control previously considered to be uneconomic. The development of corn varieties high in amino acids has improved the nutritional value of this important crop. The development of varieties resistant to diseases and insects also increases yield. Judicious collective use of the triad of crop improvement, better cultivation practices, and more effective biological control of diseases and insects may provide the necessary food without impairing the environment in the process.

THE BIOLOGICAL WORLD

The biological world is a vast interacting network of living populations in a state of dynamic equilibrium, reflecting changes in their physical environment and their relations to each other. This is an equilibrium in which the individual species follow their normal cyclic changes, or even unusual ones, without significant effect on the whole network, because compensating changes in other components maintain the balance. This is in accord with thermodynamic laws and is the result of long evolutionary adjustment. The cyclic changes of each of the factors involved is related to the balance of the biological whole as a wave on the sea is to the tides and the steady currents in the depths—a response to external forces, which does not affect the majestic sweep of the deeper movements.

Primitive unicellular bacteria and algae are known to have lived in the primeval seas more than three billion years ago. These had developed into fairly complex organisms by the time some of the evolved plants invaded the land more than 400 million years ago. Bacteria and fungi probably were parasitic on the ancestors of higher plants while still in the warm seas, before plants moved onto land and developed roots.

There have been many slow changes in the environment in the past half billion years. Those organisms that had a wide climatic tolerance (blue-green algae, bacteria) survived relatively unchanged, even to the present. Among other forms that had evolved, those that tolerated the changed conditions survived, whereas those that did not became extinct. For those that survived each major climatic change, there was a struggle for space, nutrients or substrate, light, and other factors. Organisms that produced metabolic by-products (antibiotics) that inhibited competitors for a given ecological niche had an advantage, as did those that caused some or all of their competitors to enter resting stages. Some actively occupied a site during a warm or moist period and became dormant during the cool or dry part of the year. An *ecological niche* can thus exist in either space or time. In some such manner, organisms developed an interacting system of spatial and temporal succession.

Such a situation insures maximal utilization of each niche and develops an interlocking *ecosystem,* which integrates all living and nonliving factors of the environment, and in which an alien or newly evolved organism will have great difficulty in becoming established. Such a system is said to be *biologically buffered,* and to exhibit *biological balance* or dynamic equilibrium between members of a relatively stable biotic community, and between them and their abiotic environment. A fluctuating population density of each organism is maintained within certain definite limits. That some alien plants do establish in some habitats may mean that they are better suited to a particular niche than are the residents, that they are introduced in such numbers that they temporarily or permanently swamp the residents, or that they may modify the environment in some way favorable to themselves. It usually means, however, that man has upset the natural balance in some way, making the environment more favorable to the alien than to the resident. Thus, a weed has been defined by J. R. Harlan and J. M. J. de Wet as "a generally unwanted organism that thrives in habitats disturbed by man."

The presence of a living organism in a given place and time is determined by (a) its having evolved or been introduced there, (b) the existence of a physical environment favorable to its development, (c) the presence of associated organisms (symbionts, hosts) favorable to its development, or organisms (hosts for parasites) required for its survival, and (d) the inhibition or absence of organisms (disease organisms, pests, antagonists) so detrimental to it as to cause its extinction. **An organism will increase until the limitations imposed by the biotic and abiotic environment just counterbalance the rate of increase.**[1]

Plants in the wild state were adjusted to their pathogens and pests. The very fact that they grew in mixed stands with a natural sequence or rotation, probably provided protection against root pathogens. The occupancy of plant tissues by microorganisms of differing capabilities as parasites is a very natural phenomenon and should not be construed as always harmful to the plants. If a host plant became unusually plentiful because of favorable climatic conditions, parasites probably increased sufficiently to reduce the number of susceptible plants and thus, in turn, their own numbers through a diminishing food supply. Pressure from parasites has tended to select those plants that had, through mutations and resultant variability, a measure of resistance. **The more favorable the environment to the pathogen and the more severe the selection pressure on the host, the higher the level of resistance maintained in the host.** Conversely, decreases in selection pressure favored a return of the less resistant individuals. As Odum (1971) stated, "the negative effects [of parasitism] tend

---

[1]Principles in this volume are given in **boldface.** A principle is here considered to be a concisely stated fundamental proposition or generalization which aids in the integration of related facts and has prediction value; it is a concept rather than an immutable law.

to be quantitatively small where the interacting populations have had a common evolutionary history in a relatively stable ecosystem." It is for this reason that plant breeders and pathologists, seeking resistance in a crop plant to some pathogen, obtain plants from areas of the native habitat of the host particularly favorable to the pathogen, and use these in a breeding program.

It has been suggested (Chilvers and Brittain, 1972) that strongly competing species of higher plants in natural ecosystems would be maintained in equilibrium by an ecological negative feedback system involving two sets of host-specific parasites such as fungi, insects, or nematodes. If either host increases, its parasites would probably also increase, placing the plant at a competitive disadvantage and tending to restore the equilibrium. Such a system would also help maintain the parasite at a low population.

Other balancing mechanisms also operate. The microorganisms associated with a pathogen may include parasitic plants or animals whose relationship with the pathogen, in turn, is essentially that just outlined for the host plant. Some may be competitors for available energy sources, thus preventing an increase of the pathogen. Others may tend to flourish in response to material leaking from the pathogen. Selection pressure near a fungus pathogen would favor those microorganisms that accelerate the flow of nutrients from the mycelium, either by maintaining a steep nutrient gradient outward from it (Chapter 7) or by production of metabolic waste materials (antibiotics) that cause increased leakage. Such selections would weaken the pathogen, diminish its population, and thus enforce the balance. Other microorganisms may produce substances that cause the pathogen to go into a resting or dormant state by forming chlamydospores, oospores, or sclerotia, and thus reduce the attack on the host. A pathogen, therefore, can elude its suppressors only by escaping into the tissues of a host plant or by becoming dormant.

If, as occasionally must have happened, all these checks and balances failed to reduce the activity of some pathogen, the host may have been extinguished, making the ecological niche available for some better-adapted plant. Such an event left the pathogen with a lessened substrate, an enforced saprophytic existence, or extinction. Perhaps for many organisms, the price of destroying the balance in nature is untimely death.

It is clear that the soil is more stable than the aerial environment in practically all respects, but also that it can be slowly changed by many conditions. *Slight changes in one or more factors may exert a profound effect on soil microorganisms,* perhaps because they have become adapted to this relatively stable habitat.

Organisms of a given ecosystem tend to continue the selection, through competition, for ever better-adapted individuals. Ability to change thus results in a heterogeneous variable population, from which are selected individuals best suited to the given biotic and abiotic environment. It is probable, however, that organisms rarely, if ever, become perfectly adjusted to their environment. They are undergoing constant adaptive selection. The environment, meanwhile, is also steadily undergoing gradual changes, with the result that organisms never quite catch up in adaptation. There results a resilient but balanced, complex but interlocking, and stable but fluctuating equilibrium, strongly resistant to change in composition, and generally uncongenial to alien organisms.

These minor evolutionary changes, or those imposed by environmental shifts, are generally slow acting (although in restricted areas some may be sudden) and provide opportunity for biological adjustment through selection among genetically variable lines, and hence for maintenance of a stable balance. Environmental changes induced by man, on the other hand, are major in scope and rapidly introduced, permitting little time for adjustment. This is the nub of the problem. Moreover, by selecting plants for the specialized features desired in his cultivated crops, which may have little survival value in nature, man has reduced variability and adaptability of his plants to such an extent that many are unable to survive without his protection.

Microorganisms in soil occupy such an infinite number of differing habitats that man may never be able to comprehend all the interactions. The end result—biological balance—can, however, be appraised and empirically exploited; with luck, parts of it may even be understood.

ATTRIBUTES OF A SUCCESSFUL PARASITE

Some features of a parasite apply to any successful microorganism, but not all of them are essential within any given individual if some compensating character is present.

1. An adequate number of propagules must be produced to maintain the population of the organism; because of high mortality, this usually means produced in excess. However, production of such structures as conidia, ascospores, sclerotia, oospores, and nematode eggs requires energy; too lavish an expenditure on propagules may restrict other activities, much as a person may be over-insured. That pathogens tend to produce no more spores than are necessary is shown in Table 6.1 and is discussed in Chapter 6.

2. The survival capacity of the propagules must be high, which means that they must be resistant to antagonists and to environmental extremes when dormancy is required. **In the absence of the host or adequate nutrients, a pathogen is apt to survive longest in soil under**

nonlethal conditions least favorable to germination [Park (in Baker and Snyder, 1965), paraphrased from H. Bremer, 1924]. A pathogen may be unable to compete successfully with saprophytic microorganisms for long periods in the absence of the host (which supplies a means of escaping this competition). **Parasites are at a greater competitive disadvantage than saprophytes when they are outside their host.**

3. Different levels of dormancy of propagules may insure that some are still viable if others germinate or hatch and fail to encounter a host.

4. The propagules should germinate and grow rapidly. Soil microorganisms must either be efficient saprophytes and be able to grow or move through soil until a root is encountered, or they must remain dormant until stimulated to germinate by the presence of a nearby host root.

5. Production by germinating propagules of antibiotics that will inhibit other pathogens or antagonists increases the chance of their own survival and infection.

6. Rapid infection and invasion of the host means that propagules need not provide a large energy source, because a nutrient supply is quickly reached.

7. Capacity to escape antagonists in various ways increases the chance of success of a parasite. Some are able to grow at temperatures lower than those that favor most other organisms. Others are able to penetrate and survive deep in the soil under conditions of high carbon dioxide and low oxygen contents, where there are few other organisms. Some organisms are able to grow in soil too dry for their antagonists. Still others escape antagonists by penetrating the host, or by forming resistant structures in response to an adverse environment. Some are able to attack aerial parts of the plant and thus escape antagonists in the soil.

8. Ability to withstand antagonists and to grow through field soil, as *Rhizoctonia solani* can, increases the chances of contacting a host.

9. Ability to invade and colonize organic debris may improve opportunities for infection of a host by providing a necessary food base.

10. **Possession is nine points of the law[2] for microorganisms in relation to**

[2]Devaluation seems to have occurred, for in the eighteenth century, possession was said to be eleven points of the law.

host tissue, and the organism that can get there first has an advantage. A parasite that invades a host is likely to occupy the killed tissue, depleting the nutrients required by competitors and storing them within itself, producing antibiotics that inhibit the competitor, or both.

11. A parasite should cause minimal harm to the host while completing its own life cycle. A parasite that kills its host too quickly must then contend with the myriad saprophytes that it had escaped by invading the host.

12. Ability to parasitize more than one plant species enhances the opportunities for survival of the parasite.

TYPES OF BIOLOGICAL INTERACTIONS

Organisms exhibit many types of interaction in the process of achieving biological balance. A microorganism may exert no effect on another, or it may produce one or more of the following effects: (a) It may stimulate growth or development of the associate (Chapter 7). (b) It may inhibit growth or development of the associate (Chapter 7). (c) It may stimulate the formation of resting spores by the associate (Chapter 7). (d) It may inhibit the formation of resting spores by the associate (this chapter). (e) It may enforce dormancy (as in fungistasis) of the associate (Chapter 7). (f) It may cause lysis of the associate (Chapter 2). Some of these interactions produce a harmful effect directly on a population of a given plant (weed competitors, insect pests, plant pathogens, organisms that produce phytotoxins). An indirect harmful effect is produced by others such as: organisms beneficial to weeds or plant pathogens; insect vectors of plant pathogens; organisms harmful to other organisms that are beneficial to plants, such as fungi that inhibit legume-nodule bacteria. On the other hand, many organisms are directly beneficial: the nodule bacteria on legume roots; insects that pollinate higher plants; mycorrhizal roots; a pathogen that restrains its host from producing luxuriant early growth in an environment that cannot sustain such growth (Chapter 9). Others may be indirectly beneficial: nitrifying bacteria that convert ammonium to nitrate nitrogen; organisms that decompose organic matter; pathogens of weeds or harmful insects; organisms antagonistic to plant pathogens, or those that decompose phytotoxins. An organism may thus be beneficial by interfering with the activity (for example, antibiotic activity) of a harmful one, as well as by direct assistance, and one may be harmful by interfering with a beneficial organism or by direct injury.

Daubenmire (1968) stated that "ecosystems are so fantastically complex that we can never hope to fully describe and understand any but a few of the simplest of them." This complexity has caused some plant pathologists to be skeptical of the possibility of effective biological control of plant diseases. However, such a view fails to recognize that similar complexities are involved in agriculture itself and in the successful biological control of some insect pests. We will show in this book that biological control of some plant diseases need not—indeed cannot—await the millennium of complete understanding. As Samuel Johnson observed long ago, "Nothing will ever be attempted if all possible objections must be first overcome."

Interactions Involved in Disease

An admittedly oversimplified schematic diagram of the major factor-groups acting to produce plant disease is shown in Figure 1.1 to illustrate some of the interactions involved. Each factor-group is represented by a disk, and the amount of overlapping indicates the degree of interaction. Environment is collectively used to include the several factors (soil temperature, water potential, pH) that might be operative; it could also refer to a single controlling factor. Antagonists include competitors, hyperparasites, and organisms producing antibiotics. The crop population is taken to include the many susceptible or resistant host cultivars; the pathogen includes the many strains and formae speciales that may be present, as well as the inoculum density of the pathogen. It should be understood that in each situation these three living factor-groups are interacting against background biota involved minimally or not at all in the particular interaction. Many of the specific factors in each group could be represented by other disks, since they are effective in a given disease, but this would needlessly complicate the picture.

Although factors of the physical and biotic environment may operate to some degree in each interaction, they are not equally effective in each case and may not always play the same role. For example, the resistance of the host in Figure 1.1:D is so high that no disease results. However, if resistance depends on the proper environment, it may fail under some conditions.

The occurrence of a plant disease thus indicates that some aspect of the biological balance is not in equilibrium, and the greater the imbalance, the more severe the disease is apt to be. A plant disease develops when one or more of the following conditions occurs: (a) The *pathogen* is highly virulent, in high inoculum density, or not in equilibrium with the antagonists. (b) The *abiotic environment* is especially favorable for the

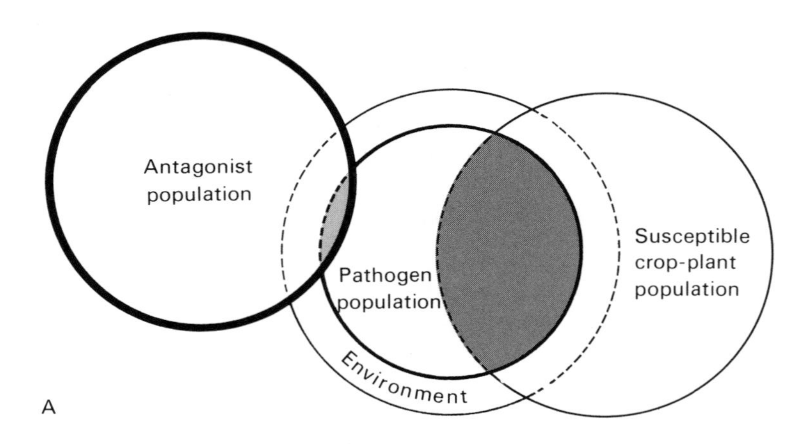

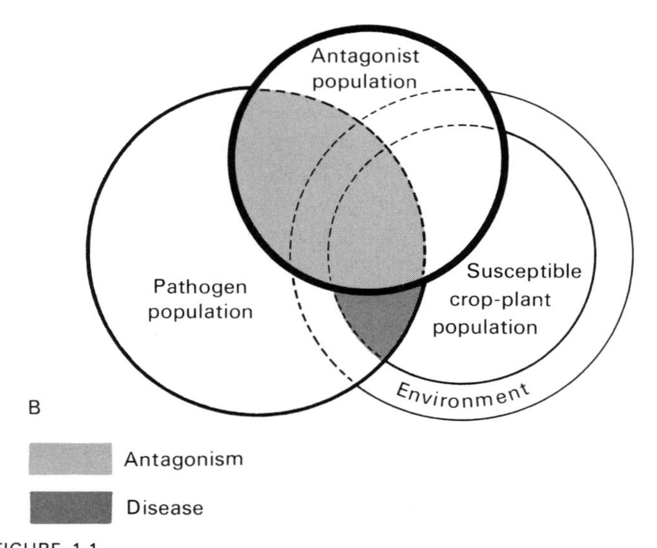

FIGURE 1.1
Schematic diagram of interactions in major factor-groups in plant disease.

A. *Severe disease loss.* Susceptible crop moderately well adapted to the environment; pathogen well adapted; antagonists not well adapted and ineffective. Exemplified by the fusarium wilt diseases in acid sandy soils.

B. *Slight disease loss.* Susceptible crop well adapted to environment; pathogen poorly adapted; antagonists moderately adapted and quite effective. Exemplified by the fusarium wilt diseases in alkaline clay soils.

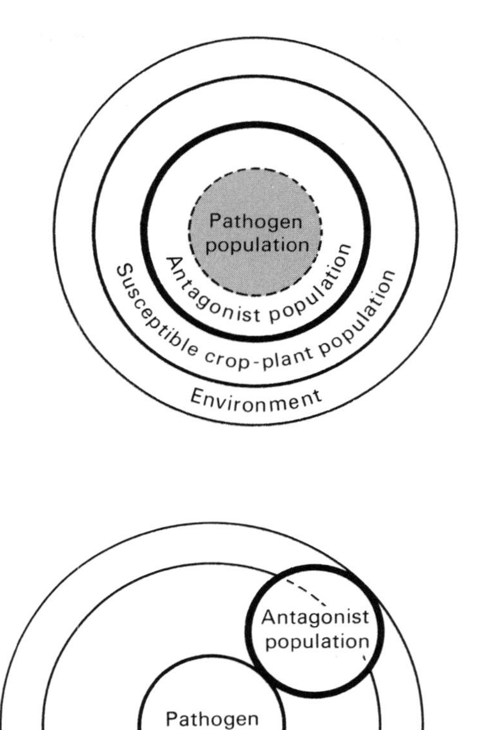

C. *No disease loss;* biological control. Susceptible crop, antagonists, and pathogen well adapted to environment. Antagonists have suppressed the pathogen. Exemplified by *Phytophthora cinnamomi* root rot of avocado in Queensland.

D. *No disease loss;* resistance. Resistant crop, antagonists, and pathogen well adapted to environment. Host resistance prevents disease. Exemplified by fusarium wilts in any soil when the crop carries monogenic resistance.

pathogen, is unfavorable to the host or antagonists, or both. (c) The *host plant* is highly susceptible, continuously grown, extensively grown, or all of these. (d) The *antagonists* are absent or in low population, lack something (nutrients, proper environment) to function as antagonists, are inhibited by other microorganisms, or the antibiotics produced are sorbed by the soil or inactivated by other microorganisms.

Conversely, absence of disease means that the pathogen is absent, the host is highly resistant, the physical environment is unfavorable part or all of the time, or antagonists are inhibiting the growth or infection by the pathogen. Biological control may thus be a restoration of a disease-inhibiting balance in nature.

Because of the multiplicity of factors involved in a disease complex, satisfactory control is rarely achieved by a single measure alone. Although fungicide applications are often thought to provide single-shot control, they usually do not. If the host is highly susceptible, the pathogen reproduces rapidly and copiously, and if the environment is favorable for extended periods, control becomes both difficult and complicated, as with *Botrytis gladiolorum* on gladiolus in southern California (Baker and Sciaroni, 1952). Similarly, *biological control should be regarded as one of the several complementary measures in an integrated program.*

In many diseases some factors are relatively unimportant, or a few factors are of overriding importance. When several factors condition the success of an organism, any one of them that approaches or exceeds the limits of maximum or minimum tolerance will be limiting and is called a *limiting factor.* This does not imply that the other factors are then inoperative; thus in Figure 1.1:B the limiting factor may be the environment unfavorable for the pathogen, but the antagonist population is also operative.

Each factor of the environment has a *minimum* below which growth of a given organism does not occur; if this is lowered still further, death will result. At the other extreme (the *maximum*), growth also ceases, and beyond that, death will ensue. At some point (the *optimum*) in between, growth is greatest. In the laboratory, the zone of optimum growth is often quite narrow, and the so-called normal growth curve (Fig. 1.2) is quite steep because other environmental variables are held constant. Under natural conditions, on the other hand, where all factors tend to vary more or less independently, this kind of situation is almost unknown.

Odum (1971) has pointed out that an organism may have a wide range of tolerance for one factor and a narrow range for another, and that if one factor is not optimal, the tolerance limits for other factors may be reduced. As Cochrane (in Horsfall and Dimond, 1959–60) observed for spores, "the limits of germination are narrower if some other factor is nonoptimal." Furthermore, because of factors that may be limiting in nature, an organism sometimes does not grow at the experimentally de-

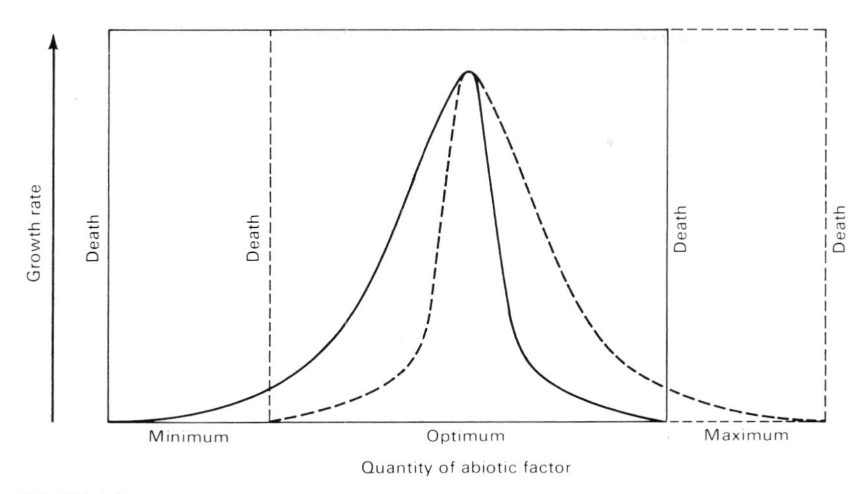

FIGURE 1.2
Theoretical relationship between growth rate of the host plant and different levels of a single abiotic factor of the environment. Broken line represents the effect of minor elements or oxygen for example, slight deficiencies of which cause drastic decrease in growth. Solid line represents the effect of moisture or temperature for example, excesses of which cause greater reduction in growth than do comparable deficiencies of these factors.

termined optimum level of another factor. The optimum and the limits of tolerance of a species may also vary geographically and seasonally through adjustment of the rate of the process involved. The optimum is not, therefore, to be thought of as an absolute, but as a range—a wandering optimum.

As noted by Daubenmire (1959), "For every change in one factor, a different optimum of all other factors comes into existence." If soil moisture is taken as the factor to be studied, as for example in the abscissa in Figure 1.3:A, growth at the lower moisture levels (curve 1) will be best under cool conditions and will fall off at higher temperatures because of excessive water loss through transpiration; however, the plants will be smaller than at the optimum in either case. Growth at the higher moisture levels (curve 5) will be possible at higher temperatures because of adequate water for increased transpiration, whereas at lower temperatures and reduced transpiration, waterlogged soil and reduced soil aeration may result. The effects would be intermediate in curves 2 and 4. As other factors are varied in the same manner, it becomes clear that the normal curve is, in fact, abnormal, and that a flatter helmet-shaped curve more nearly represents the true situation.[3] Such a curve thus represents growth under

[3]This concept was developed as a graphic analysis of the law of diminishing returns in economic theory by Peterson (1937).

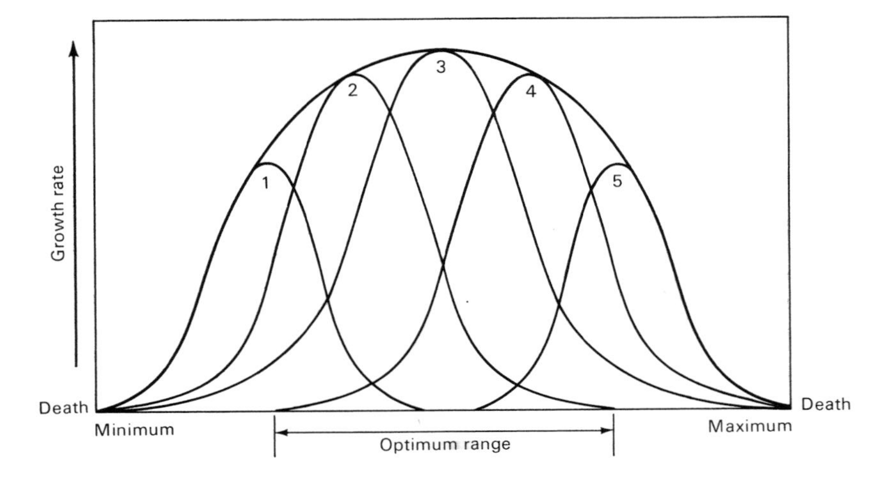

A

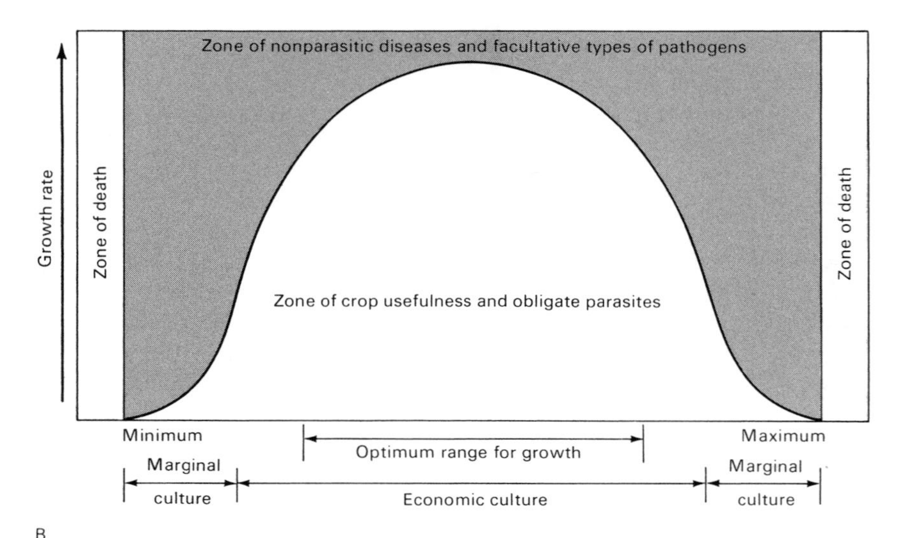

B

FIGURE 1.3

A. Growth rate of host as affected by different levels of a single factor (curves 1–5) of the abiotic environment, when other factors also vary. The helmet-shaped curve represents growth under natural conditions when the interacting factors vary independently.

B. Relationship between growth rate under natural conditions, in which all interacting factors of the abiotic environment vary independently, and different types of diseases and crop usefulness.

independently varying environmental factors in nature. A moment's reflection will show that plants do not and could not have such narrow limits of tolerance for any factor of the environment as is implied by the normal curve.

The relationship of this effect of the total environment on growth to the incidence of nonparasitic (physiogenic) diseases is shown in Figure 1.3:B. Conditions outside the curve are especially favorable to this type of disease, and thus to plant damage by facultative types of pathogens (able to grow saprophytically and parasitically), especially those that are favored by weakened or senescent tissue. **Damage by facultative types of pathogens is usually favored by conditions that weaken or stress the host.** This does not imply that infection by these pathogens is limited to senescent or weakened tissues. *The facultative and often weak parasites may establish and maintain a nonpathogenic host colonization of vigorously growing plants until the environment becomes highly unfavorable to the plant, upsetting the balance between host and parasite. Plant damage, which we recognize as disease, then appears, often rapidly.*

The zone of crop usefulness falls within the limits of the curve, with economic culture in the central area and marginal culture at either side of it; beyond the minimum and maximum limits for growth and survival, death ensues. Obligate parasites, which require a living host, generally grow best when the host grows best, and thus tend to be most plentiful and damaging in the zone of crop usefulness. They grow poorly or not at all when the host is weakened by adverse conditions. One group of obligate parasites, the viruses, however, may sometimes produce greatest damage when the plant is growing under somewhat unfavorable conditions. It is apparent that the successful grower, raising his crop in the optimum range of conditions, may still have difficulty with obligate parasites and some trouble with facultative-type pathogens (especially those that can attack healthy tissue), but he will encounter few nonparasitic diseases or facultative-type pathogens that cause damage on a stressed plant (Levitt, 1972). The marginal farmer is likely to encounter any of these several types of diseases, with perhaps less trouble from obligate parasites.

Environmental conditions relatively more favorable to facultative types of pathogens than to the host may or may not favor increased infection, but generally increase damage. Conditions most favorable for the host generally are most favorable for infection and damage by obligate parasites.

Interactions Between Microorganisms

Some of the specific types of interactions between microorganisms should be indicated, to show some of the effects that may be expected.

Sclerotium rolfsii may be stimulated to form sclerotia by certain actinomycetes. *Fusarium solani* f. sp. *phaseoli* may be stimulated by *Protaminobacter, Arthrobacter,* and *Bacillus* spp. to form chlamydospores. The cultivated mushroom is stimulated by bacteria to form sporophores, *Melanospora damnosa* by *F. roseum* f. sp. *cerealis* 'Culmorum' to form ascospores, *Phytophthora cinnamomi* by *Pseudomonas* spp. and *Chromobacterium violaceum* to form sporangia, and *Mycena citricolor* by *Penicillium oxalicum* and other *Penicillium* spp. to form sporophores (Chapter 7). *Sclerotium rolfsii* is stimulated by some isolates of *Bacillus subtilis* to produce more hyphae than in their absence. Indeed, the proposition is becoming increasingly clear that **the progression through the stages of the life cycle of a soil microorganism is determined at least as much by the associated microflora and abiotic environment as by the genes of the microorganism.** On the other hand, mycelial growth of *Phytophthora cryptogea* in culture may be enhanced by thiamine-synthesizing bacteria, and inhibited by thiamine-requiring bacteria (Erwin and Katznelson, 1961). The well-known penicillin produced by *Penicillium notatum* inhibits some bacterial pathogens of humans. Eventual death of the pathogen may result from the antagonistic effects produced.

MAN, THE DISRUPTER OF BALANCE

Plant populations growing in an undisturbed condition have a tough resistance to change. Only the well-adapted and competitive have found a niche in which to develop, and to maintain possession they have had to integrate their population, temporal sequence, and activities with those of their companions and with the climatic cycle. Those not well adjusted have simply been replaced by those that are. It is thus implicit that to be enmeshed in such an association is to gain external strength and protection. As rephrased by Marston Bates from C. S. Elton, **"The greater the complexity of the biological community, the greater is its stability."** A single type of organism in an environment is as unstable as a pyramid balanced on its apex; a complex association, by comparison, has a stability resembling the same pyramid resting on its largest face. An undisturbed flora, therefore, resists invasion by an alien plant, and an alien plant probably succeeds only when it is much better adapted than some occupant, and then only by slowly displacing it.

When man passed gradually, about 10,000 years ago, from being a simple food-gatherer to removing the competitor weeds from selected areas of cereals in the eastern Mediterranean, he began the disruption of the biological balance. When he discovered that certain seeds gave rise to food plants, and purposefully sowed them in purer denser stands, he further upset the balance. When crude cultivation of the soil, irrigation, removal of the tops to feed animals, and selection of more desirable strains of the crops began, the Era of Biological Imbalance had truly begun.

Today, man uses such a combination of fertilizers, pesticides, and tillage and seeding practices, together with specialized crop varieties, that production of food and fiber is greater than ever before. This has been necessary to feed and clothe his expanding multitudes. At the same time, however, he forfeited the biological balance that had served as the system of checks and balances against epidemics. He has usually replaced the complex community with a simple one, the variable population with a single genotype, and characteristics suited for survival with those for yield and quality. Man has become more dependent on his crops, and they in turn more dependent on him. As Ordish and Dufour (1969) stated, "farming is a most unnatural activity. Man has imposed on the environment a system of survival of what he wants to use over the Darwinian system of the fittest to survive. Consequently the farmer is engaged in a constant struggle with nature." That farmers and agricultural scientists have won many battles in this struggle is obvious when we consider the countless abundant past harvests. In America today about 5% of the people produce the food for the other 95%—hardly a testament to a failing agriculture. Moreover, the security of production in agriculture seems to be expanding both geographically and with time, as an increasing number of countries produce more food with less chance of a crop failure each year.

Failures in the efforts towards a productive agriculture have also been obvious, and the effects of some are evident to this day. The destruction of about 150 million bushels of wheat in the midwestern states in 1953 and 1954 by race 15B of *Puccinia graminis* f. sp. *tritici* is one reminder that growing vast acreages of a highly simplified biological community is not a dependable means of food production (U.S. Department of Agriculture, Cereal Rust Laboratory, St. Paul). Only a few select varieties were in use at the time, and none carried resistance to race 15B. Even better known is the potato late-blight epidemic and subsequent famine in Ireland in the mid-nineteenth century, when Ireland's uniformly susceptible potatoes were destroyed by the introduced *Phytophthora infestans.* The southern corn leaf-blight epidemic in the United States in 1970, caused by *Helminthosporium maydis,* is another example of the instability of a simple biological community; essentially every corn hybrid grown in the corn belt in 1970 carried the Texas cytoplasm for male sterility—and for leaf-blight susceptibility. Another kind of disruption in the balance has occurred when man inadvertently introduced a foreign parasite into a population of native plants. *Endothia parasitica,* imported from the Orient, destroyed the stands of American chestnut; *Cronartium ribicola* from northern

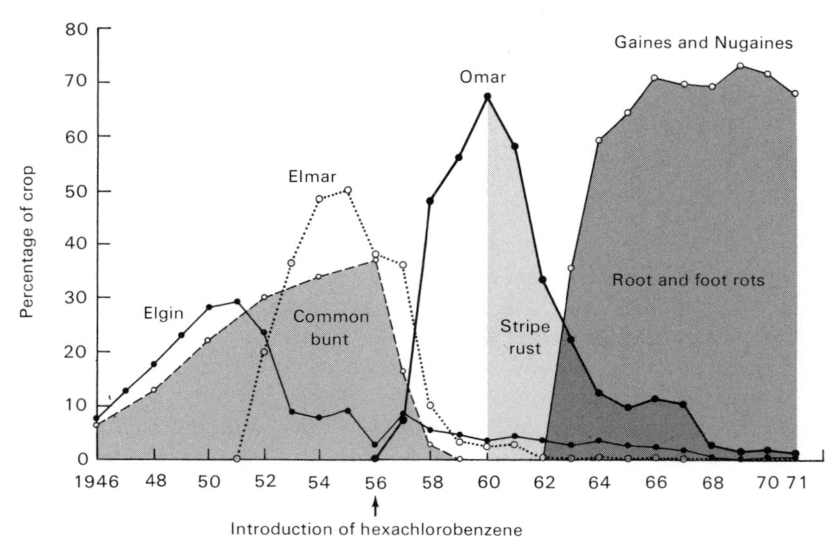

FIGURE 1.4
Production of major wheat varieties in the Pacific Northwest in the last 25 years in relation to disease incidence. Each successive variety occupies a greater percentage of the acreage—a trend toward ever greater simplification of the biological community. Plant pathogens have made continued production of each variety uneconomical. See text for details. (Data from Wheat Production Estimates of the Pacific Northwest Crop Improvement Association, Portland, Oregon.)

Europe caused widespread damage to white pines in North America; *Plasmopara viticola* from North America nearly eliminated the grape and wine industry of Europe.

Instability of the simple biological community is especially well illustrated by the rise and fall of wheat varieties in the Pacific Northwest (Fig.1.4.). Each variety has occupied a percentage of the acreage greater than its predecessor—a move toward increasing simplicity of the genetic base of the crop. For each disease problem solved by a variety, at least one new disease problem arose.

Elgin was susceptible to all races of common bunt (*Tilletia caries, T. foetida*) in the Pacific Northwest, and as its acreage increased, the incidence of common bunt also increased. Elmar, which replaced Elgin, was resistant to some races of common bunt, but helped to screen and perpetuate new and more complex pathogenic races (Kendrick and Holton, 1961). Bunt again increased until Omar, with resistance to all known races, replaced Elmar. A still newer bunt race pathogenic to Omar appeared within a year after Omar was released and probably would have eliminated it and subsequent wheats except for development of seed-treatment control with hexachlorobenzene. However, because Omar carried no resistance to stripe rust (*Puccinia striiformis*), severe losses from this disease occurred in 1961. Gaines, and later Nugaines, were released with resistance to stripe rust, as well as all known races of common bunt. They were the first commercially grown semidwarf wheats, provided a new plateau in yield potential, and initiated the era of intensive wheat-management practices to maintain high-yield capability. This produced marked changes in soil and crop microenvironment, and favored root and foot rots. Take-all [caused by *Gaeumannomyces graminis* var. *tritici* (= *Ophiobolus graminis*)] developed in irrigation districts, fusarium foot rot (caused by *Fusarium roseum* f. sp. *cerealis* 'Culmorum') in the low- to intermediate-rainfall (25–40 cm) areas, and cercosporella foot rot (caused by *Cercosporella herpotrichoides*) in the high-rainfall (50–65 cm) area of eastern Washington and adjacent Idaho. In addition, five new races of dwarf bunt (*T. contraversa*) appeared that were pathogenic to Gaines and Nugaines.

Paha and Luke, released in 1970, carry resistance to common bunt and stripe rust, and tolerance or resistance to fusarium and cercosporella foot rots. Luke carries resistance to all known races of dwarf bunt. Both are highly susceptible to flag smut (*Urocystis agropyri,* to which Gaines and Nugaines are resistant), a disease that has been important in only one area of south-central Washington and adjacent Idaho, but which could become epidemic. The simultaneous release of two varieties in 1970 is a move away from a single-variety wheat industry to a more stable multi-variety system. Nugaines will remain popular for many more years, and thus a part of the multi-variety system. There is now an intentional effort to produce wheat on an ecological basis with reference to disease distribution. The next step may be to grow the varieties as multiline mixtures rather than separately, and thereby increase the complexity and stability of the biological community. A few crops such as lima beans and oats have already been grown elsewhere as mixtures of genetic lines for this reason. The day of the pure line may be passing.

Still another kind of disruption in the balance has occurred over large acreages of one-time desert land now converted to agriculture through irrigation. In the natural state the biological community was comparatively simple, but in balance with the hot dry climate. With the sudden change in environment by irrigation, fertilization, and cultivation of specialized crops, the soil microbiological community was inadequate to prevent invasion and establishment of root pathogens. The biological balance of the soil eventually stabilized, in most places with root parasites as a major component. Crops of the western United States and inland Australia have consequently been severely affected by root diseases, and vast sums have been spent in attempted control. Moreover, as soils became unproductive because of pathogens such as *Verticillium albo-atrum, Heterodera schachtii,* and *Meloidogyne incognita,* potato, sugar beet, mint, and cotton crops were moved to virgin areas, only to have the process repeated.

Armillaria mellea is an indigenous pathogen on roots of native oaks in California, where it does relatively little damage to the host under the prevailing conditions of warm dry summers and cool wet winters. However, if a home is built among these oaks and a garden planted which requires irrigation, the oaks usually die in a few years from armillaria root rot. This is thought to be caused by the shift to warm moist conditions associated with summer irrigation, which permits the pathogen to grow vigorously, but the mechanism of operation is undetermined. In this example, man is again disturbing the biological balance, increasing disease.

Some of the forests of California have shown marked changes in composition from man's activities. According to F. W. Cobb, Jr., and J. R. Parmeter (unpublished), the dominant sugar pine in California has been replaced to some degree by ponderosa pine as a result of logging and subsequent fires. Seedlings of ponderosa pine are very susceptible to damping-off in forest litter, and survival is therefore greater in the mineral soil of burned areas. The root-invading vascular fungus, *Verticicladiella wagenerii,* is now severely attacking older ponderosa pines, killing the trees in a few years. In addition, the many stumps resulting from logging operations become infected with *Fomes annosus,* which spreads through the roots and infects nearby trees, producing root and heart rot. Neither of these pathogens was a serious problem in the original forest areas. These workers have suggested that endemic oak wilt, caused by *Ceratocystis fagacearum* (*Chalara quercina*), became important in the midwestern hardwood forests only in the 1940s, after the trees had sustained frequent injuries from logging operations and other activities of man.

Sclerotia of *Sclerotium rolfsii* are formed on plant parts at the soil surface. Under noncultivation they would remain there, subject to rapid drying, increased leakage, and microbiological destruction (Chapter 9), as well as exposure to minimal concentrations of ethylene (Chapter 6). Cultivation, however, buries these sclerotia, reducing drying, leakage, and decay, as well as exposing them to inhibitory concentrations of ethylene. Subsequent cultivation will return some of them to the soil surface where they can germinate and infect. Cultivation may thus tend to increase the severity of disease caused by this pathogen (A. M. Smith, unpublished).

There is little doubt that severe plant disease indicates that the biological balance has been destroyed in some way, directly or indirectly, and usually by man. Until agriculture began, there probably were few severe plant-disease outbreaks, primarily because there were few disturbances of

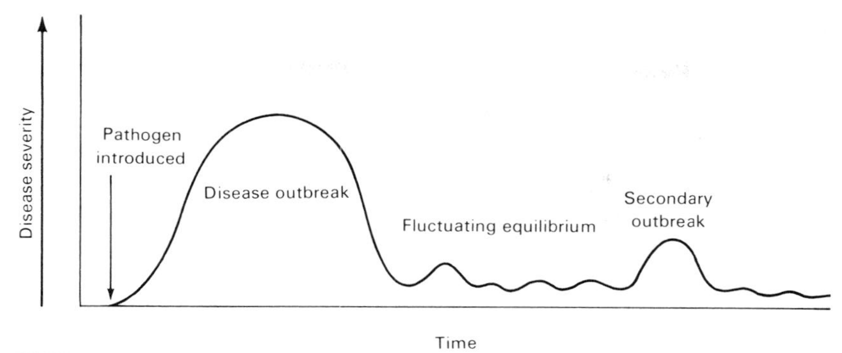

FIGURE 1.5
The grand cycle of disease. (Modified from K. S. Chester, *The Nature and Prevention of Plant Diseases.* 1947. Blakiston, Philadelphia.)

the biological balance, which in undisturbed associations tends toward ever-greater stability. It is interesting that the oldest record (Stewart and Robertson, 1968) of plant-disease organisms on crop plants is from Jarmo about 8000 years ago, in the early phases of agriculture. When a pathogen is present but does not cause significant disease, a biological balance exists. When a pathogen is repeatedly introduced into an area without producing disease, it is clear that the ecological balance has not been upset.

The serious disease epidemic that sometimes occurs when a new pathogen is introduced often follows the so-called grand cycle of disease, severe for a period, followed by less severe outbreaks, an occasional mild outbreak, and finally perhaps declining to unimportance (Fig. 1.5). This may represent the establishment of a new balance, in which the pathogen is rarely extinguished but is kept in a state of suppression. For this reason, such situations merit careful study for biological control.

Having upset the balance that held the pathogen in check, man is often dismayed at the biological backlash. He has returned to wild habitats for genetic material resistant to the pathogen in an effort to restore the balance. He has also set up quarantines and changed his cultivation practices in an attempt to avoid disease epidemics. Through use of legume cover crops or organic amendments to soil, he has attempted to restore the earlier natural condition of high organic matter and complex biological communities. To the extent that these practices have failed or have not been sufficiently rapid to satisfy his impatient nature, he has sought quicker remedies to prevent disease through fungicides. This approach has saved many an agricultural enterprise from extinction but has led to the insecticide and fungicide era, with its attendant environmental pollution and repercussions in other ecological associations, often far-removed.

Moreover, man has tended in the past to use broad-spectrum killers as quick solutions to the disease problems, and these, by their very nature, prevent a desirable biological balance, because they simplify the biological community and make it more unstable.

Although man cannot regain the original biological balance, a new one, reasonably compatible with changed components and factors, is possible and must be achieved. Ordish (1967) stated, "We are only civilised because we have upset the balance of nature to our advantage; unless we continue to do so we shall perish." It is becoming increasingly clear that if we continue to accelerate the rate of destroying the total balance we shall also perish. Daubenmire (1968) wrote, "Man must disturb ecosystems, but he should recognize that there are limits to safety in so doing, for disturbance becomes dangerous when induced conditions exceed the range of fluctuating natural conditions to which the organisms have been inured."

THE CHANGING SCENE

The small grower understandably wants a reduction of loss or control of the disease immediately, rather than in a few years. Large companies with diversified crops can better afford to take a long-term view and have tended to do so. The Del Monte Packing Corporation found that soil treatment with carbon disulfide (3300 pounds/acre; 3700 kg/ha) of land to be planted to asparagus in California gave poorer immediate control of fusarium wilt (caused by *F. oxysporum* f. sp. *asparagi*) than did chloropicrin (300 pounds/acre; 337 kg/ha), but that after several years the total control was better. For this reason carbon disulfide fumigation was used until satisfactory resistant varieties were developed.

Man has come in recent years to realize that it is better to work with natural biological forces than to ignore or override them. Rachel Carson's "Silent Spring," published in 1962, expressed this shift in man's approach to nature. The use of broad-spectrum insecticides which caused overkill has given way to subtler, more sophisticated methods. The cattle screwworm was eliminated in Curaçao and Florida by the release of large numbers of irradiated, sterile, adult males, with the resulting production of infertile eggs (Whitten, in Rabb and Guthrie, 1970). The melon fly was similarly eliminated from the island of Rota, and the oriental fruit fly from Guam. Insects have also been controlled by various attractant lures (for example, female or food odors), and by pathogenic bacteria, fungi, and viruses. On the other hand, insects have been effectively used in controlling such weeds as the prickly pear in Australia. Plant pathogens have also been used in weed control (Wilson, 1969).

More recently it has been realized that modest use of insecticides to decrease the population, followed by biological control, cultivation practices, and other methods, gives best results. It has become increasingly clear that control of plant disease is rarely achieved by a single procedure such as soil treatment. It must be supported by the use of planting material free of the pathogen, by tillage methods unfavorable to the pathogen, favorable to antagonists, or both, and by sanitation or use of fungicides. Such *integrated control* is based on the fact that different methods work best at different times and places, or under varying conditions. For example, insecticides work best on high populations, and the release of sterile insects is most effective on low populations. As Sripathi Rao wrote in 1965, "Integrated control offers the possibility of making up for the deficiencies of any single method." Biological control should not be thought of as a single-shot procedure, like spraying.

In accord with this changing attitude, there has been a shift away from single-shot overkill treatments. Entomologists have turned to more specific chemicals and to combined integrated control procedures. Plant pathologists have turned to proprietary, specific, and relatively benign materials, rather than to the broad-spectrum materials formerly used. The trend is in the right direction in both fields, but insufficient use is still made of biological control.

A number of general circumstances and misconceptions have contributed to the slow development of effective biological control of plant diseases.

1. Most insects, some diseases of plants, and many diseases of man can be controlled after they appear, by application of chemicals. This method has come to be thought of in many cases as the first line of defense rather than as a component of a more generalized system.

2. Chemical controls often are less complicated for the grower to use than are nonchemical methods, which require more intimate knowledge of the causal agent and its associates. This also affects the research worker, for it often seems more expedient for him to test a chemical than a nonchemical control.

3. Results with chemicals are faster and therefore often more spectacular than those with nonchemical methods. The grower can see the dead insects or the disease-free leaves after application of a spray. By contrast, nonchemical controls usually must be applied before the appearance of the disease, and often applied over an extended period; they are seldom spectacular because of this slowness of effect.

4. Chemical control is essentially active and provides the grower with

action therapy. If one chemical fails, another can be tried. There is satisfaction in direct engagement with the enemy, even if one wins all the battles and loses the war. Nonchemical control, because of its slower effect (often over a period of years), is essentially passive; a greater sense of commitment, of understanding of and confidence in the methods, is required for nonchemical means.

5. Large sums of money are spent in selling chemicals because it is profitable. Their use is also demonstrably beneficial for the grower; the unsprayed crop is rarely as profitable as the sprayed. As Huffaker (in Rabb and Guthrie, 1970) said, "Pesticide usage is promoted sometimes under great pressure to sell, irrespective of need or possible later consequences, and often in ignorance at least of the ecology of the ecosystem that is often disrupted." By contrast, altered cultivation practices and biological control have little profit motivation; only the grower and the public are benefited. Huffaker suggested that the alternative to the pesticide salesman is the regulated, independent, trained professional operator who sells advice rather than materials to the growers, and who is paid a retainer fee for advising the best overall control program.

6. Compartmentalization of knowledge has aided the cause of chemical control. Have too few pathologists been vitally concerned in disease control in growers' crops, and have too many justified their study of biochemical processes or laboratory minutiae because it contributes to knowledge and may provide a clue to control? Because of their reluctance to become directly involved in growers' problems, they have relinquished the field to the chemical advocates. Such direct involvement by the plant pathologist will be required to institute any major change in grower practices.

7. The decline in interest in ecology in the period 1935–60 led to diminished appreciation, by scientists and public alike, of the complex interactions between organisms, and between them and the environment. The renewed attention to the subject in the last decade, sparked by the increasing pollution problem, has caused people to challenge the advisability of continuing the barrage of chemical controls.

8. The idea has become generally accepted that large changes are necessary to produce the desired control. Chemical treatments produce large, abrupt, brief changes, and the pathogen population may fluctuate wildly. Biological control is more subtle, operates more slowly and on a smaller scale, and is generally more stable and

longer lasting. Man, an inhabitant of the more violent aerial habitat, is inclined to think that large rapid changes are necessary. With a better-developed worm's-eye view of events, he may come to see that gentle "nudging" of the epiphytic and soil microflora can be far more effective in the long run than overkill or "dynamite" treatments.

2

WHAT IS BIOLOGICAL CONTROL?

We adopt [words]. We use them for a while, and before we realize it, they direct our thinking. . . . Definitions . . . usually do not reflect this inherent fuzziness which is characteristic of all scientific measurements. . . . Thus, definitions of words used in scientific discourse should exhibit only so much precision as is commensurate with their referents.

—D. B. ZILVERSMIT, 1964

Primitive man must have observed, in very early times, the delicacy of balance in nature by noting that excessive hunting of an animal lowered its numbers and thus decreased his own food supply. Taboos evolved which attempted to maintain a favorable balance. This insight, afforded primitive man, was lost as he became civilized and had diminishing direct contact with nature. For the past several hundred years he has been learning that such a balance is difficult to maintain, and that which has been achieved smoothly in millenia is very difficult or impossible to achieve in a few years.

Slowly the idea developed that man could manipulate one organism or one aspect of the environment in such a way as to control another. The development of the entomologists' concepts and philosophies of *natural control* (maintenance of a fluctuating population density of an organism within certain limits over a period of time by the combined actions of the total environment) has been outlined by Doutt, Huffaker, and Messenger (in DeBach, 1964). Doutt cites the introduction of the mynah bird from India to Mauritius in 1762 to control the red locust as the first successful transfer of a natural enemy from one country to another. The first really successful control of an insect by another introduced species was by the vedalia beetle in California in 1888 on the cottony-cushion scale of citrus.

Such control of insects was referred to as parasitic control, bug vs. bug, use of insect enemies, and nature's remedies. In 1916 L. O. Howard referred to it as "the biological method," and in 1919 H. S. Smith called it "biological control" (DeBach, 1964). This is but one phase of natural control. Since the term biological control has now become the generally preferred inclusive one to use, it is to be expected that its meaning would gradually have been expanded. Entomologists have tended to maintain biological control for "the action of parasites, predators, or pathogens in maintaining another organism's population density at a lower average than would occur in their absence. . . . the term 'biological control' is somewhat restricted in usage . . . over the broader scope which . . . could be interpreted to include germ warfare, antibiotics in medicine, the development of plants and animals resistant to the pathogens, parasites, or predators that attack them, or . . . even the toxic effects of chemicals on organisms" (DeBach, 1964). There is little help, therefore, in turning to the original definition of the term, unless we are prepared to abandon its use in the broad sense and return to natural control. Since the latter term would be strained to include some of the unnatural things (fertilizer application, soil treatment with carbon disulfide) man may do to favor antagonists of plant pathogens, it has seemed better to adopt here the generally accepted term in the broader sense in which it is currently used. Beirne (1967) phrases the broader concept: "Any living organisms that can be manipulated by man for pest control purposes are biological control agents." He regarded biological control as an integral part of pest control rather than as a separate discipline.

Having resolved the point of original usage, it is still necessary to decide what kinds of control procedures should be included. That there is little unanimity on this point among plant pathologists was demonstrated in a seminar held after the First International Symposium on this topic in Berkeley, California, in April, 1963. After the 7 days of intensive discussions (Baker and Snyder, 1965) by world authorities, there still remained an amazing range of opinions as to what should be included under biological control; practically every procedure used in control of plant disease would have been included by someone. A term so broad would be quite useless, and there is no consistent authority to which we can appeal. Nor is it desirable to be as arbitrary as Humpty Dumpty: "When I use a word it means just what I choose it to mean—neither more nor less." Perhaps the best approach, then, is first to consider briefly the various factors and mechanisms in a natural balance involving a plant disease, and from this construct an open-ended definition. With Samuel Johnson we can say, "Every other authour may aspire to praise; the lexicographer can only hope to escape reproach. . . ."

FACTORS INVOLVED IN BIOLOGICAL CONTROL

The Host

The host population in its wild state was always involved in biological control by being a part of the biological balance that helped keep the pathogen suppressed. Exudates from roots of the individual plants served as a stimulus and a source of food for antagonists of the pathogen as well as for the pathogen itself. The remains of stems, roots, or leaves of the host served as a place of survival for the pathogen when returned to soil, but they also served as an energy source for a wealth of saprophytic microorganisms that decayed the plant tissues, sometimes destroying the pathogen in the process. Larger plant species provided shade and protection for smaller plants, which then grew, supported a specialized microbiota of their own, added to the soil organic matter, and contributed still more to the complexity of the biological community. Within the mixed stands, there were probably some plants that stimulated growth and reproduction of the pathogen of another host, and that sustained only mild or no injury themselves, but left a greater population of the pathogen to attack the more susceptible plants. Other plant species may have acted as trap plants of the pathogen, permitting penetration but not reproduction, and thus reducing the population of the pathogen. In addition to all this, individual plants with new genes or new gene combinations arose periodically within the population, and those more resistant to the pathogen gradually replaced the susceptible types.

Each of the above-described processes, and others as well, are as relevant to man's objectives of biological control today as they were to the host in the original state.

If the host or crop plant is highly *susceptible* to the pathogen, severe disease losses will occur unless the environment is highly unfavorable or antagonists suppress the pathogen; an unstable population of pathogens results. If the host is highly *resistant,* little or no disease will occur, no matter how favorable the environment or how ineffective the antagonists, and a stable pathogen population results. The host plant may simply be *tolerant* of the pathogen, that is, produce a satisfactory yield in spite of damage from the disease.

Leaves supporting a leaf-spot fungus may decompose quickly on the soil surface and in this way contribute to inoculum decline of the fungus. A legume cover crop may modify soil structure, organic-matter content, or the microbiota so as to reduce activities of the pathogen of the primary crop; or it may provide the necessary conditions of humidity and temperature at the soil surface for increased activities of a leaf or stem parasite. Root exudates of the plant may favor the pathogen, the antagonist, or both, or they may be toxic to the pathogen. Some hosts produce toxins that inhibit pathogens, thereby not only avoiding infection but reducing the pathogen population in the soil.

Finally, there is the question whether host resistance is a type of biological control. Garrett (1970) defined biological control as "being mediated by one or more organisms (excepting man himself) outside the host-parasite relationship." Host resistance is concerned with interactions between at least two organisms, the host and the parasite. It is an important factor in the biological balance and reduces injuries by pathogens through biological means. Resistant varieties may starve a pathogen quite as much as does crop rotation; both result in the absence of a suitable host, and both are forms of biological control.

The mechanism of resistance is known in only a few instances. In some the host may be resistant because it stimulates antagonists to grow in its rhizosphere. Resistant varieties may be rendered susceptible by products of decomposition of organic matter in soil. Thus, Linderman and Toussoun (1968a) found that cotton varieties resistant to *Thielaviopsis basicola* became susceptible in the presence of hydrocinnamic acid, a naturally occurring compound in decomposing organic matter in soil. It was thought that this compound removed or prevented formation of inhibitory materials, increased exudation of stimulatory materials, or both. In the interest of an open-ended interpretation not restrictive to future research, resistance and tolerance are considered to be types of biological control. However, because of the voluminous literature and extensive and effective work on breeding for resistance in crop plants, it is treated minimally in this volume. Wood (1967) has presented a thorough treatment of the subject.

An obvious point frequently overlooked in analyzing disease situations is that the host is itself a biological system, a unit consisting of interacting parts. Photosynthesis in the tops supplies the necessary carbohydrates for growth of the whole plant; the roots absorb water and minerals from the soil for the whole plant. Roots of many plants are also the site of formation of amino acids, utilizing carbohydrates from the tops and nitrogen from the soil (Bollard, 1960). It is thought that growth-controlling substances are also formed in the roots (Chapter 8).

Tops that have been weakened from insect attack, disease, or other injury may, therefore, develop weakened roots and increased susceptibility to disease. Susceptibility of oaks to root rot caused by *Armillaria mellea* has been increased by defoliation from gypsy moth or leaf rollers, stem-boring insects, or drought (Boyce, 1938). Susceptibility of balsam fir to this pathogen has recently been shown to increase from 2.5% to 4, 20, 32, or 88% as the infestation of balsam woolly aphis increased from none, through light, medium, severe, and lethal (Hudak and Singh, 1970). Although a tropical *Armillaria mellea* (*Armillariella elegans*) normally forms large brown rhizomorphs on citrus roots at Somersby, New South Wales, infection is infrequent. However, when vitality of the trees is reduced by a disease of unknown cause (called "sudden death"), the incidence of infection by *Armillaria* is greatly increased (P. Broadbent, unpublished). Stripe rust on wheat reduces the extent of roots and hence the amount of water removed from the soil (Martin and Hendrix, 1967).

Louis and Nightingale (1937) showed that pineapple crowns with high starch and moderately low organic-nitrogen content rooted far better than those with low starch and higher organic nitrogen, and formed larger roots.

It was found in 1943 (K. F. Baker, unpublished) that complete removal of the tops of belladonna plants (from which the drug was to be extracted) greatly increased the number of plants killed by *Phytophthora* and *Pythium* over that in the untrimmed checks. If a single branch per plant was left, the root decay was reduced nearly to that of the checks. A similar situation was observed in petunia plants kept over for a second year in commercial seed fields. Apparently carbohydrate storage in these two plants is in stems rather than roots. It had been found earlier that removal of or severe injury to cotyledons of American elm seedlings also increased damping-off caused by *Rhizoctonia* and *Pythium*. D. T. Hartigan (unpublished) showed that the sapwood starch content of *Eucalyptus saligna* in moist valleys near Gosford, New South Wales, was inversely correlated with severity of attack by psyllids (*Glycaspis* spp.), and that low starch appeared to be associated with the incidence of *Phytophthora cinnamomi,* to which *E. saligna* normally is quite resistant. Defoliation of apple trees also increased susceptibility of their roots to attack by *Xylaria mali* (Cooley, 1944), and defoliation of sugar maple increased injury from *Armillaria mellea* (Wargo, 1972). However, Rossetti and Bitancourt (1951) found the opposite effect from defoliation of citrus trees on enlargement of trunk cankers caused by *P. citrophthora.*

Similarly, Chapman (in Baker and Snyder, 1965) reported that "anything which tends to devitalize the tops of trees, such as climatic extremes, insect infestation, insecticidal damage, and nutrient deficiencies and excesses, will lower the vitality of citrus roots and thus open them to destructive influences by both pathogenic and nonpathogenic soil organisms."

Injury to roots will produce injury to the tops from water deficiency, and often from translocated toxins produced by root pathogens. For example, injury from club root of crucifers and root-knot nematode of tomato is greatly increased by the frequent invasion by low-grade pathogens that decay the galls. The degree of injury to the tops from root rot is often considerably greater than that produced by excision of a similar root area, even though water will continue to enter such infected roots for a time.

Wilhelm (in Holton et al., 1959) emphasized the interdependence of the shoot and root systems, noting that "the plant is essentially a continuous single mass of protoplasm partitioned only partially by cell walls. . . ."

Newhook (in Toussoun et al., 1970) demonstrated this interdependence in the control of *Phytophthora cinnamomi* on *Pinus radiata* in New Zealand. Aerial application of 560 pounds/acre (628 kg/ha) of superphosphate to P-deficient infected forests caused a remarkable recovery of the trees without diminution of the pathogen. Elimination of this deficiency improved crown density, which led to improved root growth and mycorrhizal development—and to a better root-shoot ratio. Increased top growth removed more water from the soil through the more extensive root system, making soil conditions less favorable for *Phytophthora* (Fig. 2.1). The greater abundance of mycorrhizae on roots perhaps further protected them from infection by the pathogen (Marx, 1972). Effective control was thus provided for 16 years by a single application of superphosphate. Probably phosphate was the limiting factor in tree growth in this area, and correction of it led to spectacular improvement. In other instances, however, some other factor(s) may be limiting. When a limiting condition is widespread, as in the New Zealand example, the deficiency may not be recognized, and indeed may not be very significant, unless it increases damage by a pathogen.

Eradication of an alternate host (barberry for stem rust of wheat) or of a companion host (weeds susceptible to the spotted-wilt virus of tomato) is regarded as a form of biological control because the direct effect on the pathogen is through the removal of its host. The eradication of an introduced plant pathogen from an area (the citrus-canker bacterium, *Xanthomonas citri,* from Florida, Texas, and Darwin, Australia) through destruction of infected plants would be a similar type of biological control.

Sugar beets in areas of the western United States, where curly top virus can be severe, may escape disease if sown in winter. This enables the plants to reach a size that covers much of the soil surface before flights of the infective beet leafhopper from the desert vegetation occur. The leafhoppers will not settle on fields where about 80% of the soil is covered by vegetation. Plants infected while very young remain small and sickly, create conditions favorable to leafhoppers, and thus lead to increased inci-

FIGURE 2.1
Indirect control of littleleaf disease of *Pinus radiata* in New Zealand (caused by *Phytophthora cinnamomi*) by a single application of superphosphate at 560 pounds/acre 12 years previously. Left: Unfertilized area. Right: Fertilized area. Population of pathogen remained high, but improved growth conditions for the host and mycorrhizal associates caused greater root and top growth, increased transpiration loss, and reduced soil moisture. [From F.J. Newhook (Toussoun et al., 1970).]

dence of disease. Early planting is thus a form of biological control operating through the host and affecting the insect vector of the virus. A similar control of common bean mosaic of beans in central Washington is provided by early planting.

The Pathogen or Parasite

Many special reference works are available on types of soil microorganisms. Soil fungi are treated by Barron (1968), Domsch and Gams (1972), and Gilman (1957); soil bacteria by Gray and Parkinson (1968); soil actinomycetes by Waksman (1967). Soil fauna are covered by Kevan (1962), Thorne (1961), and Wallwork (1970).

A parasite (an organism living in or on another living organism and obtaining organic nutriment from it) may or may not be a pathogen, that is, produce symptoms of disease. *Endogone* spp. may thus locally invade cortical cells of roots (Mosse, 1973), and *Deuterophoma tracheiphila* may systemically invade chrysanthemum (K. F. Baker, unpublished) without producing significant damage. Some organisms, on the other hand, are able to produce disease without infecting the plants, by producing toxins, as in *Penicillium oxalicum* on corn seedlings (Johann et al., 1931), *Aspergillus flavus* on corn (Koehler and Woodworth, 1938) and citrus (Durbin, 1959b) seedlings, and *Bacillus cereus* and *Aspergillus wentii* as possible causes of tobacco frenching and yellow strapleaf of chrysanthemum (Steinberg, 1952; Woltz and Littrell, 1968). These are pathogens but not parasites. Other pathogens (*Periconia circinata* on milo maize) infect the host only after the toxins have practically killed it (Leukel, 1948). In these cases the pathogen is fully exposed during all or most of its life cycle to both the physical environment and the antagonists.

The specialized metabolism of parasites and pathogens, which enables them to invade the host and reduce their environmental vulnerability, is not an unmixed blessing for the microorganism, but can be advantageous to man in its control. Pathogens and parasites are generally more sensitive to unfavorable abiotic factors than are saprophytes (Chapter 7). This makes possible selective treatment of soil by heat or chemicals to eliminate pathogens while leaving many antagonists (Chapter 5).

The plant-parasitic habit has developed in many types of organisms; among these, we are here concerned with fungi, bacteria, algae, seed plants, nematodes, viruses, and mycoplasmas. As would be anticipated from this array, there are many variations in life histories which affect biological control of the pathogens.

Most pathogens invade the host early in the disease and, being internal, are generally protected from antagonists. Secondary organisms may, however, invade diseased tissue and rot it, as clubroot (*Plasmodiophora brassicae*) galls of broccoli are invaded by saprophytes that greatly augment the injury to the plant. Application of pentachloronitrobenzene (PCNB) will control the secondary microorganisms, but only partially the *Plasmodiophora,* permitting approximately normal plant growth (Wilhelm, in Holton et al., 1959).

Possession of host tissue after it has been killed by the pathogen is maintained by depleting available substrates or by producing antibiotics. In addition, the hyphae may be thick-walled and resistant. Similarly, the egg masses of the root-knot nematode (*Meloidogyne* spp.) may be deeply embedded in the roots. Tobacco mosaic virus is said to survive as long as the tissue in which it occurs remains undecomposed. All these internal structures are better protected than those exposed on the host surface, such as powdery mildew mycelia. That time of survival of many of them is approximately that required to decay the organic matter shows that they are better protected than similar mycelia in soil. Therefore, most successful soil fungi produce resistant thick-walled mycelia or resting structures (sclerotia, chlamydospores, oospores), which may remain viable after decomposition of host tissue. Pigmentation of these structures commonly occurs and has been associated by some workers with resistance to antibiosis and enzymatic attack (Chapter 6). Thin-walled mycelia, conidia, germ tubes, or zoospores are more vulnerable to attack by antagonists.

A successful microorganism has the features necessary for its survival under the prevailing conditions, and this situation is enforced by the total environment in the ecological niche it occupies.

Vascular pathogens (*Verticillium albo-atrum, Cephalosporium gramineum,* formae speciales of *Fusarium oxysporum*) are very well protected in the interior of the plant. Such pathogens have the specialized metabolism required for such invasion. However, specialization of pathogens also generally leads to increased vulnerability to the physical environment and to antagonists.

Garrett (1944) suggested that **the least specialized soil microorganisms are the soil inhabitants,** that is, members of the general soil microflora whose parasitism of a wide range of hosts is incidental to their saprophytism, and which are able to survive indefinitely as soil saprophytes. In addition, these soil inhabitants are generally viewed as the most primitive of root-infecting fungi. Alternatively, they may be highly advanced microorganisms, having acquired the ability to invade many kinds of plants as well as dead substrates. This concept is supported by the ability of most nonspecialized soil inhabitants to invade their hosts under a wide range of environmental conditions, but to produce little disturbance except under conditions unfavorable to the hosts. In the wild state, plants are necessarily adapted to the prevailing environment (those not adapted are eliminated, particularly in years of environmental extremes) and are not stressed, and the internal microorganisms may not, therefore, be injurious. Evolution in the parasite has, therefore, been toward the ability to infect but cause minimal damage to the host under near-optimal conditions, a situation associated with the facultative, rather than obligate, types (Chapter 1). This would suggest that the facultative types are physiologically more advanced than obligate parasites, but less advanced than obligate saprophytes.

Soil inhabitants are, in general, more refractory to biological control than are *root inhabitants* (Chapter 6). Garrett (1956) observed that root-inhabiting fungi were able to survive more or less indefinitely in living hosts, but that they declined during the saprophytic phase because of the activity of antagonists. Fortunately, most of the important soilborne plant pathogens are root inhabiting and are therefore susceptible to biological control. Obligate parasites either produce dormant resting stages on or in the host before it dies, or they themselves die.

Durbin (1959a) found that *Rhizoctonia solani* isolates that attack roots as much as 0.3 m below the surface are slow-growing, tolerant of carbon dioxide, and that few of them produce sclerotia, even in culture—possibly because sclerotia-forming isolates are not suited to a subterranean environment. Isolates from the soil surface, on the other hand, are moderately fast growing and tolerant of carbon dioxide, and usually produce sclerotia. Isolates that attack leaves and other aerial parts take advantage of brief favorable periods by rapid growth, and most of them quickly produce many sclerotia; they are comparatively intolerant of carbon dioxide.

Dormant structures in the soil may resume growth when conditions become favorable, and the pathogen may grow through soil until it makes contact with a host plant. Such organisms are, of necessity, quite tolerant of antagonists. The dormant structures or cysts of other pathogens (*Verticillium albo-atrum, Heterodera schachtii*) may remain dormant until a suitable root grows near them and stimulates germination or hatching, and infection follows. Such organisms may be quite sensitive to soil antagonists. A few fungus propagules probably germinate from time to time under conditions of nutrition and temperature suboptimal for most of the population. This may be a mechanism of attrition of a pathogen's propagule population.

Organisms vary greatly in their ability to invade soil at different depths. Thus, *Sclerotium rolfsii* occurs at the soil surface, whereas *Pythium* and *Phytophthora* may occur considerably deeper. The latter type escapes the majority of antagonists, which do not thrive at these depths; this also places these organisms beyond the reach of most economic soil treatments. This escape mechanism conferred by the ability of a pathogen to persist deep in the soil was dramatically illustrated in a field of garden stocks in southern California (K. F. Baker, unpublished). A soil-surface treatment with Vapam effectively killed *Phytophthora* in the upper 15–20 cm. The stocks did well for a time, but were eventually destroyed when the pathogen grew up the tap roots from below. Antagonists restricted to the upper 15–20 cm of any soil might be ineffective as a disease-control measure for the same reason.

The survival of some sensitive nematode-transmitted viruses depends on a complex set of conditions (Cadman, in Baker and Snyder, 1965). The tomato black ring and raspberry ringspot viruses are spread in England by the root-feeding needle nematode (*Longidorus elongatus*), but are not transmitted through eggs to the progeny. A fallow of 9 weeks until infective nematodes die will break the virus cycle, therefore, if no infected plants are available to infect the larvae. However, since the viruses are seed-transmitted in many weeds, the nematodes again become infective when these seeds, planted or volunteer, germinate.

Biological control through the pathogen itself may be possible with avirulent strains. There is increasing evidence that inoculating the host with such strains may prevent subsequent infection by a virulent strain (Chapters 7, 8, and 10). This approach may be analogous to the control of an insect by radiation-sterilized males.

Quarantines are sometimes regarded as a type of biological control, since the spread of the pathogen is prevented. This is, however, a direct activity of man against the pathogen, which therefore should not be included. The natural barriers to spread, such as oceans, extensive deserts, or high mountains, are types of natural control effective prior to man's influence but are not types of biological control.

The Physical Environment

Control of a disease through the inhibitory effects of the physical or chemical environment directly on the pathogen would not be a type of biological control, because it is not then operating through another organism. When, however, the environment favors the host and causes it to maintain its resistance to facultative types of microorganisms (Fig. 1.3:B), this is biological control. Such a case is that of fusarium (*Gibberella roseum* f. sp. *cerealis* 'Graminearum') seedling blight of corn and wheat at high and low temperatures, respectively (Chapter 9). Bliss (1946) showed that the optimal temperature for root growth of peach, casuarina, pepper tree, apricot, and geranium was 10–17° C, and that most root damage from *Armillaria mellea* occurred at 15–25° C. The figures for both citrus and rose were approximately reversed (17–31° for root growth and 10–18° C for damage).

Soil temperature may be adjusted to some extent by time of planting. Thus, postponing planting beans (a high-temperature crop) in coastal valleys of California until the soil temperature has reached about 18° C decreases hypocotyl infection by *Rhizoctonia solani,* but increases that by *Fusarium solani* f. sp. *phaseoli* (W. C. Snyder, unpublished).

Tillage practices that modify the environment so as to favor antagonists are certainly part of biological control.

Conditions of the environment may be made unfavorable to the pathogen or vector. Garden stocks do not flower unless exposed to 10–15° C for extended periods, and thus are grown in gardens and for cut flowers during cool seasons. They usually escape injury from fusarium wilt (*Fusarium oxysporum* f. sp. *mathioli*) in the cool soil. However, production of seed requires nearly a year, and growth necessarily extends into the season of warmer soils; the disease may therefore cause severe losses in seed fields (Baker, 1948). In growing the plant under conditions suited to it but not to the pathogen, man is practicing biological control of the pathogen.

Soil water may vary with field topography or may be modified to some degree by altering culture such as fertility, planting date, row spacing, and raised beds. Cook and Papendick (in Toussoun et al., 1970) reported that wheat in eastern Washington sustained losses from *Fusarium roseum* f. sp. *cerealis* 'Culmorum' on side hills but not in the lowlands, although populations of chlamydospores of the pathogen were about equal. Bacterial populations and activity were greater in the moister soils and presumably lysed the germ tubes of chlamydospores there (Chapter 6). Infections that did occur remained restricted if soils were moist but resulted in severe foot rot if soils were dry. This was apparently caused by plant water stress and predisposition (Chapters 6 and 9). This is a type of biological control because the soil water has operated through the medium of the host, soil bacteria, or both to reduce foot rot.

Other instances of soil-water manipulation for reducing plant disease apparently affect the pathogen directly and therefore are not forms of biological control. Thorne (1942) showed that in Nevada and California planting potatoes and tomatoes on raised beds between deep irrigation ditches, and where they received frequent shallow applications of water so that the ridges were never saturated, permitted production of a fairly good crop in soil heavily infested by root-knot nematode. The roots were able to function at moisture levels too low for nematode movement. Roots extending into moister soil were severely galled, and a single soaking of the ridge nullified the method by permitting heavy infection. Similarly, raised beds may greatly reduce damage from *Phytophthora* and *Pythium* spp.

By planting nonirrigated native plants around native oaks infected with *Armillaria mellea* in California gardens, man exploits the fact that the fungus does little damage except under moist conditions in summer. By this method man is enabling the host to grow under conditions where it can develop but where the pathogen is inhibited; the effect may be directly on the pathogen, and if so, is ecological, not biological, control.

Soil pH may be modified by the addition of lime, sulfur, fertilizers, or organic matter, and on a microscale, by the presence of host roots and by microorganisms. It is well known that liming of soil will in many instances reduce the severity of club root of crucifers, and that addition of sulfur will reduce the severity of common scab (*Streptomyces scabies*) of potato (Walker, 1969). It is suspected that the first is caused by an effect on spore germination of the club-root pathogen, and the second by reduced growth of *Streptomyces*. If the effect is directly on the pathogen, without affecting host susceptibility or the activity of antagonists, it would not be an example of biological control. If the effect of environment on the pathogen is mediated through the host or some microorganism, it would be considered to be biological control. It should be remembered, however, that such practices work, and from the practical standpoint, the mechanism is probably not important.

Soil aeration is an important factor in parasitism by root pathogens and involves at least three types of effect.

1. Oxygen level. It is well known that a diminished oxygen supply greatly reduces root growth. Chapman (in Baker and Snyder, 1965) showed that susceptibility of citrus roots to decay in culture solutions increased as aeration was decreased, especially with low calcium. Some roots, such as rice, have specialized tissues that permit internal gaseous exchange with the atmosphere through the tops and are able to grow under waterlogged conditions. Most plants, however, sustain rapid root injury under these conditions, perhaps because of host predisposition to pathogens. Root-pathogenic fungi are generally aerobic, but the water fungi, *Pythium* and *Phytophthora,* are able to grow at oxygen levels injurious to roots (Stolzy et al., 1965). Oxygen requirements of several root pathogens were reported by Mitchell and Mitchell (1973).

2. Carbon dioxide level. The carbon dioxide content of the atmosphere is 0.03%, and in soil is commonly 0.5% in temperate areas, but it may reach 5–10% after rain, even in porous soil. The carbon dioxide content increases with the fineness, compaction, and impaired drainage of the soil, and with depth. The variability of pathogens probably enables many of them to develop in habitats differing in carbon dioxide level. *Phymatotrichum omnivorum* tolerates fairly high levels, *Rhizoctonia solani* may tolerate high levels (depending on the strain), and *Sclerotium rolfsii* cannot tolerate appreciable levels (Mitchell and Mitchell, 1973). Fungi growing in organic matter generally tolerate high carbon dioxide levels. Durbin (1959a) commented, "the more tolerant a pathogenic fungus is of CO_2, the greater may be its advantage in relation to the host's root system." High carbon dioxide levels may also favor the pathogen over less tolerant microorganisms. Furthermore, carbon dioxide at concentrations above 10% inhibits root growth and may check water absorption, causing the top to wilt, another form of predisposition.

3. Other volatile materials. It has recently been demonstrated that volatile materials (acetaldehyde and others) produced by decomposing alfalfa in soil stimulate germination and growth of sclerotia of *Sclerotium rolfsii* and *Verticillium dahliae* (Owens et al., 1969). Some bacteria and actinomycetes, and a number of other fungi, were also stimulated. At higher concentrations, *V. dahliae* was apparently

killed, possibly by the increased number of antagonists. These materials diffused through at least 5 cm of soil. Similarly, King and Coley-Smith (1968) showed that volatiles from onion and garlic stimulated germination of sclerotia of *S. cepivorum*. There is also evidence now that volatiles are involved in the widespread fungistasis phenomenon (Hora and Baker, 1972; Smith, 1973; Chapter 6).

The formation of fruiting caps of the cultivated mushroom provides an interesting example of microorganism interaction controlled by volatile emanations. Mycelial spawn growing in the prepared compost produces volatile materials (ethanol, acetaldehyde, and two others) which stimulate *Pseudomonas putida*, and perhaps others, to greater activity (Hayes et al., 1969). Apparently these volatile materials inhibit other bacteria that might be antagonistic to *Agaricus*, as shown by Eger (in Baker and Snyder, 1965). Thus the fungus stimulates the bacteria in the surface casing soil, which in turn, stimulate, in an unknown manner, the formation of caps and suppress antagonists at the same time. If this work is confirmed, it will provide a remarkable example of microbial interaction and balance. The layer of casing soil applied to the surface of the spawned compost provides the medium for these interactions, which certainly provide commercial biological control of fruiting.

Available organic matter for a carbon source is necessary for pathogens and antagonists alike. A source of inorganic nutrients is also essential. The competition for these energy sources is intense, and manipulation of them may provide a means of biological control of a pathogen. However, some methods apparently operate directly against the pathogen and would not be examples of biological control. Thus, the plowing under of surface organic matter to remove it from the area favorable to *Sclerotium rolfsii* (Chapter 4), as a means of preventing infection of peanuts is a form of biological control only if the buried sclerotia are destroyed by microorganisms.

The Antagonists

Any disease control in which antagonists are involved is biological control. Several examples have already been presented, and many more will be. Antagonism is now considered to include three types of activity:

1. Antibiosis and lysis. *Antibiosis* is the inhibition of one organism by a metabolic product of another. Although it is usually an inhibition of

growth, it may be lethal. The metabolite may penetrate a cell and inhibit its activity by chemical toxicity.

Lysis is a general term for the destruction, disintegration, dissolution, or decomposition of biological materials (Lamanna and Malette, 1965). Because of the variety of ways it can be produced, and the number of its effects on plant cells, confusion has resulted in the literature.

Although partial enzymatic digestion of the walls of living cells by other organisms external to them has been termed lysis by some workers, it is obviously a different effect than (b) below, and is here called *exolysis.* It is actually a form of parasitism or pathogenesis, and is considered in (3) below.

Endolysis is here considered to be a dissolution of the cell protoplast without prior or concomitant digestion of the wall, whether self-induced or initiated by outside agents. It can result from:

a. Internal metabolic changes produced by aging or senescence, nutrient deficiency, inability to use nutrients because of some unfavorable environmental condition such as deficient oxygen, or the accumulation of self-generated toxic metabolic products. These changes have been referred to as *autolysis* (Dean and Hinshelwood, 1966). It can also result from microorganisms, especially bacteria, that commonly proliferate around mycelia, particularly when leakage of contents is increased (Chapters 6 and 8). This is commonly referred to as *heterolysis. Microorganisms may be purely saprophytic on the surface of dead mycelia, or may be harmful to living mycelia by producing toxins, by increasing exosmosis, or by using the exogenous supply of oxygen, nutrients, or both.* Much of the endolysis observed in soil results from these processes.
b. Exposure to toxic materials, such as those produced by other organisms or from decomposing organic matter, or from fungicides applied by man. The effect of low dosages of carbon disulfide or of exposure to mild heat on mycelia of *Armillaria mellea* (Chapter 4) in weakening them and increasing susceptibility to *Trichoderma viride* are specialized examples. The effect of antibiotics is of this type, since they may penetrate and cause protoplasmic dissolution, collapse, plasmolysis, or rupture of the cell (Fig. 2.2). Huber et al. (1966) referred to this type of endolysis of bacteria as "bacterial necrosis." Bacteria may be inhibited by antibiotics produced by other bacteria (bacitracin), actinomycetes (streptomycin), or fungi (penicillin).

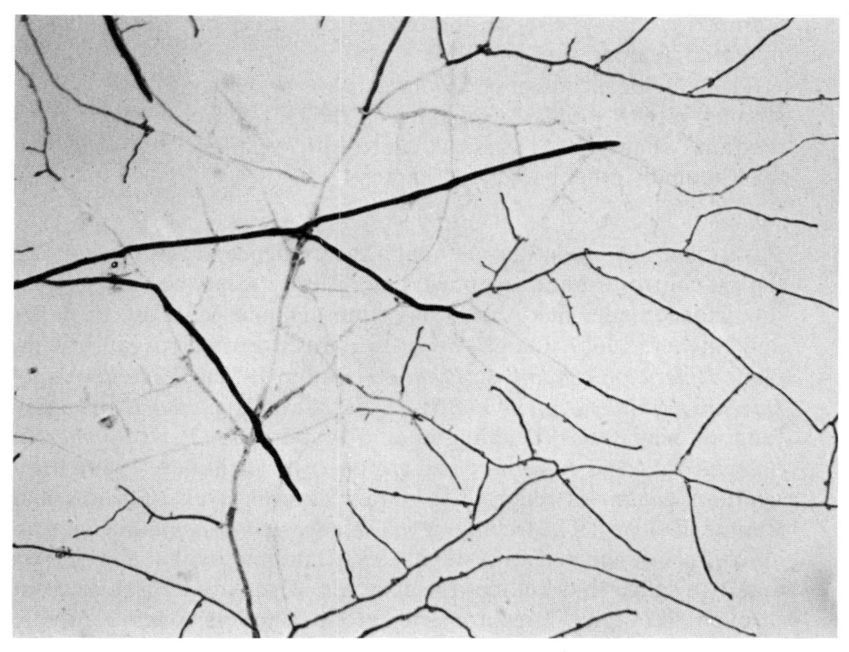

FIGURE 2.2
Lysis of mycelium of *Mycena citricolor* by *Trichoderma* sp. on agar plate, showing effect beyond area of contact. × 220. (Photo from Salas, 1970.)

c. Immunological reaction. Bacteria that have combined with antibodies produced by animals as a defense reaction to them may have the permeability of their surface membranes altered to such an extent that the contents leak out. This occurs with pathogens of animals and may be involved in the lethal cytoplasmic incompatibility ("killer reaction") between different strains of fungi in culture (Flentje, in Baker and Snyder, 1965; Hankin and Puhalla, 1971).

2. Competition. Competition was viewed by Clark (in Baker and Snyder, 1965) as "the endeavor of two or more organisms to gain . . . the measure each wants from the supply of . . . a substrate, in the specific form and under the specific conditions in which that substrate is presented . . . when that supply is not sufficient for both." In essence, competition is for nutrients (particularly high-energy carbohydrates, but also nitrogen), and possibly certain growth factors.

Competition may also be for oxygen or space, but not for water potential, temperature, or pH.

The biological control of *Fomes annosus* by inoculating freshly cut stumps with *Peniophora gigantea* is thought to result from competition (Chapter 10), as is the control of *Pseudomonas tolaasii* on mushroom by other bacteria (Chapter 4).

3. Parasitism and predation. Although the existence of this type of biological control is not questioned, there is uncertainty about its actual importance under field conditions (Boosalis and Mankau, in Baker and Snyder, 1965). Fungi known to parasitize other organisms include *Rhizoctonia solani* on *Pythium* (Butler, 1957), *Trichoderma viride* on *Armillaria mellea* (Bliss, 1951), several genera of trapping fungi on nematodes (Duddington and Wyborn, 1972), and *Tuberculina maxima* and *Fusarium roseum* on rusts (Kimmey, 1969). *Verticillium dahliae* is reported to parasitize even itself (Griffiths and Campbell, 1970). The free-living nematode (*Aphelenchus avenae*), an ubiquitous fungivore, thrusts its stylet into a hypha and injects digestive saliva that liquifies the contents, which are then sucked out through the stylet. Predatory nematodes such as *Seinura* rapidly paralyze other nematodes by injecting saliva and later sucking out the digested contents.

   Still a different type of parasite is the virus (phage) that infects a bacterium by penetrating the wall, multiplying within the protoplast, and then rupturing the wall, again liberating newly formed virus. This phenomenon is an example of endolysis. Phages are specific and common, and undoubtedly reduce the number of bacteria, but man thus far has been unable to make use of them for control of pathogens.

   The partial digestion of the cell walls of fungi, frequently referred to as lysis, is actually a type of pathogenesis by an external organism, and is here referred to as exolysis. Mitchell and Alexander (1962) reported that the addition of chitin to soil infested by *Fusarium oxysporum* f. sp. *cubense* stimulated *Bacillus* and *Pseudomonas*, which gave control of the pathogen by digesting its walls, which also contained chitin. Okafor (1970) failed to confirm this.

The boundaries between the three types of antagonism are quite blurred. Bacteria may hasten the demise of mycelia in a series that ranges from accelerated death, which may involve competition for the substrate, through endolysis of cell contents by antibiosis, to parasitism and exolysis, in which the cell walls are digested.

This chapter has developed most of the features of biological control and has excluded some factors that are often included. An important additional exclusion remains. Man cannot act as the antagonist. *Although man alone of the organisms involved can consciously control the environment, he is not an active participant in the interactions involved.* Furthermore, if man were to be included in biological control, it would expand the subject to include practically all control measures against plant disease, such as spraying to kill the pathogen, soil sterilization to produce a biological vacuum, and chemical seed treatment.

A PLANT PATHOLOGIST'S DEFINITION OF BIOLOGICAL CONTROL

Taking into account the preceding inclusions and exclusions, a unified concept of biological control may be stated:

Biological control is the reduction of inoculum density or disease-producing activities of a pathogen or parasite in its active or dormant state, by one or more organisms, accomplished naturally or through manipulation of the environment, host, or antagonist, or by mass introduction of one or more antagonists.

This definition may be paraphrased for two specific examples: Biological control of pine-stump infection is the reduction of disease-producing activities of *Fomes annosus* in its active state, by *Peniophora gigantea,* accomplished through competition from the mass introduction of the antagonist. Biological control of wheat seedling blight is the reduction of the disease-producing activities of *Fusarium roseum* f. sp. *cerealis* 'Graminearum' in its active state, by the wheat plant, accomplished through manipulation of soil temperature.

It will be noted that this concept is necessarily broader than that used by entomologists for insect control, or by many plant pathologists.

The objective of biological control of plant pathogens is the reduction of disease by:

1. Reduction of inoculum of the pathogen through decreased survival between crops, decreased production or release of viable propagules, or decreased spread by mycelial growth.
2. Reduction of infection of the host by the pathogen.
3. Reduction of severity of attack by the pathogen.

COMPARATIVE APPROACHES TO BIOLOGICAL CONTROL OF PLANT PATHOGENS AND INSECTS

Because of the traditionally close association of plant pathology and entomology, comparison of the approaches to biological control in these fields may be fruitful, although the objectives and methods are quite different.

Biological control of insects has been largely by predators and parasites, which are actively mobile, seek their prey, and operate on the basis of a single predator or parasite against a single prey. Biological control of plant pathogens is largely through antibiosis and competition, apparently with infrequent effective hyperparasitism. Antagonists are largely passive and nonmobile, make accidental contact with the pathogen, and operate in mixed groups more than as individuals. Because of these differences, the concept of density dependency (repressive environmental factors intensify as the population increases and relax as it declines, stabilizing the population) may have limited application in biological control of plant pathogens.

Entomologists have tried a predator or parasite, and then analyzed the reasons for its success or failure. Plant pathologists have studied individual pathogen-antagonist interactions, but often have not tried them in the field. Entomologists freely introduce foreign predators or parasites when native ones have not been effective; pathologists have depended on native antagonists and attempted to strengthen their effectiveness by manipulation of the environment. They are just beginning to introduce antagonists into soil successfully.

The above points suggest that plant pathologists should: (a) reverse their approach, that is, try the antagonist(s) without awaiting complete knowledge; (b) try introducing known antagonistic microorganisms into soils where they do not normally occur; (c) try mass transfer of the whole microflora into soils.

Over the years, plant pathologists, faced with the inherent difficulty of using microscopic agents, have developed biological controls empirically by emphasizing tillage practices that manipulate the environment. They have made effective and widespread use of pathogen-free seeds and planting stock, as well as sanitation practices, and breeding for resistance to pathogens has been extensively utilized. Comparatively, entomologists have made much less use of these methods.

Pathologists have, however, tended uncritically to attribute the successes of nonchemical controls to direct action on the pathogen rather than to biological control, denying the role of antagonistic microorganisms.

Many successful nonchemical controls, however, depend on resident antagonists (Chapter 4).

A promising but largely unexplored area of biological control is the combined use of insects and bacteria or fungi in the control of weeds (Wilson, 1969). The sensational effectiveness of the moth, *Cactoblastis cactorum*, in the control of *Opuntia* cactus in Queensland in 1920–40 is an example (Dodd, 1940). Larvae of this insect tunnel into the cactus pads, which then develop soft rot, much as does *Carnegiea gigantea* attacked by *Cactobrosis fernaldialis* and *Erwinia carnegieana* in Arizona. In each case, the insect acts as an efficient vector of the bacteria, as well as producing direct injury.

APPLYING BIOLOGICAL CONTROL

Biological control rarely eliminates a pathogen from the site, but rather reduces its numbers or its ability to produce disease; such control may be achieved with little or no reduction in population of the pathogen, or perhaps without preventing infection. If an antagonist eliminated the pathogen, it might itself be eliminated from the ecosystem, particularly if the pathogen was a preferred substrate. If the pathogen should then return, it could be destructive because of the absence of antagonists. For this reason drastic treatments tend to produce wildly fluctuating disease levels, whereas biological control tends toward stability at a low level of disease. The resulting losses may be so slight as not to require special control procedures.

When suitable antagonists are already present in the soil but do not provide a satisfactory level of disease control, it is desirable to intensify their activity, at least during the period when such control is needed. This may be accomplished by one or more of the following methods:

1. Crop rotation is the oldest and best known example of biological control (Chapters 4 and 8). It generally lowers inoculum density of the pathogen, but in some cases it may actually have the long-term effect of prolonging disease loss by delaying return to a natural balance, as in the take-all disease of wheat.

2. Adding amendments that selectively stimulate the antagonists (Chapters 4 and 8).

3. Altering the pH of soil to one favorable to antagonists, unfavorable to the pathogen, or both, by adding sulfur or lime or by the selection of proper fertilizers (Chapter 9).

4. Employing tillage methods that modify soil structure, crumb size, or aeration (Chapter 9).

5. Selecting a planting date that will expose the principal infection courts to the pathogen at a time when the environment is more favorable to the antagonists or the host than to the pathogen (Chapter 9).

6. Applying organic amendments in such a way as to reduce, by their decomposition, the available nitrogen at the infection site (Chapter 8). Fertilizer placement should be outside the infection court so as not to nullify this effect.

7. Employing irrigation practices that will maintain a soil-water potential favorable to antagonists at the infection site (moist for bacteria, drier for actinomycetes), but will insure water available for the host (Chapter 9).

8. Using trap and inhibitory plants to reduce injury from nematodes and some fungus pathogens (Chapter 8).

9. Selective treatment of soil to reduce or eliminate the pathogen as well as reduce competitors of the antagonists, but leave as many antagonists as possible (Chapters 5 and 9).

10. Adding supplemental antagonists, probably inoculated on a substrate selectively favorable to them (Chapter 5). Such antagonists may also be introduced by inoculation of the seed (Chapter 4).

11. Maintaining a favorable environment for the antagonist by applying to the foliage a translocatable material that may be exuded into the rhizosphere. This method might be a combination of (2) and (10) above, but would primarily stimulate activity against pathogens in their aggressive state.

12. Selecting a method of culture that favors the antagonists at the soil depth where infection by the pathogen occurs (Chapter 6). Thus, *Sclerotium rolfsii* and some strains of *Rhizoctonia solani* infect very near the soil surface, but *Phymatotrichum omnivorum* and *Pythium* and *Phytophthora* may be active at depths of 1–3 m, as well as at the surface.

13. Manipulating the environment to produce maximal resistance of the host, as discussed below.

Utilization of the Host

A common objective in breeding for resistance is prevention of infection of the pathogen, often by a hypersensitive host reaction (Chapter 8). This may be the proper approach with obligate parasites, which usually cause damage once they are established in a plant. For many plant-microorganism associations, however, establishment within the plant tissues is not in itself harmful. The association may be beneficial (mycorrhizae), but for most, no positive or negative effect can yet be assigned. Hollis (1951) thus showed that bacteria are common occupants of the internal stem tissues of potato but cause no apparent damage (Chapter 7). *Fusarium* species are very common occupants of cortical tissue of roots (Cook, 1969) and may cause no visible harm, or the slight necrosis causes little or no reduction of crop yield. Many such cases could be cited. When does a microorganism become a pathogen? The relationship between host and occupant may be commensal, perhaps even symbiotic. The biological balance described in Chapter 1 for the total biota may apply on a reduced scale to situations where one organism lives within another. Damage may occur only when the host is stressed by an unfavorable environment upsetting the balance between plant and occupant (Chapter 9).

The development of varieties best adapted to a given environment is a common objective of plant breeders. The tillage practices used by growers commonly are those that modify the environment to favor the given variety. In both these practices, an underlying objective may be to produce plants under minimal stress, and therefore least susceptible to damage from facultative types of microorganisms. This is actually an effective method of applying biological control.

3

BIOLOGICAL CONTROL IN PLANT PATHOLOGY

Relationships may be found long before they are understood, but in all cases it is desirable to be assured of their truth before attempting to explain them.

—SIR RICHARD GREGORY, 1916

. . . biological control is much neglected, not because it does not work but because not enough research is done on it. Biological control has worked, is working and can, if we desire it, greatly extend its successes. . . .

—G. ORDISH, 1967

Because there is demonstrable biological balance in nature (Chapter 1) maintained through a network of intricately interacting organisms and the abiotic environment, biological control, which involves the same mechanisms, must be efficiently functioning under natural conditions. Some level of biological control must have been operating in nature or the many wild plants susceptible to various pathogens would not have survived. Consistent with the principles of biological control, the pathogens have not been eliminated, although their activity has been curbed. When man briefly restrains his constant efforts to change nature, a new balance may be established, as in the diminishing severity of take-all disease under continuous wheat cropping, or of common scab under continuous culture of potato (Chapter 4).

An infinite number of associations occurs in nature before a viable parasite-host-environment combination appears. That this is indeed a rare event is shown by the small number of disease-producing combinations relative to the number of possibilities that must have occurred in the last million years. Perhaps it is only when the biological balance implicit in this situation is upset in some way that significant disease loss is produced. It would be easier, therefore, to reduce disease loss by environment alteration if the soil were not so well-buffered that lasting significant changes are difficult to achieve.

We know of these successful disease complexes because of the diseases they produce—a positive observation. It is reasonable that a completely suppressive antagonist-pathogen relationship may be equally rare. Because such a successful antagonistic effect would be evident only by the absence or reduction of a disease, there is nothing to see—a negative observation. Man is too busy coping with the problems he has to spend much time or money discovering those he has not. It is rather like seeking a saprophytic virus that can only be detected by a disease that it cannot produce. Probably for this reason it is seldom realized that effective biological control *is* working. Man perhaps attributes a relatively small disease loss to favorable environment or superior cultivation practices rather than to obscure organism interactions difficult to observe.

Biological control must be common, because if all biotic and abiotic environments were favorable to pathogens, it is doubtful that successful agriculture would be possible. Such protection is obviously imperfect, because if all environments were unfavorable to pathogens, there would be no plant diseases. We need only recall the familiar example of pathogenicity tests in "sterilized" versus nontreated soil to be reminded of the influence biological control has on the activities of root pathogens. The increased destructiveness of facultative-type organisms on a crop plant under environmental stress (Chapters 1 and 9) illustrates the influence on disease of active biological processes within the plant. **Pathogenicity** (the production of disease) **is the exception rather than the rule. Parasitism appears to be the norm for some microorganisms, but pathogenicity is abnormal** and perhaps occurs only when natural biological control is diminished by environmental disturbance. Most pathogens are restricted to a few host species, and without biological control, the roster of plant pathogens and their host ranges probably would be much larger than they are.

THE STATURE OF BIOLOGICAL CONTROL OF PLANT PATHOGENS

There are numerous examples (Chapter 4) of effective biological control of microorganisms, either natural or from man's activities. Furthermore, there are abundant examples of biological control of pathogens of large organisms (man and other animals), as well as of the large organisms themselves (insects, wildlife, weeds). Nevertheless, some plant pathologists have a hopeful skepticism that biological control of plant disease will ever be more than a novelty studied by petri-dish biologists, with little chance of success under field conditions. It has been referred to as "empirical, unreliable, and inextensible," and it was stated in 1962 as a principle of plant pathology that "Biological control plays a minor role in the control of plant disease." "Biological Control—Mission Impossible?", the title of a symposium at the 1971 annual meeting of the American Phytopathological Society, fairly expresses this viewpoint.

There are a number of explanations for undervaluation and neglect of biological control by plant pathologists:

1. There has been a tendency to equate biological control with the complicated application of antagonistic microorganisms, overlooking other more successful methods. The success of entomologists in mass-rearing and releasing predacious and parasitic insects (DeBach, 1964; Huffaker, 1971; Ordish, 1967) may have led to overemphasis of this approach by pathologists. Greater success has been attained with, and more emphasis placed on, host resistance, crop rotation, cultivation practices, and crop management by pathologists than by entomologists. However, pathologists have tended not to attribute any significant part of these successes to biological control.
2. Plant pathologists have worked under the notion that fairly complete comprehension of each ecological situation is necessary before biological control by antagonistic microorganisms is possible. The widespread view that "there are no shortcuts to biological control" has undoubtedly discouraged many from working in this field. To this end they have collected incidental facts in the hope that when this information attained a critical mass, biological control would appear. They have disregarded H. Poincaré's comment that "Science is built up of facts, as a house is built up of stones; but an accumulation of facts is no more a science than a heap of stones is a house." Very few pathologists have attempted either to analyze instances where biological control is naturally working or to apply biological control under field conditions. It has largely been made a laboratory and glasshouse exercise. A biological-control entomologist, D. A. Chant, commented at the 1963 International Symposium in Berkeley, California, "One thing that worries me is how this wealth of complex, precise, quantitative information can be brought together in any practical way to achieve biological control of specific diseases in specific circumstances."

Entomologists have operated on the philosophy that the only way to determine whether a parasite will be effective in a given location is to try it there. As Huffaker stated, "we cannot wait for the extensive long-term studies of life systems . . . envisaged . . . as necessary before we can develop new pest-control procedures . . . certain key factors can be discovered qualitatively, if not quantitatively, by analytical, perceptive, experienced ecologists, which can be exploited and control programs built around them" (Rabb and Guthrie, 1970). Although plant breeders consider host resistance to pathogens as a form of biological control, they have concentrated on the simpler direct host-parasite effects and have rarely studied host genes that may influence interactions among microorganisms (Chapter 8). Similarly, disease control through application of organic amendments or tillage practices is tried, and promptly put to use if it works; the mechanism involved may or may not then be studied. Western Regional Project W-38, "Nature of the Influence of Crop Residues on Fungus-Induced Root Diseases," (Cook and Watson, 1969) thus investigated this subject in 1956–71. On the other hand, biological control of plant pathogens, particularly by introduced antagonists or by purposeful management of resident antagonists, has been approached cautiously by attempting to understand the mechanism before testing the practice. As Fox (in Baker and Snyder, 1965) commented, "Since problems of biological control of root diseases are so intricate, it is inevitable in our present state of knowledge that chance or accidental observations may still lead to solutions where refined and methodical approaches fail."

3. Antagonist populations have often been sought in areas where both the pathogen and disease occur, rather than where they do not, or where the pathogen is unable even to persist. A fertile source of antagonist populations is an area long in monoculture where the disease has declined to a low level. In some instances the continued presence of the host-stimulated pathogen may provide a food base for antagonists, increasing their numbers and reducing the amount or severity of disease (Chapter 6). Antagonists should be sought in the rhizosphere rather than in the soil mass, as their effective activity probably will be on the root surface. Better still, they may be isolated from the surface of hyphae (particularly of lysed hyphae) of the pathogen on the root surface. A change in method of investigation may lead to increased effectiveness if it takes interactions into account. The recent upsurge of promising systemic fungicides was thought (Bent, 1969), for example, to result from the change

from screening fungicides on spores or mycelia of a few fungi in artificial media, to their use on specific pathogens on their hosts.

Antagonist populations to a given pathogen or disease occur only in soils biologically suppressive to the pathogen, but individual antagonists may occur in many soils. A biologically suppressive soil probably can not be explained in terms of a single antagonist. The area sampled for antagonists will depend, then, in part on whether multiple or individual antagonists are sought.

4. Emphasis has sometimes been placed on unpromising antagonists. The mycological heritage of pathologists has often caused them to emphasize fungi, although many of these tend to occupy a restricted area of the soil, often extending only a few centimeters from the sparsely distributed germinated spores. Bacteria and actinomycetes are better distributed through the soil (Jones and Griffiths, 1964), perhaps because the cells are carried downward more efficiently by percolating water and moved more readily by the soil microfauna. Furthermore, fungi multiply much more slowly than do bacteria or actinomycetes. Fungi grow at lower water potentials than bacteria, but most of them not as low as actinomycetes. A combination of compatible bacterial and actinomycete antagonists might compensate for the individual deficiencies. Finally, spore-forming bacteria tend to be more heat resistant than most fungi, permitting use of selective thermotherapy (Chapter 5).

5. Although soil microorganisms rarely stage solo performances, many studies have assumed that a single antagonist introduced into soil should be sufficient (Chapter 11). It appears from present data that individual antagonist species may be effective in nearly sterile soil or when applied to seeds or propagules favorable to their growth, but that a complete microflora or multiple antagonists are best for nontreated soil. The use of a suppressive soil as a "starter" to inoculate soil of near-sterility (this chapter) has not been sufficiently exploited.

The same situation obtains for aerial pathogens. Etheridge (in Nordin, 1972), reviewing work with wood-decay fungi, suggested the application of two or more antagonists that (a) are able to intensify antagonistic activity of each other, (b) possess different antagonistic mechanisms, (c) have different substrate preferences, and (d) are insensitive to antibiotics produced by the pathogen. These differences undoubtedly contribute to the stability of protection afforded by mixed populations of antagonists. Direct application of a single specific antagonist to a wound was thought, in contrast, to be unstable and inconstant over the protection period. These same relationships exist for soil microorganisms.

6. In searching for individual antagonists, too few (often 1–10) candidate microorganisms have commonly been used. Because of the many characteristics required for success (wide-spectrum effect against pathogens, nontoxicity to the host and other antagonists, adaptation to the given environment, ability to survive in the rhizosphere, ease of commercial handling), it is necessary to work with a large number of candidates. Plant breeders, for example, expect to screen thousands of progeny to find a few with the desired characteristics. Pesticide manufacturers screen 10,000 compounds at a cost of about $10,000,000 to find one that meets specifications (Johnson, 1971). S. A. Waksman indicated that from 10,000 cultures tested, only one that produced streptomycin was selected. Available 1969 figures for the United States indicate that of 485–520 parasites and predators of 77–80 crop pests introduced since 1884, 3.8–19.6% effected some degree of control. Up to 1969, worldwide attempts to import natural enemies of 223 insects provided substantial control for 40.3% of them and partial control for 13.5% more (DeBach, 1971). Searching for antagonists is no less complicated than these examples, and large numbers of organisms must be screened. If these candidate organisms can be simultaneously screened against several pathogens instead of a single one, greater economy of effort will be achieved.

7. Some studies on antagonists have been conducted only on agar media and are not carried further in soil tests. Agar tests supply presumptive assays for antagonism but have little value unless followed by tests in soil (Chapter 5). Dilution plates of nontreated soil may give an erroneous idea of the kinds and numbers of antagonists present, depending on the kind of media used. Selective heat treatment of soil before plating will permit a fair appraisal of *Bacillus* spp., and use of modified King's medium B (Sands and Rovira, 1970) gives an approximation of *Pseudomonas* spp. present.

8. The conditions of the soil test may be excessively drastic. The inoculum density of the pathogen may be raised to unnaturally high levels by addition of cultures when antagonists are tested in soil, "swamping" the biological control and giving the erroneous conclusion that the antagonists are ineffective. **Inoculum density of the pathogen in experiments should not exceed that found under severe natural disease conditions.** Dormant propagules or hyphae in soil or

plant tissue rarely exceed 10,000 per g of soil (Chapter 6), and are not often more than a fraction of a percent of the soil mass. Environmental conditions of the test may also be unfavorable to the antagonists. Soil temperature may be too low, or the soil too dry or too high in nitrogen for the antagonists to be effective (Chapter 9). Antagonists should be applied to nontreated soil in high population numbers, and with amendments favorable to their growth, or with the crop in whose rhizosphere they will develop.

We suggest, in view of the above considerations, that a successful approach to biological control of plant diseases would be a direct and somewhat empirical one: *Seek antagonists (either individuals or populations) where the disease does not occur, declines, or cannot develop. Try transferring antagonist populations, or mixed isolates of them, to conducive soil. Screen large numbers of candidate antagonists and test the more promising ones in laboratory and field trials as pure or mixed cultures.* There are ample precedents for suggesting this approach (see 2 above).

9. The test soil may lack the necessary substrate for production of effective quantities of an antibiotic by the antagonist, or the antibiotic may be sorbed or inactivated by other organisms (Chapter 7).

10. The emphasis on killing microorganisms by application of chemicals has been notable in plant-disease control. The idea of disease control through killing microorganisms by broad-spectrum methods such as chemicals or heat was developed after 1867. By 1900 the dominant idea of soil treatment was destruction of microorganisms by carbon disulfide (1869), steam (1893), or formaldehyde (1900), and this concept of overkill, unquestioned until recently, still largely prevails.

    The effect of a similar changed viewpoint is illustrated in the field of medicine. The idea that antiseptic chemicals had to kill microorganisms in order to prevent human infections persisted until the 1930s. It was then shown that sulfanilamide inhibited growth of bacteria and thus cured infections, but did not itself kill the organisms. This was referred to by Hare (1970) as "one of the most important medical discoveries that has ever been made, for it . . . showed us how wrong had been our ideas about how a successful chemotherapeutic agent was likely to act." The widespread use of antibiotics followed shortly thereafter.

    The trend in chemical control of plant disease has been from broad-spectrum to more selective materials, and from killers to inhibitors. The replacement of mercurials in the 1950s with the more specific hexachlorobenzene for cereal smut control is one example. Other examples are dichloropropene-dichloropropane (D-D) soil fumigant (nematodes), Dexon (*Pythium, Phytophthora*), PCNB (*Rhizoctonia, Sclerotinia, Sclerotium*), benomyl, thiabendazole, oxycarboxin, and carboxin.

The idea of total control also dominated breeding for disease resistance in crops for a time. High-level, specific, monogenic resistance was sought until it became clear that for many pathogens, a lower-level, broader-spectrum, general, polygenic, stable resistance was more effective. Today it is realized that this polygenic ("horizontal") resistance, because of its effectiveness against more strains of the pathogen, often gives less total loss from disease than does monogenic ("vertical") resistance. The use of genetic multilines is another evidence of this trend.

11. There has been insufficient purposeful research to provide a foundation for understanding biological balance and methods for manipulating it. For example, there has been little attempt to intensify or direct the rhizosphere flora or propagule dormancy in soil. There have not been enough attempts to produce slight but effective changes in soil microflora that will reduce disease incidence along lines shown for common scab of potato (Chapter 4), take-all disease of wheat (Chapters 4, 8, and 9) or *Verticillium albo-atrum* in potato (Chapter 7).

Biological control is rarely spectacular, and usually goes unnoticed when it does operate. Almost all plant pathologists have, however, observed unexplained situations in which a pathogen has failed to establish in a field soil or declined rapidly or over a period of years after its introduction, or in which a disease failed to appear in a few rows or part of a field previously planted with mixed crops.

RESIDENT ANTAGONISTS

One school of thought considers that field soils contain the necessary antagonists to control almost all root pathogens, and that occurrence of disease merely shows that some necessary environmental condition is lacking. We do not now, and probably will not for a long time, have the necessary information to confirm or refute this idea, but present evidence provides little or no support for it. For example, although legumes are both numerous and ubiquitous, it is necessary to inoculate seed with the proper strain of nodule bacteria (*Rhizobium* spp.) before planting. Indeed, some soils have microorganisms that may prevent or restrict nodulation, as in Western Australia (Holland and Parker, 1966). However, many soils may have very effective antagonists to some pathogens. Indeed, one of the theses of this book is that such localities should be sought out and exploited.

The discovery of penicillin in 1929 by Fleming, its purification in 1940 by a group at Oxford University, and the demonstration that it was useful against bacteria causing disease in animals (Hare, 1970), probably provided the single greatest stimulus to studies of antagonists against plant pathogens. These discoveries were quickly followed by the isolation of gliotoxin in 1936 and viridin in 1945 from *Trichoderma viride,* of streptomycin in 1944 from *Streptomyces griseus,* of chloramphenicol in 1947 from *S. venezuelae,* and of chlorotetracycline in 1948 from *S. aureofaciens.* More than 100 antibiotics were described between 1929 and 1949, 75 of them in the final 5 years. Significantly, most of the commercial antibiotics have been obtained from saprophytic soil microorganisms (Goldberg, 1959). If similar effort and expense were lavished on finding and using antagonists for controlling plant diseases, there is little doubt that success would be attained.

INTRODUCED ANTAGONISTS

Biological control through introduction of antagonists to the soil apparently was first attempted by Hartley in 1921 against damping-off of coniferous seedlings, Millard and Taylor in 1927 against common scab of potato, and Sanford and Broadfoot in 1931 against take-all of wheat (in Baker and Snyder, 1965).

Nontreated Soil

Biological control by introduction of antagonists into nontreated soil is difficult to achieve, because it attempts to establish an alien antagonist in a biologically buffered community. Although this is difficult, it can be done when the right organisms are obtained and are properly used.

To have achieved the biotic balance found everywhere in nature, countless new organisms must have invaded each site, some successfully, some not. One that fitted into the association better than some resident gradually displaced it. No association is forever. This is according to the competitive-exclusion principle that **different species having identical ecological niches** (that is, ecological homologues) **cannot coexist for long in the same habitat.** An analogy is provided by a flock of perching birds crowded on a power line: A new arrival attempting to alight will occasion much flapping and shifting about; if he is successful in alighting, some other bird must leave, if not he perches elsewhere. If a strong wind disturbs the balance of the birds, some may be dislodged, increasing the opportunity for a new arrival to find a place. An association of microorganisms in virgin soil must similarly change with time, and with man's disturbance of the environment (Chapter 1), this must occur frequently. The "right" microorganisms must continually be successfully invading, even as weeds appear in man's cultivated areas. All introduced plants do not become weeds, but the success of the occasional one refutes the statement that establishment cannot occur.

Our inability to see invasions by alien soil microorganisms does not reduce the probability of their occurrence. Perhaps we are observing the conseqences of it in the reduction of a disease under continued monoculture (Chapter 4). It is well documented that such a population shift occurs in soil recently treated, during mushroom composting, and in the decomposition of fallen leaves. That plant pathogens can establish and flourish in new ecosystems shows that alien species can invade an established biological community, particularly when favored by a host as a selective substrate. There is no valid reason why the same thing cannot occur with antagonists. It is as difficult to introduce a pathogen into nontreated soil without its host as it is to introduce an antagonist. Successful introduction of antagonists into nontreated soil simply requires extensive trials with many varied types under a wide range of conditions, that is, it should be done the way it occurs in nature. In view of the extraordinarily limited attempts to introduce antagonists into nontreated soil, it is understandable that complete success has not yet been achieved.

In attempting to introduce antagonists into nontreated soil, microorganisms isolated from soil have usually been screened for antagonistic properties, and selected individuals have been grown in mass culture. If they are reintroduced into the same ecosystem from which they came, they decline in numbers because the prevailing biotic and abiotic environment did not support the natural increase of the antagonist to such levels. Moreover, nutrients carried over with the organisms from the culture (the medium itself should not be added to the soil), and released by the death of many of the cells, may cause changes in the resident flora that may even hasten the return of the introduced species to its original place in the community.

Treated Soil or Freshly Cut Host Surfaces

In each of the examples (Chapters 4 and 10) of biological control by an introduced antagonist, the infection site is occupied by the antagonist before arrival of the pathogen, or seeds provide a favorable food base for antibiotic production. Such biological control is obviously easier to accomplish than if the antagonist is introduced some distance from the infection site. An even easier approach may be the introduction of antagonists into soil with a simplified microflora (for example, steamed, fumigated, desert, or polder soils) or onto freshly cut host surfaces. Since there is no established microflora, the antagonist luxuriates and may even become the dominant organism.

MANAGING THE BIOLOGICAL BALANCE

It should now be clear that the key factor in nearly all aspects of biological control is the biological balance achieved. Pathogens flourish in man's crops because his fertilizer, tillage, seeding, and weed-control practices, together with genetic stocks selected for production characters or for resistance to some other disease (Day, 1973), have combined to simplify the biological community and thus reduce its stability. On the other hand, an antagonist introduced into soil usually will not establish or function because of the existing biological balance. Crop rotations, organic amendments, and certain cultivation practices lessen the impact of a pathogen because they help intensify or shift the biological balance. The simplest place for establishing biological control is a situation that approaches a biological vacuum, for example, Rishbeth's stump inoculations against *Fomes annosus* (Chapter 10), and the introduction of antagonists into nearly sterile soil or medium in which plant propagules are to be grown (Chapter 4). However, the proper introduction of a suitable antagonist into field soil has been shown to be possible when the conditions created are sufficiently favorable. In short, development of an effective biological control depends on learning to manipulate the biological balance.

The microbiological balance of the soil can be altered by some major change in the environment, an ecological shock to stimulate, inhibit, or destroy some element(s) of the population. Treatment with some mild chemicals as fumigants (carbon disulfide) or drenches (PCNB) may upset the balance. Treatment with aerated steam may destroy the pathogens and many saprophytes, while leaving a flora of effective antagonists (*Bacillus* and *Streptomyces* spp., ascosporic *Penicillium* and *Aspergillus* spp.) (Baker, 1962; Warcup and Baker, 1963). Addition of a specifically favorable amendment (such as soybean green residue for *Bacillus subtilis* against potato scab) may have a similar effect.

Treating with high dosages of a potent fumigant (chloropicrin) or with regular steam will induce a condition of near-sterility, into which a microorganism (antagonist or pathogen) may be introduced without difficulty. Unless a mixed population of well-adapted organisms is used to inoculate such a soil (for example, using a quantity of the same soil which has been treated with aerated steam to free it of pathogens), a stable microflora will not result. When only one or a few antagonists are added, a gradual shift in populations will inevitably result, with random contaminants from air, tools, planting material, and so forth, playing a prominent role.

Large acreages of land are now treated annually with chemical fumigants, and steam or aerated steam is used routinely in glasshouses. Since chemical and regular steam treatments may lead to rapid recolonization by pathogens, and are largely used on high-value crops (strawberries, flowers), treatment is usually on an annual basis. Introduction of suitable antagonists into soil after treatment might protect plants in the field. That this can be done under glasshouse conditions has been demonstrated many times. If field treatments could by this means be decreased, for example, to an alternate-year basis, it would probably be economically feasible. What seemed impossible some years ago is now probable, even as the irrigation and fertilizing of wheat, considered economically infeasible 30 years ago, have now increased yields in eastern Washington manyfold to a world record of 209 bushels/acre (182 hl/ha). These yields make economically possible such procedures as soil treatment, formerly not even considered. Rising lumber values have also greatly expanded the range of feasible practices in forestry and forest pathology. An increase in production per acre or in the value of a crop may thus make possible the future application of biological control in areas where it is now considered impossible. As W. R. Whitney said, "The impossible is that which we have not yet learned to do."

The management of soil microorganisms through soil treatment for root disease control is more difficult to accomplish for long-term than for annual crops. Thus, soil to be used for fruit trees is generally treated with low-potency fungicides, and that for annual crops by high-potency materials.

Where the biological balance cannot be directly modified, perhaps we can play a numbers game. The more complex the biological community, the greater its stability. *The more varied and numerous the soil microorganisms, the greater the chance of biological control of a pathogen.* No microorganism, be it pathogen or saprophyte, dominates the scene; competition for nutrients is intense, oxygen may be in short supply in a microsite, toxic metabolites accumulate, and antagonism is widespread. Soil fungistasis is an expression of such a situation. The rapid lysis of fungus germ tubes is nonspecific and probably caused by the total intense microbial life around them. These microorganisms are stimulated by the same nutrients that induced the spore to germinate, and they replace the supply of nutrients and oxygen with metabolic wastes. The eventual death of propagules of pathogens (particularily hyphae in host tissue, but also sclerotia, chlamydospores, and other resistant structures) from prolonged exposure to microbial activity is also implicit in this concept. These points will be considered further in Chapters 4, 6, 7, and 9, but it should be noted here that some of the best examples of commercial biological control are of this complex-community approach.

Stability of the biological community through management should be selectively reinforced by tillage practices disadvantageous to the pathogen, or advantageous to the antagonist or the host, and by introducing new genes into the host for resistance and enhanced survival.

The trend of the last 70 years suggests that control procedures of the future will be ever more subtle and involve greater specialization in an integration of biological and chemical control. As suggested in 1959, "it is increasingly clear that mastery of the soil microflora will come only when we understand the nonpathogenic organisms, even though the pathogens have seemed the logical point of attack. We must see that the whole soil flora is in a state of fluctuating equilibrium, and that, in the worm's-eye view, no organism is more important than another" (Baker, in Holton et al., 1959). At the very least, we will strive to leave the nonpathogenic microorganisms as undisturbed as possible. More likely, we will more actively use the microbiota to control disease and increase production, and will learn to manage second-, third-, and fourth-order interactions. This will best be learned by empirically trying control methods and analyzing the findings, rather than by analyzing those variables of which we are aware and by trying to formulate controls from these limited data.

4

EXAMPLES OF BIOLOGICAL CONTROL

. . . the end result of an outstanding example of biological control is not spectacular and is likely to go unnoticed and unappreciated because the formerly abundant organism has been reduced to a rare species. . . . It is easy to overlook the results and to forget a problem when it has disappeared. Many striking examples of 'invisible' biological control are everywhere around us, and can be demonstrated experimentally. . . .

—P. DEBACH, 1964

To provide the necessary background before considering the methodology of biological control, we shall describe some examples in which it is functioning in agriculture.

BIOLOGICAL CONTROL BY RESIDENT ORGANISMS

There are numerous examples in which biological control is being accomplished by microorganisms that occur naturally in field soils.

Pathogen-Suppressive Soils

The inhospitality of certain soils to some plant pathogens is such that either the pathogens cannot establish, they establish but fail to produce disease, or they establish and cause disease at first but diminish with continued culture of the crop. These categories are used for convenience and do not imply that the three phenomena are regulated by distinct biological principles.

Ironically, in the search for practical systems of biological control, plant pathologists have tended to ignore pathogen-suppressive soils and the clues they present. Indeed, there has been some tendency to regard such soils as a nuisance to research, and to seek elsewhere for soils that do not present the problem of hostility to the pathogen. Obviously, not all such instances involve biological control, but many apparently do. Because pathologists have tended to view pathogen-inimical soils as an interesting but bothersome phenomenon, data on causes of the inhospitality are rarely available. However, growers intentionally capitalize on certain cases favorable to their crop production.

Failure of Pathogen to Establish. Soils of this type, referred to as resistant, long-life, immune, intolerant, antagonistic, or suppressive, have been noted for years for species of *Fusarium, Streptomyces, Rhizoctonia, Phytophthora, Gaeumannomyces, Fomes,* and probably others. Some examples have been published, but others are known only through personal communication, since negative results are rarely published.

East of the Cascade Mountains of the Pacific Northwest, fusarium root and foot rot of wheat, caused by *F. roseum* f. sp. *cerealis* 'Culmorum,' is destructive to wheat in sandy low organic-matter soils of low rainfall (25–40 cm or less). By contrast, large populations of this fungus (and hence the foot rot) are rare in the finer-textured, higher organic-matter soils of the Palouse area along the Washington-Idaho border, where rainfall reaches 50–65 cm per year. This distinction is so remarkable that wheat growers of this latter region generally ignore Culmorum when making decisions on varieties or farm-management practices (R. J. Cook, unpublished). An exception occurred in 1966, when a field near Pullman, cropped to oats in 1964 and 1965, developed a high inoculum density of Culmorum. A wheat crop in 1966 showed severe fusarium foot rot, and the Culmorum population in the field immediately after the crop approached 2000, 3000, and in some areas, 10,000 propagules per g (Cook, 1968). Wheat grown in this field in 1968 and 1970, however, showed no effect from Culmorum. Population counts indicated that during the two years after 1966, the fungus had gradually disappeared from the field. Similarly, in a field plot at Pullman, a population of 5000 Culmorum propagules/g, introduced experimentally to a depth of 15 cm in September, 1965, gradually declined to undetectable levels over a 24-month period, in spite of a wheat crop grown on the plot in the interim. In the sandy soils of the drier areas, Culmorum populations seem to persist indefinitely (R. J. Cook, unpublished).

Differences in amounts of rainfall can be correlated with the differences in fusarium root and foot rot in these two regions. In addition, soils of the dryland area, and particularly those of the Ritzville series, are more favorable for Culmorum under controlled conditions in the laboratory and glasshouse than those of the Palouse series. Chlamydospores of the fungus germinate in lower percentages and lyse more rapidly in Palouse silt loams than in Ritzville silt loams. Even the effects of higher rainfall may be through biological factors, perhaps through increased germ-tube lysis (Chapter 6) or greater host resistance (Chapter 9). The significant point here is that differences in soil type effect a form of biological control on Culmorum, to the extent that one of the world's most fertile wheat-growing areas, the Palouse, can grow wheat with little concern for this *Fusarium*.

The best examples of failure of a pathogen to establish or function probably are the so-called wilt-resistant soils. Walker and Snyder (1933) noted the phenomenon in Wisconsin for pea wilt caused by *F. oxysporum* f. sp. *pisi.* Wilt spread rapidly and established in Carrington and Miami loams, but made little progress in a Colby silt loam, and was never observed, nor would it establish, in Superior red clay. Climate was not responsible for the differences in wilt severity, because Superior red clay would not support wilt development even if transported to and tested in field sites adjacent to the Carrington or Miami silt loams. This fusarium wilt has been found in California only in two small coastal areas, even though susceptible pea varieties have been widely grown, and although many hundreds of tons of pea seed carrying spores of *pisi* have been brought in over the years and planted for canning, freezing, and seed production. If soil from the large suppressive areas is steamed, *pisi* can then be established in them (W. C. Snyder, unpublished). Apparently there is a somewhat similar field situation with this disease in Illinois and in New South Wales, Australia.

Reinking and Manns (1933) showed the clear-cut existence of soils "tolerant" and "intolerant" to *F. oxysporum* f. sp. *cubense* in Guatemala, Honduras, Costa Rica, and Panama. One hundred percent of the banana plants would succumb to Panama disease within 3–4 years of planting on some soils, yet on others often only a few yards away, plants would remain healthy for up to 20 years. Soils of so-called short-life (wilt in 3–10 years), intermediate (wilt in 20 years), or long-life (more than 20 years required for severe wilt) types have long been important to the banana industry in Central America in selecting fields for new plantings of susceptible varieties.

Other examples of soils suppressive to the wilt pathogens and other fusaria are also known. Wensley and McKeen (1963) showed that wilt of muskmelon in Ontario, Canada, is consistently more severe in Fox sandy loam than in a Colwood loam. Smith (in Cook, 1969) noted that *F. oxysporum* f. sp. *melonis* produced hyphal growth in "wilt" but not in "nonwilt" soils in California. *Fusarium oxysporum* is prevalent on pine seedlings in forest nurseries of California, but disappears from the seedlings when transplanted to native forest soils; it is also conspicuously absent from the rhizosphere of pine seedlings in natural stands (Toussoun et al., 1969). Fusarium foot rot of bean (caused by *F. solani* f. sp. *phaseoli*) occurs in some but not in other soils of the irrigated Columbia Basin in Washington. The pathogen quickly forms chlamydospores from conidia and persists in one soil, but makes extensive mycelial growth, does not readily form chlamydospores, and endolyses in the other soil (Burke, 1965a,b).

Soils of the coastal area near Castroville, California, are conducive to fusarium wilt of peas, crucifers, and other wilt-susceptible crops grown there, but soils in the Salinas and San Joaquin Valleys are suppressive to fusarium wilt of tomato, peas, sweet potatoes, or other wilt-susceptible crops grown in those areas. Smith and Snyder (1971, 1972) showed that soils suppressive to one forma specialis of *F. oxysporum* may be suppressive to other formae speciales, but conversely, wilt-conducive soils are favorable to many different formae speciales. A soil from near Castroville, tested in the glasshouse, was thus conducive to development of wilt in sweet potato caused by *F. oxysporum* f. sp. *batatas,* even though the Castroville area itself is climatically unsuited for and does not produce sweet potatoes. Wilt development in the conducive soil was possible with only 50 propagules of *F. oxysporum* f. sp. *batatas* per g, whereas one-third of the plants remained wilt-free in suppressive soil from other areas when 5000 propagules/g were used. In addition, the pathogen multiplied in conducive soil, but remained at a steady population level in suppressive soil, even after two consecutive plantings of sweet potato cuttings.

Causes of the so-called *Fusarium*-resistant and -tolerant soils are about as poorly known today as they were 40 years ago, when the first observations were made. Certain physical and chemical properties of the soil have been shown to correlate with this tolerance to *Fusarium,* but these are only correlations and do not hold in every instance.

The work on soil influences on banana wilt has been reviewed by Stover (1962). Although fusarium wilt is generally associated with acid soils, some alkaline soils permit rapid development of the disease. As with texture, the proportion of clay, particularly montmorillonitic clay (Stotzky and Rem, 1966–67), generally correlates directly with intolerance to *Fusarium.* However, exceptions occur, in that some soils conducive to banana wilt are high in clay, including the montmorillonitic type.

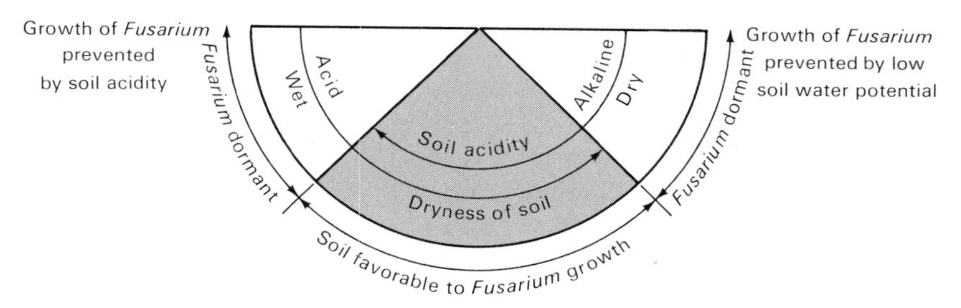

FIGURE 4.1
Schematic diagram showing the favorableness of low water potentials and of soil acidity to growth of *Fusarium*. Soil acidity may compensate to some extent for a high water potential, and a low water potential for an alkaline soil, but there are upper and lower limits for each factor.

It seems that key soil factors, such as pH and texture, are important, but that their effects are subtle and may be masked if other soil factors dominate. Stover stated, "In local areas in which soil texture varied, other soil characteristics being similar, texture was important." Moreover, it is interesting to note, not only for fusarium wilt of banana, but for many other fusarium diseases as well, that alkaline clay soils are least favorable and acid sandy soils most favorable. The tendency may be that a sandy texture of soil offsets the suppressive nature of an alkaline soil, and that acidity offsets the unfavorable nature of a clay coil (Fig. 4.1).

Alkaline soils would be relatively more favorable to soil bacteria than to the fusaria. Sandy soils, on the other hand, have a low water-holding capacity, are quick to drain, dry, and aerate, and thus may help the fusaria escape from, or at least compete with soil bacteria. Conversely, clay soils would remain favorably moist for bacteria over longer periods, and thus relatively unfavorable to fusaria unless some advantage, such as soil acidity, was provided them. It is well known that germlings of *Fusarium* can grow in dry soil where soil bacteria generally cannot multiply or compete effectively (Chapters 6, 7, and 8; Cook and Papendick, 1970). Thus, it may be more than coincidence that in the studies of Toussoun et al. (1969) with pine soils intolerant to *F. oxysporum,* Smith (in Cook, 1969) with soils suppressive to *F. oxysporum* f. sp. *melonis,* and Burke (1965b) with soils suppressive to *F.* solani f. sp. *phaseoli,* lysis of germ tubes of the respective fusaria was greatest in the suppressive soils. Moreover, Smith and Snyder (1971, 1972) showed by plate counts in California soils that bacteria multiplied more rapidly in suppressive than in conducive soils.

This does not imply that all soils intolerant to *Fusarium* are so because they harbor specific soil bacteria. It seems more likely that the total bacterial activity contiguous to the germ tubes is detrimental to *Fusarium,* possibly because of preempting of nutrients and oxygen, elaboration of metabolic wastes, or both. P. W. Brian referred to this lysis of germ tubes as "early selective lysis." It may be endolysis triggered by the activity of soil bacteria (Chapters 2 and 6). Any adverse effect of soil alkalinity and heavy texture would thus be indirect, which would explain why these soil characteristics were not always adverse; the effects are mitigated by at least second-order interactions. Also, being nonspecific, the effect could be measured only in terms of a final outcome after many seasons of interaction between the fusaria and soil bacteria. Some individual germlings of each *Fusarium* undoubtedly succeed in situations highly favorable to bacteria, and fail in situations unfavorable to them.

The soil environment influences the *relative competitive advantage* between these two groups of organisms and thus determines which is most successful. A soil factor unfavorable to growth of *Fusarium,* but even more unfavorable to that of soil bacteria, obviously would be more to the advantage of *Fusarium* than to the bacteria. Conversely, a factor more unfavorable to *Fusarium* than to bacteria would have an inverse long-term effect and be more to the competitive advantage of bacteria. **An organism that is relatively more favored than another by a given environment will be the more successful in a specific ecological niche.** This applies to interactions between crop plants and pathogens (particularly the facultative type), and between microorganisms in general.

Garrett (1970) gave a similar explanation why colonization of wheat straw by four cereal foot-rot fungi may be maximal at 10° C, when their growth in pure culture is optimal at 25° C: "a soil temperature of 10° C depressed the overall inoculum potential of the competing saprophytes more than it depressed that of any of the four fungi. . . ." The effect is also similar to that reported by Leach (1947) for seedling blight of different crops caused by *Pythium* and *Rhizoctonia,* and by Dickson (1923) for seedling blight of wheat and corn caused by *Fusarium roseum* f. sp. *cerealis* 'Graminearum'; a temperature that retards the rate of seedling emergence more than it retards the growth rate of the pathogen will result in increased preemergence damping-off (Chapter 9).

Nelson (1950) reported for *Verticillium albo-atrum* var. *menthae* on peppermint in Michigan that "some soils resist invasion by the wilt fungus while others apparently do not. . . . This resistance was destroyed by steaming." The level of suppression by some soils apparently was such that, even when mint rhizomes from fields with severe wilt were planted in such areas, it was several years before the disease appeared. In other fields the disease developed promptly and severely.

A strain of *Rhizoctonia solani* from the Eyre Peninsula of South Australia, where it causes an important bare-patch condition in wheat fields, was grown in cornmeal-sand culture and inoculated into a field at the Waite Agricultural Research Institute, Adelaide, at time of seeding (Baker et al., 1967). The pathogen grew abundantly and rotted the roots and stunted the wheat seedlings for a time, but within 4 months disappeared from the soil, and the plants recovered. This performance was repeated in another year. A crucifer strain of *R. solani* is, however, commonly present in both this soil and that of the Eyre Peninsula. The wheat isolate grew copiously and formed sclerotia in the Waite soil in containers covered with gas-permeable plastic for a month, and then died; other strains survived under these same conditions. The Waite soil thus showed selective antagonism to a strain of this fungus. Similarly, attempts to establish the lima-bean strain of *R. solani* from the Ventura area of California in field soil at Riverside proved unsuccessful. An attempt to introduce *Streptomyces scabies* from the Shafter, California, potato fields into field soil at Davis was also unsuccessful.

Pathogen Becomes Established but Fails to Produce Disease. A south Queensland 34-year-old avocado grove has beautiful, large, high-yielding trees growing in clay soil infested with virulent *Phytophthora cinnamomi* under extremely favorable climatic conditions (152 cm average annual rainfall; more than 254 cm in 1972). Root rot has been rare there in most years and was insignificant even in 1972. Trees in nearby groves have sustained slight to severe root rot during this period. This situation has been under study since 1969 (Broadbent et al., 1971; Broadbent and Baker, 1973a,b). These investigations and extensive surveys by K. G. Pegg (unpublished) have shown that suppressiveness of these soils to *P. cinnamomi* is associated with: (a) high content of organic matter, sometimes with a dark humus layer 23 cm deep; (b) high calcium levels (often 3000–4000 ppm), with the calcium tied up in the organic cycle; (c) pH of 6.0–7.0; (d) high levels of ammonium and nitrate nitrogen, also involved in the organic cycle; (e) adequate levels of phosphate; (f) very high populations of many types of microorganisms—a really "live soil"; (g) reasonably well-drained red basaltic soils; (h) relatively high magnesium content. Feature (g) resulted from the flocculated clay soil, enhanced by (a) and (b). Feature (f) undoubtedly reflected conditions (a) through (e), as well as the abundant, well-distributed rainfall.

The suppressiveness of these soils apparently results from complex microbiological interactions that provide biological control, and from a well-drained soil improved by 40 years of intensive cover-cropping and application of calcium, chicken manure, and superphosphate. Neither factor alone is sufficient to suppress the pathogen, but collectively they have prevented significant root rot for 34 years. The pathogen is not, however, eliminated from the soil.

This suppressiveness may be overcome by: (a) experimentally waterlogging the soil (this also occurs in an area of a drainage sump in the field); (b) addition of subterranean clover meal; (c) addition of excessive quantities of *Phytophthora cinnamomi* inoculum. Susceptible seedlings planted in soil that had been steamed (100° C/30 minutes) to destroy antagonists and inoculated with *P. cinnamomi* had severe root rot, despite the good drainage. Soil that was nontreated, or treated at 60° C and inoculated with the pathogen, gave no root rot. This situation therefore strongly involves biological control.

Phytophthora cinnamomi requires stimulation from soil bacteria (*Pseudomonas* spp., *Chromobacterium violaceum*) for profuse zoosporangial formation (Zentmyer, 1965). The stimulatory effect is eliminated from soil by aerated steam treatment (40–50° C/10 minutes) or by Millipore filtration of the soil extract. The bacterial cells normally increase, probably on the surface of *P. cinnamomi* hyphae, where the stimulatory material reaches effective concentrations. The concentration in the steamed or filtered soil extracts is apparently too low for stimulation. The antagonistic bacteria (*Bacillus subtilis* type) survive treatment at 60° C/30 minutes. Mycelia of *P. cinnamomi* added to extracts from the suppressive soil treated at 60° C was lysed by antagonists, and formed few chlamydospores and no zoosporangia. These antagonists cause mycelia to leak, and hyphal tips may swell and burst.

Zoospores are the principal infective structures of this pathogen, although mycelia may also infect, apparently mainly through wounds. Ayers (1971) showed, and this has been confirmed by C. L. Schoulties (unpublished), that the material that stimulates zoosporangial formation is nonvolatile, heat-stable, water-soluble, and effective in high (10^{-9}) dilution. Schoulties showed further that the molecule could be released from bacterial cells by treatment with EDTA and lysozyme, or by using old cells. The number of zoosporangia formed is directly related to the number of stimulatory bacteria present. Stimulators are abundantly present in all soils, including suppressive ones, regardless of the presence of *P. cinnamomi*. There are inhibitory agents in suppressive soils that apparently either inhibit stimulatory bacteria or break down their stimulatory compound(s); these agents are also inactivated at 60° C and seem not to occur in conducive soils.

The few zoosporangia formed in suppressive soils are attractive to antagonistic bacteria, which attach and cause lysis and discharge of the undifferentiated contents (Broadbent and Baker, 1973b). Soil suppressiveness to *P. cinnamomi* appears then to involve (a) antagonists that lyse hyphae and sporangia or perhaps "smother" mycelia in a bacterial sheath, (b) inhibitory agents that inactivate sporangial stimulators or their active metabolite(s), and (c) excellent soil drainage attained by tillage practices that also favor biological control (Broadbent and Baker, in Bruehl, 1974).

The suppressive soil also inhibits sporangial production by *P. citrophthora,* but not by *Pythium ultimum,* in laboratory studies, although mycelial growth of both was inhibited (Broadbent and Baker, 1973a).

Evidence now seems to favor the view (Anonymous, 1971) that *P. cinnamomi* is indigenous to at least the northern half of the eastern coast of Australia for the following reasons, although some (Newhook and Podger, 1972; Podger, 1972) favor its fairly recent introduction: (a) Trees and shrubs in Queensland have more indigenous species with a high order of resistance than do those in Western Australia, where the pathogen (a recent introduction) is causing a major epidemic in many forest species (Wallace, 1969). (b) The pathogen occurs in isolated forest areas where its spread by man is unlikely (L. R. Fraser, unpublished; Pratt et al., 1973; Shaw et al., 1972). (c) Both mating types of the pathogen occur in eastern Australia and New Guinea; the rare A1 type is not uncommon there (Pratt et al., 1972; Shaw et al., 1972). (d) The only known soils truly suppressive to *P. cinnamomi* are in south Queensland (Broadbent and Baker, 1973a).

An ecological balance between the various hosts, the pathogen, the antagonists, the stimulators and their inhibitors, and the abiotic environment seems to have evolved in the rainforests of coastal Queensland. The high organic matter, calcium, and ammonium and nitrate nitrogen content (a) favor antagonists and inhibitors nutritionally and by soil pH, as well as (b) improve soil aeration and drainage. (c) The pathogen would, however, maintain selective pressure on the vegetation, particularly in years of extreme rainfall when even well-drained soils become waterlogged, necessarily leading to development of resistance or tolerance in the flora. Such a triple means of preserving the biological balance would be a reasonable development, markedly superior to dependence on a single mechanism for achieving a vital objective.

Under these rainforest conditions, there would be a rapid recycling of nutrients. When the "bush" is cleared and crops planted, this favorable balance may quickly be lost. Organic matter is quickly lost in soil exposed to a tropical sun; calcium is then leached out (frequently down to 200 ppm), and nitrogen is lost. The pH drops to about 4.5, a level inhibitory to bacteria. Suppressiveness to *P. cinnamomi* is thus systematically diminished. Tillage practices maintained in the suppressive soils since they were first cultivated have aimed, consciously or unconsciously, toward maintenance of the soil conditions of the original vegetation (Broadbent and Baker, 1973a). The question whether it will be possible to apply these practices for effective control of the pathogen in areas where *P. cinnamomi* is not indigenous (and the antagonists and inhibitors are possibly lacking) remains to be determined.

Consideration of the relationship of continental drift to plant dispersal in Australasia (Schuster, 1972) affords as yet no evidence whether *P. cinnamomi* migrated north to, or south from, Asia after the Australian-New Guinea plate contacted the Eurasian plate in the late Tertiary (within the last 10 million years).

Another example of the effect of tillage practices in restricting activities of the pathogen is afforded by *Sclerotinia sclerotiorum* on leafy vegetable crops in the Santa Maria Valley, California (K. F. Baker, unpublished). Because of equable climate and high land rental fees, fields are continuously cropped, and are plowed and almost immediately replanted after each harvest. Masses of green debris briefly left in the fields after a lettuce, celery, or cabbage harvest are often overrun with *S. sclerotiorum* and form large numbers of sclerotia in various stages of maturity before being plowed under. The disease has, however, not become severe on the crops, although present in each. The rapid turning under of the green succulent debris limits the number of mature dry sclerotia that are formed. Perhaps this, the rapid decomposition of debris because of the high fertility, moisture, and warmth of the soil, and continuous planting, have favored a microbial population that decomposes sclerotia or prevents their effective germination. Whatever the explanation, the pathogen is there kept at a low level, contrary to pathologists' expectations.

An unusual use of suppressive soil developed by Ko (1971) in Hawaii was reported to give successful biological control of root rot of papaya seedlings caused by *Phytophthora palmivora*. Planting holes about 30 cm in diameter and 10–20 cm deep were dug in *Phytophthora*-infested field soil and filled with pathogen-free virgin soil. Healthy papaya seedings were then planted in the "islands" of virgin soil. The seedlings became sufficiently resistant to *P. palmivora* with age that little disease developed when the roots spread into the infested soil. The introduced virgin soil thus protected the young seedling during the most susceptible stage. Ko suggested that *P. palmivora* was prevented from crossing the virgin soil barrier because of nutrient deprivation and other forms of antagonism imposed by microorganisms in the virgin soil.

Pathogen Diminishes with Continued Monoculture. In this situation the pathogen becomes established in a soil and initially causes severe disease. With continued planting of the same crop, the disease declines in severity, and crop yields recover. This pattern has been referred to as the grand cycle of disease (Chapter 1).

The control of take-all, caused by *Gaeumannomyces graminis* var. *tritici (Ophiobolus graminis),* with a 2- or 3-year rotation to nonsusceptible crops has been so successful around the world that few growers attempt to grow wheat or barley continuously. Yet reports have for years showed that severity of take-all reaches a maximum in 2 or 3 years of monoculture, and subsequently declines.

Fellows and Ficke (1934) reported that with continuous wheat in Kansas, patches of take-all appeared the first year, enlarged the second, decreased in size the third, and disappeared by the fourth year. At about this same time, Glynne reported an almost identical observation in the annual incidence of take-all in continuous barley or wheat plots at Woburn, England (in Baker and Snyder, 1965). Fellows and Ficke (1939) casually stated, "It is rather common, in take-all-infested fields in central Kansas, for the disease to disappear from fields continuously cropped to wheat." The appearance of take-all and its subsequent disappearance with wheat or barley monoculture has since been repeatedly observed in England, The Netherlands, Switzerland, Denmark, Yugoslavia, Australia, the United States, and undoubtedly in many other wheat-growing areas. This recession has been reported in France, and Lapierre et al. (1970) suggested that it might result from an increase of virus infection in the pathogen (Chapter 7).

Shipton (1972) studied take-all decline in England in the 1960s in fields that had been in pasture and then cropped continuously to wheat or barley (Fig. 4.2). Gerlagh (1968) studied take-all decline in The Netherands in polders recently reclaimed from the North Sea, planted first to reed grass and then continuously to wheat. Take-all increased in severity during the first 2 or 3 years in both soils, then decreased to a uniform and economically tolerable level. Wheat developed severe take-all in glasshouse tests in polder subsoil or grassland soils to which live *G. graminis* inoculum had been added experimentally. By contrast, the same quantity of inoculum added to a soil that had entered the decline phase resulted in only mild or very little take-all.

A similar phenomenon has been observed in Washington. In the fall of 1967, R. J. Cook (unpublished) began to establish a high population of *G. graminis* in a 1-acre (0.4 ha) plot at Lind. The site selected had a long history of dryland wheat in rotation with fallow. Approximately 500 pounds (227 kg) of inoculum of *G. graminis* as severely infected stubble from a field 5 km away was distributed uniformly and disked into the surface 12–15 cm of soil. The field that supplied the stubble inoculum had been recently converted from native sagebrush-bunchgrass vegetation to agricultural use, and had been cropped from 1965 to 1967 to wheat under sprinkler irrigation, which greatly favored take-all. To duplicate these conditions at Lind, winter wheat of the variety Gaines, and later Nugaines,

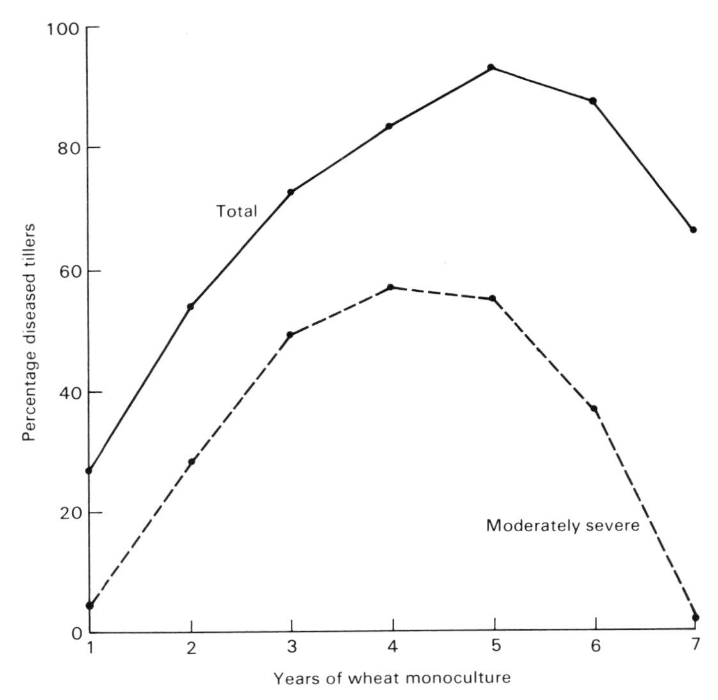

FIGURE 4.2
Incidence and severity of take-all (*Gaeumannomyces graminis*) in winter wheat at time of heading in field plots in monoculture wheat during the years 1963–69, inclusive. (Data from Shipton, 1969.)

was grown annually on the plot with sprinkler irrigation of up to 76 cm per growing season in addition to the 20–25 cm received annually as natural precipitation. By 1972 (the fifth consecutive irrigated crop) the soil, by wheat-seedling bioassay in the glasshouse, was uniformly infested with *G. graminis,* but severe take-all under field conditions still had not developed in most of the plot, and where present was erratic and did not reappear in the same place each year. Moreover, in glasshouse tests with soil from the plot, the bioassay plants showed small lesions that remained restricted.

Other fields of long-term wheat monoculture, but without take-all, were subsequently identified in other areas of the state, including the Columbia Basin irrigation district, where this disease can be severe. The record now leaves little doubt that Columbia Basin soils newly reclaimed from the native sagebrush-bunchgrass vegetation are highly conducive to take-all, but eventually acquire suppressive properties with continued wheat monoculture (Shipton et al., 1973). If this suppressive soil is treated by fumigation, irradiation, or steam, however, it again produces severe take-all when inoculum is added. Gerlagh (1968) and Shipton et al. (1973) showed that heat treatment of the soil at 40° C diminished, and at 60° C essentially eliminated the antagonistic factor (Fig. 4.3), which indicates that this antagonism is not caused by spore-forming bacteria or thermal-tolerant actinomycetes. Pugh and von Emden (1969) suggested that cellulose-decomposing fungi are the principal antagonists in Dutch polder soils.

Gerlagh (1968), Pope (1972), Shipton (in Bruehl, 1974), and Vojinović (1972) distinguished between a general antagonism to *G. graminis* present in many soils "which moderates the potential pathogenicity of the fungus," and a specific antagonism developed during monoculture. The general antagonism was thought to be due to fungi and the specific to bacteria (Pope, 1972). Bacterial populations in "take-all decline" (TAD) soil exceeded those in nondecline soil by 60–75%, actinomycete by 12%, but fungus populations did not. Bacteria and actinomycetes antagonistic to *G. graminis* were three times more abundant in TAD than in nondecline soil.

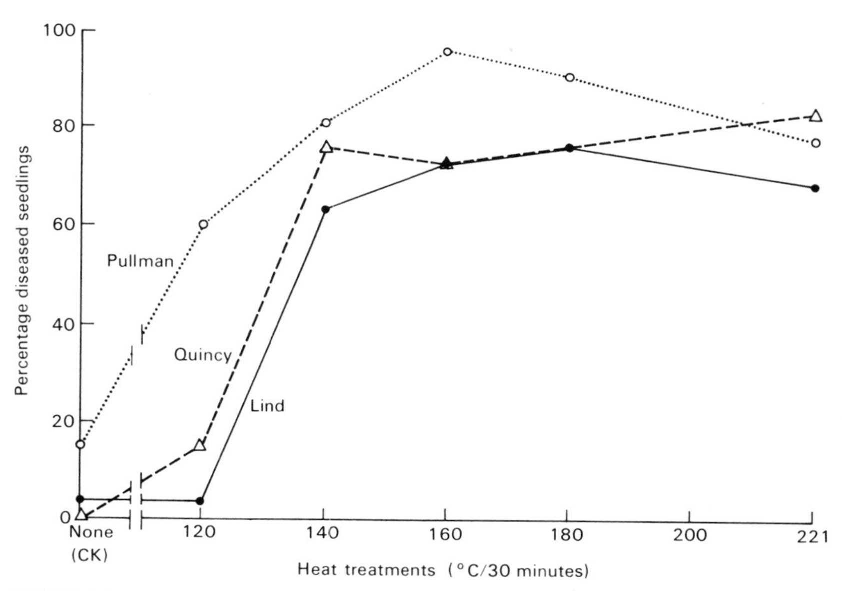

FIGURE 4.3
Effect of heat treatment on suppressiveness of three long-term wheat field soils to take-all disease of wheat seedlings 3 weeks after sowing. Glasshouse test with heat-treated suppressive soil (1%) introduced into methyl bromide-fumigated soil (99%) and inoculated with 1% of a *Gaeumannomyces graminis* culture. Suppressive soil was treated at indicated temperatures with aerated steam for 30 minutes. Values are percentages of the plants with severe disease as indicated by presence of base-plate mycelia on the stem. (From Shipton et al., 1973.)

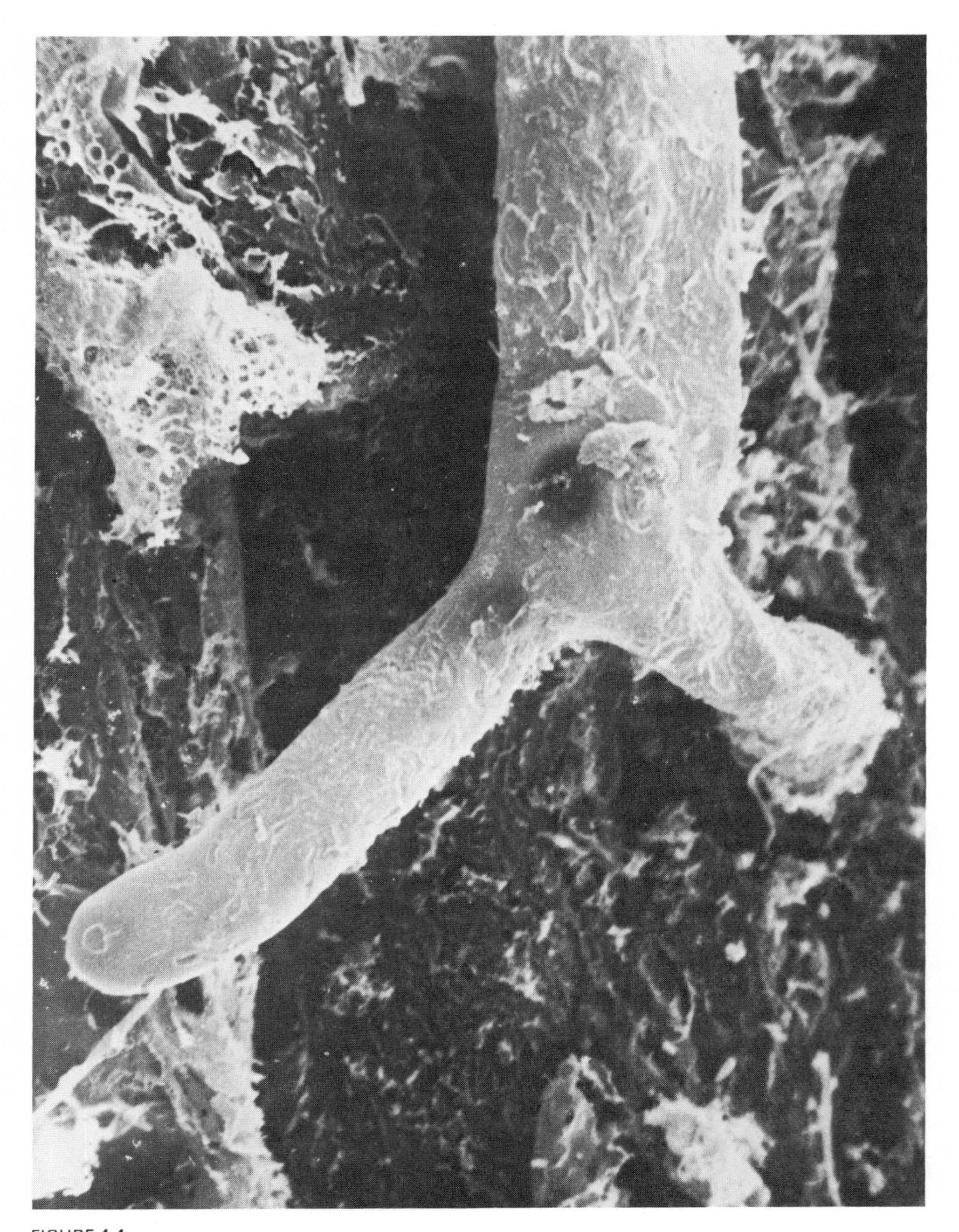

FIGURE 4.4.
Runner hypha of *Gaeumannomyces graminis* var. *tritici* on a young wheat root, showing feeder hypha (*right*) penetrating the host. The hypha has an irregular roughened surface, possibly of partially eroded mucilage. The dark areas in the hypha (*left*) are electron-transparent areas with thinner walls that may be points of incipient lysis. The circular pit at the tip of the hypha (*left*) may be an early stage in development of a hole in the wall. Such holes were observed in walls of old hyphae. At upper and lower left are fragments of hyphae almost completely digested. Scanning electron photomicrograph, × 2300. (From Rovira and Campbell, 1974.)

Vojinović (1972) found that after a fourth wheat crop, 90% of the microorganisms on the roots were actinomycetes (33% of them antagonistic to *G. graminis*), 9% were bacteria (1% antagonistic), and 1% fungi (10% antagonistic). After several crop cycles the wheat rhizosphere had abundant gram negative, nonspore-forming, motile, rod-shaped bacteria difficult to culture, and highly specific and antagonistic to *G. graminis*. Both the fungus and the bacterial antagonists involved would be killed by heat treatment at 40–60° C. Endolysis of mycelium on a root by bacteria is shown in Figure 4.4.

Pope and Jackson (1973) and Vojinović (1973) suggested that the microflora in TAD soils affected the growth rate and tropic response of *G. graminis* hyphae to wheat roots. Is it possible that the specific rhizosphere microflora increases the production of ethylene there (Chapter 6), and thus indirectly influences infection by this pathogen? Shipton (1969), however, found no decrease in number of root infections in TAD as compared with conducive soil (Chapter 6).

Apparently, the inability of *G. graminis* to tolerate or perhaps compete with other organisms in soil and in the host rhizosphere has had a major impact on losses caused by take-all in cereals and grasses around the world. The evidence suggests that the microflora selectively enriched through several years of wheat or barley monoculture may eventually exhibit antagonism to *G. graminis*. In eastern Washington and southern Idaho, the disease is severe mainly in desert soils recently reclaimed and planted to irrigated wheat, or after an 8- to 10-year-old alfalfa stand that had been largely replaced by grass. In contrast, the disease does not occur on wheat in the long-established wheat-fallow area of the low- to intermediate-rainfall (25–40 cm) area of Washington, nor does it occur in the annually cropped wheat fields of the high-rainfall (50–65 cm) Palouse area, even under supplemental irrigation.

Gaeumannomyces patches occur in turf in western Washington primarily in new development areas cleared of forest, and where the soil is fumigated with methyl bromide for weed control prior to seeding grass. Older turf is not nearly so severely affected, and the patches eventually disappear from new turf (C. J. Gould, unpublished). A. M. Smith (unpublished) has made similar observations in Australia. The severe take-all of 40 years ago reported by Garrett (1944) for the mallee desert soils of South Australia, and by Fellows for relatively new prairie soils of Kansas, provide additional examples. The fact that this disease is one of the easiest to control with organic amendments (Fellows and Ficke, 1934) would suggest that amendments change the quantity and quality of the microbiological makeup in a way comparable to that accomplished by long-term monoculture. In the long-established cereal-producing areas of the world, growers have enjoyed the benefits of a highly significant biological control that we are only now beginning to appreciate.

Awareness of the decline of another disease can be traced to the 1916–18 observations of Scofield (1919) in Texas on the appearance and disappearance of cotton-root-rot spots caused by *Phymatotrichum omnivorum*. After a 3-year study of the location of spots in a single plot planted annually to cotton, Scofield concluded simply, "It seems clear from the evidence here presented that these rootrot spots do not carry over from year to year."

This conclusion was confirmed by King (1923) for alfalfa in Arizona, and by McNamara et al. (1931) for cotton in Texas. King noted that the enlargement of root-rot spots in a perennial stand of alfalfa could be likened to that of a fungus fairy ring, "becoming free from the disease after the active mycelium has passed on. . . ." In a classical demonstration, King showed not only that infected alfalfa plants recovered from the disease in the centers of the spots, but also that if the centers were replanted to alfalfa, healthy plants developed. He further showed that *P. omnivorum*-susceptible cowpeas planted in the centers of so-called root-rot spots would also grow luxuriantly during the entire season. Likewise, in cotton in Texas, McNamara et al. (1931) noted that disease diminution usually appeared first in the centers of the spots and then progressed outward, as had the fungus during its initial invasion of the soil 2 or 3 seasons earlier. These workers also noted that root-rot spots would "break-up" in one area of the field, whereas in other spots often only a few yards away, cotton root rot remained severe. Their statement of explanation still can hardly be improved upon: "It remains to be seen whether the advance and retreat of infection of the spots can be associated with features of the life history of the organism, or possibly with an accumulation of its by-products in the soil which might inhibit its vigorous growth, or with some environmental factor not yet understood."

The fact that *Phymatotrichum* in root-rot spots may allow crop plants to recover from the spot center outward, or may disappear entirely has, unfortunately, been of temporary practical value at best. Repeated observations have shown that spot disappearance is generally followed within 2 or more seasons by its reappearance. According to McNamara et al. (1931), large spots of uniform infection give way to numerous smaller spots within the original boundary. Recolonization of the recovered areas originates from live mycelia or sclerotia deep in the soil in remnant areas. In other instances, recolonization of the center of the spot may originate with inward growth of live mycelia from the periphery, a reversal of the usual outward radial growth of the fungus. How long this cyclic appearance and disappearance of disease spots can continue is unknown.

King (1923) indicated, however, that the largest or oldest spots showed the greatest tendency toward "immunity" from reinvasion, which could suggest that the amplitude of the cycle decreases with time and approaches a state of equilibrium, possibly at a low level of disease. Since phymatotrichum root rot is still a serious disease of the southwestern United States 50 years after the work of Scofield and others, we may safely conclude that the disease is not self-eliminating. On the other hand, a modern study of the cyclic nature of the disease would provide helpful clues to biological control.

Common scab of potatoes may also decline with monoculture. Goss and Afanasiev (1938) recorded scab severity on irrigated potatoes in Nebraska from 1929 to 1936 on plots that had been subjected to different rotations since 1912. Only small percentages of severe scab, and essentially no pitted scab, were recorded in the plots of continuous potatoes. In contrast, severe and pitted scab were recorded in high incidence in all short-term rotations, including sugar beet-potato, oats-potato, and corn-potato. They described their observation as "somewhat at variance with popular opinion." On the other hand, results of the long-term rotations of sugar beet (or oat)–alfalfa (2 or more years)–potato were in more agreement with popular opinion, in that these plots had less scab than those with continuous potatoes. Moreover, Werner et al. (1944), who studied scab incidence under dryland conditions in Nebraska using various rotations, reported that scab was most severe with continuous potatoes.

The long-term irrigated potato plots of Weinhold et al. (1964) at Shafter, California, with potatoes in various rotations, and with or without green manures, also recorded a decline of scab with monoculture. Beginning in 1949, when the plots were established, scab on tubers from the continuous potato plots increased each season until the eighth year, and then generally decreased in severity. Scab was distinctly less severe with rotation than with continuous potatoes, but rose steadily with each potato crop. Thus, in 1961, scab severity was actually less with continuous potatoes than with any of the 2- or 3-year rotations, including potato–cotton, potato-sugar beet, potato-barley, or potato-sugar beet-cotton. Of even greater interest, however, is the suggestion by Figure 4.5 that scab incidence on potatoes in rotation with other crops rises, then levels off as under continuous potatoes, but does so more slowly, as if rotation served to dampen and delay a process that occurred within 8 years with continuous potatoes. It would be interesting to have followed these various 2- and 3-year rotations for several additional years to determine whether scab would continue to decline with rotation as it did with potato monoculture. Potatoes after 3 years of alfalfa gave less scab than did 13 years of continuous potatoes, but as with the other rotations, there was a steady but

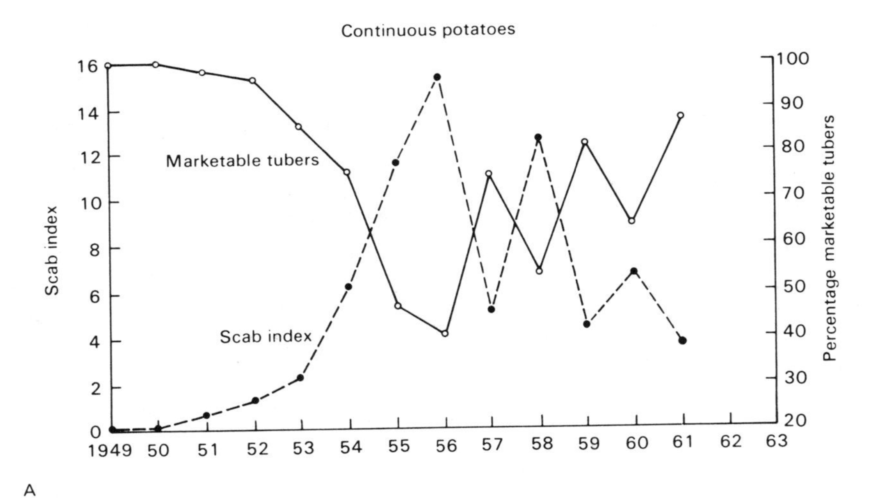

A

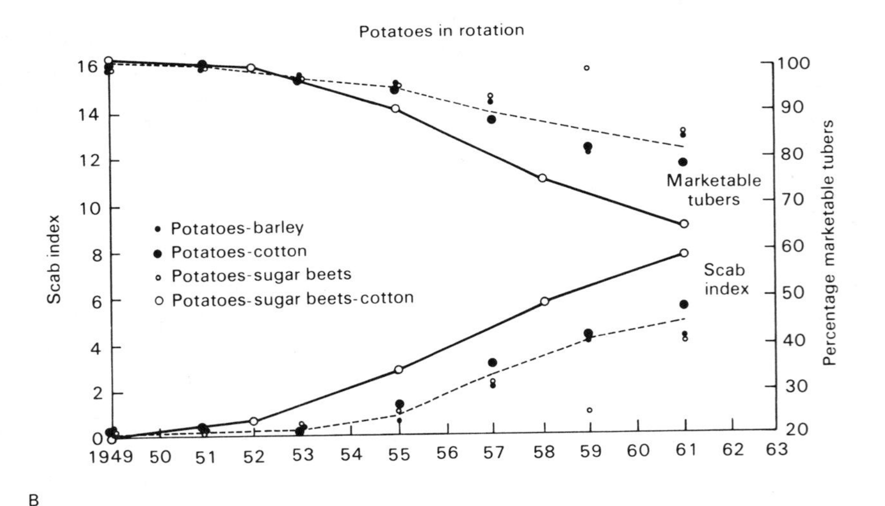

B

FIGURE 4.5.
Occurrence of common scab on potatoes grown in Shafter, California, from 1949 to 1961: A. In continuous monoculture. B. In 2- or 3-year rotation with cotton, sugar beet, or barley. Scab increased under monoculture for 8 years and then declined; under rotation it slowly increased for thirteen years. (Data from Weinhold et al., 1964.)

slow rise in scab each year. Goss and Afanasiev did not begin their observations until their plots were 18 years old (1912–29), and scab in the irrigated plots of Nebraska may thus already have declined to the steady low level.

Irrigated fields of the Yakima Valley and Columbia Basin in Washington may also become suppressive to scab with intensive potato cropping. Menzies (1959) showed that potatoes in soil just reclaimed from the desert developed severe scab, whereas the disease was mild in nearby or adjacent fields of long-term cultivation, referred to as "old scab land" (Chapter 7). Similarly, R. D. Watson (unpublished) observed in southern Idaho that scab is commonly most severe in lands recently converted from sagebrush to agricultural use. He attributed the severe disease to a lack of soil biological buffering. Menzies showed that the responsible factor could be eliminated by soil sterilization, or transferred by mixing as little as one part suppressive soil to nine parts conducive soil. A practical biological control of potato scab apparently has thus been accomplished in three ways: by a rotation of 4 or more years to reduce the population of the pathogen; by a soybean green-manure crop to favor *Bacillus subtilis* (Chapter 7); simply with continuous cropping to potatoes. With this many means of control, one may ask why scab is a problem at all. More properly, we should remember that, but for these methods of biological control, natural and contrived, scab would undoubtedly be much more serious than it is. The same is true of take-all disease of wheat, which can also be controlled by three comparable methods. Dissatisfied as we are with less than complete control, science must continue to work toward still better biological methods to protect potato and wheat crops that still become diseased.

An instructive example of biological control of the grape leafhopper in California vineyards is perhaps analogous to the above effects of monoculture. The pest was observed not to be troublesome in untreated vineyards adjacent to streams where wild evergreen blackberries grew. An efficient parasite of the leafhopper overwintered only in eggs of another leafhopper on the blackberry. Planting small patches of such blackberries adjacent to vineyards has extended this protection far from the streams and reduced the leafhopper population in about 4 years to a tolerable level without the use of insecticides. In addition, spider mites, which were a secondary problem associated with insecticide application, were not aggravated (Huffaker, in Rabb and Guthrie, 1970).

Crop Rotation

Farmers have known for thousands of years that yields are improved if crops are grown in some type of rotation, and few practices have had a greater impact on agriculture (Curl, 1963). The trend from diversified to more specialized farming, predicted years ago as the way for the future, has been subverted because no satisfactory substitute practice has been found to perform the functions of rotation. Some types of agriculture continually migrate to new or virgin soils because crop rotation is not or cannot be practiced. Yet even with these pathogen-infested new lands, we can rationalize that the contamination is temporary and that reclamation is still possible, given sufficient time with production of nonsusceptible crops.

Soilborne pathogens of cereals are foremost among those for which crop rotation is our best method of control. Control of speckled snow mold of wheat caused by *Typhula idahoensis,* and especially of pink snow mold caused by *Fusarium nivale,* can be striking in northeastern Washington if but one spring-wheat crop is used in the usual winter wheat–fallow–winter wheat rotation. Spring wheat is susceptible, but avoids the snow and cold environment favorable to snow mold. A fallow–spring wheat–fallow combination thus results in a 3-year period without hosts for snow-mold fungi, during which time their populations may sink to relatively harmless levels. Cephalosporium stripe of wheat, caused by *C. gramineum,* is likewise favored by winter environment, and thus attacks the winter wheats. Rotation to spring wheat or peas for a single year in the wheat–fallow–pea or wheat–fallow–wheat areas of the Pacific Northwest controls the disease. Take-all, caused by *Gaeumannomyces graminis,* can be controlled by a 3- or even 2-year rotation to crops other than susceptible cereals or grasses (in addition to its control in long-term wheat monoculture). Cercosporella foot rot caused by *C. herpotrichoides,* and leaf spots caused by *Rhynchosporium secalis* and species of *Septoria, Selenophoma, Ascochyta,* and *Helminthosporium,* are adequately controlled by 2 or 3 years away from cereals or grasses. In all cases, soil microorganisms decompose the wheat refuse containing these pathogens, whether roots with *G. graminis,* stems with *C. herpotrichoides,* or leaves with leafspot fungi.

Longer rotations of 4, 5, or more years away from susceptible crops are needed to control some pathogens, particularly those that survive by means of sclerotia or thick-walled spores. Four years away from cotton and other suscepts will reduce losses caused by *Phymatotrichum omnivorum,* a recommendation that apparently dates back to observations by B. M. Duggar in 1888. Since the host range of *P. omnivorum* is extremely wide, and the fungus can persist for years in deeply buried infected woody tissue and other residues, rotation has never worked satisfactorily in control. On the other hand, a 4- to 6-year rotation has proved effective in the control of scab (caused by *Streptomyces scabies*), black scurf (caused by *Rhizoctonia solani*), and verticillium wilt of potato, particularly when the rotation included 2 or more years of clover or alfalfa. Certain crop rotations are also partially effective against *Sclerotium rolfsii, Sclerotinia sclerotiorum,* and species of *Pythium, Phytophthora,* and *Fusarium;* since rotations are generally insufficient control measures in themselves, they are thus best used in combination with other disease-control methods.

Sclerotium rolfsii on peanuts is controlled by plowing under the surface refuse before sowing. This deprives the fungus of a food base, since it can grow only on or near the soil surface. The effect, therefore, is very much like that of crop rotation.

This does not imply that whenever crop rotation is effective, the mechanism is one of biological control. Bacterial wilt (Moko disease) of banana, caused by *Pseudomonas solanacearum,* is controlled by disking the surface 25 cm of soil five times during the dry season, followed by 9 months fallow, apparently because the causal organism cannot withstand desiccation (Sequeira, 1958). A short-term rotation away from tobacco apparently controls *Phytophthora parasitica* var. *nicotianae,* because the sporangia germinate and discharge zoospores in response to water and are lost unless a host is present.

Many, if not all, pathogens would probably die of starvation or old age, irrespective of any biotic environment, without access to a host or other substrate. Soil microorganisms nevertheless use food that otherwise might serve to perpetuate pathogens, including those that persist as resistant spores. As pathogenic propagules age, weaken, or approach extinction, surrounding organisms would naturally hasten their death (Chapter 6). Even during a supposedly good state of health, the energy requirements and hence the rate of exhaustion of reserves in propagules is undoubtedly enhanced by the presence of surrounding and potentially competitive microorganisms, particularly in soil. A microscopic living propagule resists decay in soil, whereas a piece of organic matter of identical shape, size, and chemical makeup, if added to the same soil, will decay within hours. Soil organisms shorten an otherwise very long process. Without their help, propagule death would probably take more time than man has patience and more seasons than he has crops. Fungus propagules survive much longer in sterile than nonsterile soil.

Organic Amendments

The early attempts to use soil microorganisms in practical biological control under field conditions were through organic amendments. Early work of Sanford in Canada on control of potato scab by plowing under green rye, of King, Hope, and Eaton in Arizona on control of phymatotrichum root rot with barnyard manure and certain other organics, and of Millard and Taylor in England on control of potato scab with grass cuttings have been discussed by Garrett (in Baker and Snyder, 1965). Other early workers were Fellows and Ficke (1934) in Kansas; after 6 years of testing they stated "chicken manure always gave perfect control of take-all," the effect was "on the causal organism rather than on the wheat plant."

Even before these early scientific reports, people were unquestionably aware of the value of organic manures to plant health. Indeed, the use of manure to improve crop production is nearly as ancient as farming itself. We now know that manures stimulate high populations of soil microorganisms, and in so doing, limit the germination of pathogenic spores or the growth of hyphae, or hasten the microbial digestion of propagules and pathogen-infested remains. The digestion of sclerotia of *Phymatotrichum omnivorum* by soil organisms after treatment of the soil with chicken and barnyard manures illustrates this (Chapter 6).

Increase in dormancy of propagules and their digestion by soil microorganisms are among the more important disease-control processes initiated by organic amendments. Adams et al. (1968b) described this effect as an increase in soil fungistasis. Whereas propagules are normally dormant in nonamended soil, they are generally quick to germinate in response to root exudates and various other nutritive substances. By contrast, in soil amended with organic residues, wastes, or manures, propagule germination by the pathogen may not be possible, even if very rich mixtures of nutritive substances are supplied. Available evidence suggests that the effect is relatively nonspecific in origin and involves the sum of intensified activity of the complex microbial community, with all concomitant features, including increased liberation of toxic metabolites and competition for nutrients (Chapter 7). As microbial activity increases, the expenditure of propagule energy during dormancy presumably increases as a protection mechanism, the net result being an increase in the frequency of propagule exhaustion and death.

The use of cover and legume crops, particularly green legumes plowed under, has been an especially effective means of biological control of plant pathogens. A crop of green peas or dry sorghum plowed under before planting cotton in the southwestern United States apparently also provides an excellent field control of phymatotrichum root rot (Chapter 8). The effectiveness of legume cover crops for the control of take-all has been frequently demonstrated. Germinability, and possibly viability, of sclerotia of *Typhula idahoensis* is greatly reduced in Idaho fields where alfalfa is introduced into the rotation with wheat, as shown by Huber and McKay (1968).

Leguminous residues are rich in available nitrogen and carbon compounds, and they also supply vitamins and more complex substrates. Biological activity becomes very intense in response to amendments of this kind and may increase fungistasis and propagule lysis. In addition, some propagules are known to germinate in response to leguminous amendments, only to terminate in germ-tube lysis. There is now good evidence (Owens et al., 1969) that certain volatile substances released from leguminous tissues during decomposition in soil may trigger germination of propagules, but do not support their continuing growth. This results in their death. In attempted control of *Phytophthora cinnamomi* on avocado with alfalfa meal in California, saponins released reduced root rot (Zentmyer, 1963), but subterranean clover meal increased zoosporangial formation in Australia (Broadbent and Baker, 1973a). Ammonia is also released from legume substrates during their decomposition, temporarily increasing the soil pH. These pH changes, as well as possible accumulation of nitrites or availability of mineral nitrogen, are known to effect significant complex shifts in biological balance and disease severity.

One noteworthy example of biological control with green manuring is that of potato scab with a soybean cover crop in southern California (Weinhold and Bowman, 1968). In field trials of continuous potatoes over a 13-year period, scab was prevented from increasing if soybeans were grown annually as a cover crop and incorporated green into the soil each year before planting potatoes. Scab increased when the cover crop was green barley. With continuous potatoes without the soybean amendment, however, disease increased for 8 years, then decreased. A strain of *Bacillus subtilis* antagonistic to *Streptomyces scabies* was prevalent in all plots, with or without the soybean amendment. Moreover, glasshouse studies showed that the strain of *B. subtilis* multiplied equally well in pots of soil amended with either green soybean or green barley tissue. Although multiplication by the bacterium was similar with the two amendments, the antibiotic effective against *S. scabies* was produced by the bacterium in much greater amounts on soybean than on barley. Nutrition of the antagonist and antibiotic production may thus be significant factors making one green manure superior to another.

The specificity of the green-manure effect was further emphasized by Nelson (1950), who found that a 1- or 2-year planting of soybean, or a single year of carrot, increased verticillium wilt of peppermint more than did 11 other rotations involving 18 nonsusceptible crops and fallow.

Bean growers in the Salinas Valley of California observed that root rot caused by *Fusarium solani* f. sp. *phaseoli* was less severe or nearly absent if beans were preceded by a crop of barley. A similar observation was made by Burke (in Cook and Watson, 1969) in bean fields of the Columbia Basin in Washington. Population studies of the pathogen revealed that numbers of propagules of the fungus are as high, if not higher, following barley, although the disease caused by the fungus is less. Laboratory studies in California and Colorado (Cook and Watson, 1969) suggested that mature barley straw incorporated into soil immobilizes the inorganic nitrogen in the soil needed by the fungus for growth and parasitism. The N-immobilization theory has been substantiated by the fact that in glasshouse tests, sawdust, cellulose, cellobiose, and even sugar, added to soil may control the bean root rot, and nitrogen added to the high-carbon amendments negates the control (Maurer and Baker, 1965). Because straw plowed under occupies the same layer as the pathogen (the tillage layer), nitrogen starvation of a pathogen, but not of the host, is possible. The plant roots presumably penetrate and absorb below the zone of nitrogen immobilization. It is significant that control of bean root rot with mature straw is nutritional (competitive) in action, whereas fresh soybean tissue apparently controls potato scab through antibiotic activity (Cook and Watson, 1969).

Addition of abundant organic matter to ornamental plantings in home yards has been reported to reduce the incidence of and injury from phytophagous nematodes. The mechanism involved is unknown but may well involve various trapping and pathogenic fungi and other predators, as found by Linford and co-workers (1937, 1938) in laboratory tests in Hawaii, and by Duddington and Wyborn (1972) in England (Chapter 7).

Fomes lignosus, Ganoderma pseudoferreum, and *F. noxious* cause, respectively, white, red, and brown root diseases of plantation rubber. The basis of control is as follows (Fox, in Baker and Snyder, 1965; Fox, 1966; Fox, in Toussoun et al., 1970): Old jungle trees in a prospective planting site, or old rubber trees in an area to be replanted, are poisoned before new ones are planted. Their rapid death allows quick invasion of these potential nutrient sources by saprophytes, and thus inhibits the extension of the pathogens in already infected roots. Creosoting the cut surfaces of stumps decreases the number of potential new disease centers by stump colonization from airborne spores. A mixed cover of creeping legumes is planted between the new rows of trees, on either side of which a clean-weeded strip 3 feet wide is maintained. Quarterly foliage inspections are made to detect infected trees, which are removed and their neighbors treated as explained below. Only if an old stump repeatedly causes new infections is it isolated by a trench; otherwise it is left to rot beneath the creeping cover plants, where conditions generally favor rapid decay. These conditions also encourage luxuriant rhizomorph formation by *F. lignosus* in and on the litter layer at a time when there are but few rubber roots to infect, and with all three pathogens, these conditions enhance development of fructifications, thus depleting the food bases of the pathogen. Actinomycetes and other saprophytes in the soil are also favored by the cover plants.

Using the buried-slide technique in the field, R. A. Fox (unpublished) observed dissolution of the cross walls of the hyphae of *F. lignosus* by actinomycetes in soils where legumes were grown. Mycelia in tests in which this pathogen was grown in tubes of sand wetted with a sterilized soil extract that had been inoculated with nontreated soil (1 g/100 ml) were sparser and more fragile than those in sand wetted with noninoculated soil extracts. No wall erosion was noted in this laboratory system, but strands, which were dotted or encrusted with colonies of bacteria, were greatly weakened. It was thought that their continuity, as with ectotrophic growth in the field, was interrupted, as it may also be by mycophagous nematodes and other members of the soil microfauna which multiply beneath the cover plants. Soil actinomycetes have also been shown to inhibit growth of *F. lignosus* by toxins apparently formed on bits of plant debris. Autoclaving inhibitory soil from under cover crops, or treating it with propylene oxide, eliminated this effect, even if glucose was subsequently added. As regards *G. pseudoferreum* and *F. noxious,* Fox stated that "the beneficial effect of cover plants in reducing losses caused by these two pathogens is more by encouraging antagonists that contain them within infected roots, where they slowly waste away, and by encouraging saprophytes that deny potential food sources to them, rather than by early dissipation of their food reserves by enhanced vegetative or reproductive growth"—factors more effective with *F. lignosus* (Fox, in Baker and Snyder, 1965). Fox (1964) has summarized this postplanting system as "biological control with limited manual assistance."

Infected trees are removed at each quarterly inspection. Neighboring trees are inspected by exposing the collar of each, until noninfected or mildly infected trees are reached. Trees in areas infected with *F. lignosus* are painted with PCNB in emulsified bitumen or wax around the area where surface lateral roots join the tap root (Fox, 1966; Fox, in Toussoun, et al., 1970). The epiphytic rhizomorphs of this pathogen, by which new infections are started along the root, are thus prevented from spreading for up to 2 years.

Organic amendments have played a long and significant role in prevention of plant diseases, often unknown to the user, probably through as many mechanisms as there are potential amendment-disease interactions. This does not imply, however, that such amendments are without their limitations in plant-disease control, or that they hold the key to the future of successful biological control. Experience has shown that amendments, whether green manures, crop residues, or other form of organic materials, can be frustratingly unreliable in plant-disease control, and that much more research is needed before these materials can be dependably used.

Soil Treatments

The control of *Armillaria mellea* in orchards is a classical example of combined soil fumigation and biological control and may be the first instance of this type of integrated control in plant pathology. Effectiveness apparently comes from killing the more exposed mycelia and rhizomorphs by carbon disulfide, weakening the rest in some undetermined manner, which enables *Trichoderma viride* to kill them. The Del Monte Corporation has for years been controlling *A. mellea* in its California orchards by such fumigation.

Bliss (1951) found that *A. mellea* in roots was not killed immediately after soil fumigation with carbon disulfide at moderate dosage, but was killed in 24 days. The pathogen survived at least 6 years in citrus roots 2.5 cm in diameter in nontreated soil that contained *Trichoderma.* Carbon disulfide did not kill the pathogen in roots in sterilized soil, but did when *Trichoderma viride* was present. Bliss concluded that carbon disulfide would kill *Armillaria* at a high dosage, but that at a lower dosage (302 gallons/acre; 2825 l/ha), *Trichoderma* was the effective agent. Darley and Wilbur (1954) and L. O. Lawyer (unpublished) found that a high dosage killed the pathogen in 2–3 days in roots 0.9 m deep in the field, but a sublethal dosage of carbon disulfide killed the pathogen in roots at that depth only after 30–50 days. *Armillaria* could be recovered for 26 days after treatment; bacteria then dominated for 4 days and were in turn replaced by *Trichoderma.* These times increased as the soil temperature was lowered, and *Trichoderma* could not then enter roots until after *Armillaria* had been killed. If *Armillaria*-infected roots were buried in soil 18 days after fumigation, no control resulted, perhaps because the pathogen had not been weakened, or because microorganisms that inhibited *Trichoderma* had not been destroyed by the treatment. The fungus in these California orchards apparently did not invade the wood unless it was infected by *Armillaria.*

Benomyl has an effect opposite to that of carbon disulfide, as it controls *Trichoderma* and permits *Armillaria* to grow. Methyl bromide, however, acts like carbon disulfide on the soil flora in the field (L. O. Lawyer, unpublished; Ohr et al., 1973). All these workers concluded that the fumigant weakened *Armillaria,* and that *Trichoderma* killed what the fumigant missed.

Heating soil to 33° C for 7 days weakened *Armillaria* in a manner comparable to that of carbon disulfide; heating to 36° C for 7 days or 43° C for 2 hours killed the fungus. Living citrus roots withstood 43.9° C for 2 hours and 36.5° C for 5 hours without injury. Peach roots showed slight injury at 48.9° C for 6 hours (E. F. Darley and W. D. Wilbur, unpublished). The effectiveness of these sublethal chemical and thermal treatments suggests that it might be possible to treat with some chemical or with heat at some level noninjurious to the host, but which would weaken *Armillaria* or diminish antagonists of *Trichoderma,* and perhaps eventually control the pathogen. Mild heat treatments of the root system of valuable ornamental trees could, perhaps, be used to control *Armillaria* in urban areas where carbon disulfide would not be permitted.

Saksena (1960) found that *Trichoderma* was only moderately tolerant of carbon disulfide, but that its rapid growth rate in soil with a reduced microflora enabled it to rapidly occupy fumigated soil.

Green-spored trichodermas were lumped in 1939 by G. R. Bisby under *T. viride,* but were again separated into nine species by M. A. Rifai in 1969. As would be expected from work on specificity described in Chapter 7, species, and isolates within species, vary widely in antagonism to *Armillaria* and *Rhizoctonia* (Mughogho, 1968). It was thought that some of the conflicts in reports on the effectiveness of *Trichoderma viride* as an antagonist were caused by this variation. However, most of the tests for antagonism have been run in the laboratory, where mutants and variants tend to be preserved. In nature, there is probably a natural selection for the virulent strains of *Trichoderma* which could account for the difference between laboratory data and commercial field results. Garrett (1970) commented about this situation that "our hopes of mediating biological control of root-disease pathogens by *Trichoderma* species through soil fumigation . . . have received a set-back. My account . . . is merely a progress report; it is not an epitaph." In the meantime, the sublethal dosages of carbon disulfide continue to provide economic control of *A. mellea* in fruit orchards.

BIOLOGICAL CONTROL BY INTRODUCED ORGANISMS

Trap Plants

The early literature on control of the important root-knot nematodes (*Meloidogyne* spp.) suggested that a highly susceptible "catch crop" be grown and then carefully removed from the soil before egg masses were deposited. Because of the obvious difficulties of timing, and the hazard of thus increasing, rather than decreasing, the population, this method was never popular. This practice reflected the generally accepted idea that nematode larvae penetrated roots of only susceptible plants. Field tests in Georgia and Florida had shown by 1937 that *Crotalaria* spp. reduced root-knot damage in tobacco.

Barrons (1939) found that as many larvae entered roots of resistant as susceptible bean varieties, unexpectedly raising the question "whether resistant plants may not be superior to susceptible plants as trap crops. . . ." A rotation of the resistant *Crotalaria spectabilis* with cucumber was found by C. W. McBeth in 1942 to provide excellent control of root-knot, greatly reducing the soil infestation. The same effect was observed in home yards. McBeth and Taylor (1944) showed that growing this plant as a cover crop in infected peach orchards increased, over a 5-year period, the average annual yield per tree from 10.4 kg under a susceptible cover crop to 60.6 kg under *Crotalaria.* The tops of the trees were about twice as large in the latter plots, and trunk diameter increase was 6.3 cm and 11 cm. Female larvae penetrated roots, and because giant cells were not formed there, the immobilized nematodes starved, or at least did not lay eggs.

A field in Ventura County, California, heavily infested with root-knot nematode was planted to *C. spectabilis* for one summer and replanted to sugar beets the following year. A 200% yield increase over the checks resulted, with a proportional decrease in the number of galls (K. F. Baker, unpublished).

Although this method of biological control of root-knot nematode had great promise, it was dropped because the inexpensive soil fumigant, D-D, came on the market at that time. However, the use of a *Crotalaria* cover crop in salvaging nematode-infested orchards still seems to have great promise, as chemical treatment of living trees is expensive and injurious.

Trap plants have also been suggested for use as biological barriers against spread of *Radopholus similis,* the burrowing nematode, in citrus groves (Ford, 1968). Several rows of Milam rough lemon rootstock are planted around the infested area. The roots of other adjacent varieties must be cut every 6 months to prevent nematode spread along their surfaces; injections of D-D were suggested for this purpose.

Inhibitory Plants

African and French marigolds (*Tagetes erecta* and *T. patula*) were observed by G. Steiner in 1941 to be resistant to root-knot nematodes; larvae penetrated roots as in *Crotalaria* but did not lay eggs. Others observed that planting *Tagetes* as a cover crop and turning it under increased the yield of narcissus 141% over the control, as against 202% by D-D soil treatment. This was caused by a reduction of population of the nematodes, *Pratylenchus* and *Haplolaimus.* M. Oostenbrink, K. Kuiper, and J. J. s'Jacob showed in 1957 that infested soil near *Tagetes* roots had fewer *Pratylenchus penetrans* nematodes than did that near other kinds of plants. Certain other nematodes (*Heterodera rostochiensis* cysts) were unaffected. This inhibitory effect was produced by growing plants but required 3 to 4 months to become evident. The inhibitory material was released from growing roots in sand, peat, or silt soils, even without root decay. The effect was observed on four *Pratylenchus* spp. and on *Tylenchorhynchus dubius*. Uhlenbroek and Bijloo (1960) isolated highly nematocidal polythienyls from the roots of *T. erecta*. One of the best of the many derivatives of 2, 2'-bithienyl was α-terthienyl, which was found to be 0.1% of the air-dried weight of roots. This low concentration accounted for the requirement of prolonged cropping for effectiveness.

Daulton and Curtis (1963) found that mixing 200 ppm of α-terthienyl into soil did not control *Meloidogyne javanica*. Larvae that penetrated the roots of *Tagetes*, however, did not develop beyond the second larval stage. Hijink and Winoto Suatmadji (1967) showed that other Compositae (*Helenium, Gaillardia, Eriophyllum*), as well as *Tagetes*, suppressed *Pratylenchus penetrans*. These studies have been reviewed by Winoto Suatmadji (1969).

C. M. Olsen (unpublished) showed in 1960 that α-terthienyl completely inhibited germination of conidia of *Fusarium oxysporum* f. sp. *callistephi, Septoria tageticola,* and *Helminthosporium sativum* at concentrations above 50 ppm.

S. Wilhelm (unpublished) planted *T. minuta* as a cover crop in an olive grove infected with *Verticillium albo-atrum* in the San Joaquin Valley, California. Because annual infection apparently is necessary for continuance of this disease, factors that inhibit this reinfection may provide control. The planting of *Tagetes* greatly improved the condition of the trees in the second year, when the experiment was concluded because *T. minuta* was declared a noxious weed. With 33 species in this genus, there is reason to believe that a nonnoxious species can be found that will be useful in biological control.

African savannah grass (*Hyparrhenia* spp.) produces a thermolabile, partially water-soluble toxin in soils of Rhodesia and Ghana, which is thought (Meiklejohn, 1962) to inhibit *Nitrobacter* and especially *Nitrosomonas* in high-veld grasslands that are very low in available nitrogen but not in nitrogen-fixing bacteria (*Clostridium, Azotobacter,* and *Beijerinckia* spp.). There is usually an inhibition of nitrification in grassland soils in many areas of the world (Clark and Paul, 1970). Forest soils of the African areas, in which *Hyparrhenia* is lacking, are abundantly supplied with nitrifiers and nitrogen-fixing bacteria. Because *Armillaria mellea* does not produce rhizomorphs in Rhodesia, spread of the pathogen depends on root contact. The usual formation of rhizomorphs in agar culture is inhibited by addition of soil extracts sterilized by filtration. Autoclaving the soil destroys the inhibitory effect, but treatment with

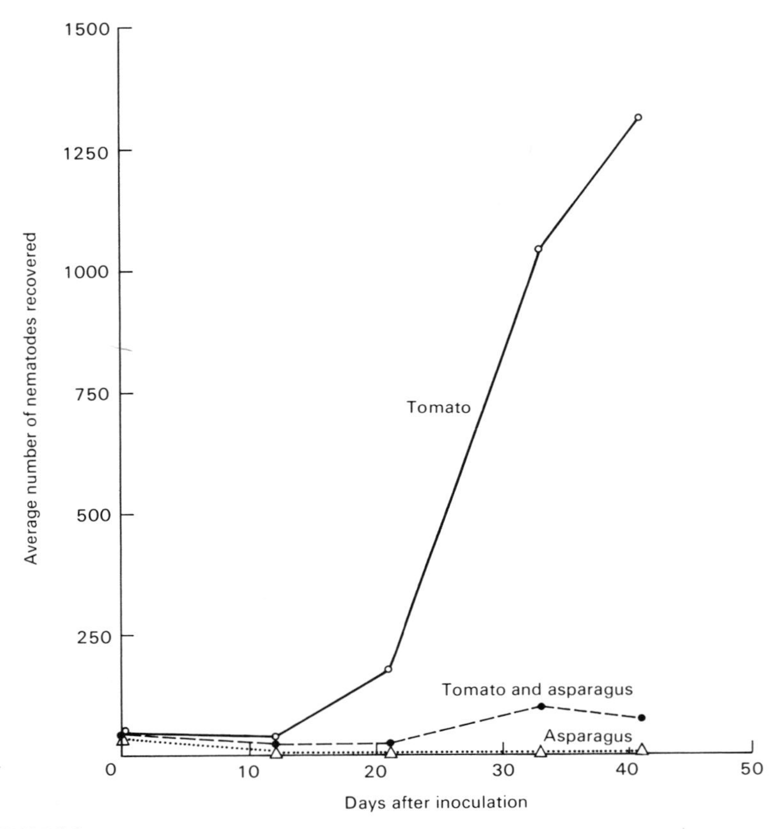

FIGURE 4.6.
Populations of stubby-root nematode (*Trichodorus christiei*) on tomato and asparagus plants grown separately and in the same pot. Separate curves are the mean of five replicates. (Data from Rohde and Jenkins, 1958.)

propylene oxide does not. The toxic material thus appears to be fairly stable in soil apart from living roots. Swift (1968) concluded that "the absence of rhizomorphs in the field in Rhodesia may be primarily due to the presence of a toxic substance or group of substances in the soil." It was thought (Boughey et al., 1964) that the toxic effect on nitrifiers and *A. mellea* may be related in some way.

Rohde and Jenkins (1958) found that the stubby-root nematode, *Trichodorus christiei,* multiplied rapidly on tomato roots but did not feed on asparagus roots. When tomato and asparagus were grown together in pots, nematode populations were substantially smaller than those in pots with tomato alone (Fig. 4.6), showing that the toxic material, a glycoside, diffused through the soil. It could also be leached from soil, and to some extent translocated in plants, because when it was sprayed on tomato tops it restricted nematode increase in the roots.

Phenyl isothiocyanate given off by roots of mustard plants prevented hatching of the cysts of *Heterodera rostochiensis*, the golden nematode of potato, when mustard was grown in pots with potato (Ellenby, 1945). Allyl isothiocyanate, which occurs in the seeds, was more effective because of its lower vapor pressure; mixed with peat and added at planting time, it gave good control of the nematode. The effect was only inhibitory, as transfer of the unhatched cysts to soil with potato caused them to hatch.

Antagonistic Microorganisms

Inoculation of Seeds. The success of inoculation of legume seed with *Rhizobium* spp. prior to field sowing is well known. It is also possible to inoculate pine seeds with spores of *Rhizopogon luteolus*, which will establish as a mycorrhiza on the roots in field soil (Theodorou, 1971). In nontreated soil, 56% of the roots from inoculated seed were mycorrhizal, and as would be expected, more (80%) were obtained in methyl bromide-treated soil. There was a 36% increase in plant growth from seed inoculation in nontreated soil. The principles and energy relationships involved in the movement of microorganisms from inoculated seeds onto and along roots have been examined by Bowen and Rovira (1973).

The inoculation of seeds with microorganisms before planting in nontreated soil has long been studied in Russia, and has apparently resulted in increased yield and control of some root pathogens. These results have been partially confirmed by workers at Rothamsted Experimental Station in England (Brown, 1972, 1974; Brown et al., 1964), and in the Commonwealth Scientific and Industrial Research Organization (CSIRO) in Australia. They provide evidence that selected organisms can be introduced into a new habitat. The increased growth may arise from (a) control of harmful microorganisms and pathogens in the rhizosphere, (b) altered balance of rhizosphere microflora, producing an indirect effect on the crop, (c) production of indoleacetic acid or gibberellin (Brown, 1972) by the inoculated microorganism (however, the organism used in the Victoria tests below apparently forms neither of these materials), (d) possible nitrogen fixation by the inoculated microorganism (nitrogen was not fixed by the Victoria organism below), (e) possible production of vitamins, or conversion of materials to a form useful to the host, or (f) release of nutrients from soil or organic matter.

Brown (1972) showed that bacteria (mainly *Pseudomonas* and

Achromobacter spp.) isolated from the rhizoplanes and rhizospheres of 6-day-old pea or wheat seedlings produced three or more substances that inhibited elongation of pea internodes and lettuce hypocotyls. These organisms were stimulated during early stages of plant growth. At later stages of growth, different bacteria (mostly *Nocardia* and *Arthrobacter* spp.) were present; these produced gibberellins and indolyl-3-acetic acid, which increased plant growth. Thus, root exudates may stimulate rhizosphere microorganisms that may, in turn, affect plant growth. This provides a unique opportunity to improve plant growth by influencing the rhizosphere microflora through inoculation of the soil or seed with specific bacteria.

Field experiments were conducted in Victoria, Australia, in 1969 and 1971 (Price et al., 1971), in which *Bacillus subtilis* A 13 was used to inoculate barley, wheat, and oat seed, which was sown in areas in a field affected by "bare-patch" caused by *Rhizoctonia solani, Pythium* sp., and *Fusarium* sp. There were 10^6–10^7 bacterial cells applied per seed in a water suspension. Wheat and oats showed increased tillering and dry weight, and barley and oats gave up to 9% increase in yield and about 2 weeks earlier maturity than plants from nontreated seed. The suspected pathogens were present both inside and outside the bare patches, but there were only about one-fourth as many pseudomonads inside as outside the patches. If confirmed, this would suggest that the slowly enlarging areas result from receding antagonist populations rather than from advancing pathogens. This may be effect rather than cause, however, since there were fewer roots in the patches and less available nutrient for pseudomonads. Weeds and volunteer barley in the areas previously planted with inoculated seed still showed as much as a fourfold increase in size 14 months later, and the effect was still visible after 24 months. The bacteria thus showed a direct or indirect relatively prolonged effect on a wide range of plants in nontreated soil.

In other field tests in Victoria, Australia, in 1971, Merriman et al. (1974) applied *B. subtilis* A 13 and *Streptomyces* sp. in water suspensions to oat seeds at, respectively, 10^6 to 3×10^7 and 5×10^4 to 3×10^5 per seed. Grain yield was increased in one locality by 40 and 45%, respectively, highly significant increases over the water checks. Number of tillers was significantly increased by 28 and 50%, respectively, and *Streptomyces* sp. increased the dry weight of foliage by 25%. The differences in plots with wheat and barley at two other localities were largely nonsignificant. Numbers of *B. subtilis* and *Streptomyces* sp. on roots indicated that these microorganisms had spread from the seed and persisted for a least 5 weeks.

The effect of a microorganism introduced in an early critical stage of plant development may have an influence on the plant that extends beyond the survival period of the organism. The effect on other microflora may also last longer than the survival of the introduced microorganism, as apparently was the situation here. Hameed and Couch (1972) found that inoculation of axenic *Tagetes* with *Penicillium simplicissimum* also increased plant size and weight, and produced earlier flowering than that in the uninoculated checks.

Merriman et al. (1974) conducted careful field trials in Victoria in 1972–73 with Royal Star carrots, applying four different bacteria to the seed, which was sown in nontreated field soil. The greatest increase in growth was from seed pelleted with *B. subtilis* A 13 in Prillcote, which gave 33.5 tons/acre (83.8 metric tons/ha) of marketable roots, followed by A 13 liquid dip with 25.4 tons/acre (63.4 metric tons/ha); pelleting without bacteria gave 22.7 tons/acre (56.8 metric tons/ha), and water dip alone gave 21.5 tons/acre (53.8 metric tons/ha). The increase in the first treatment was highly significant (48% more than pelleting without bacteria); the other differences were not. *Streptomyces griseus* did not give a significant increase in this or in a second trial. *Bacillus subtilis* A 13 failed to give significant increases in the second trial, perhaps because of a more than 4-month delay between pelleting and planting. However, nearly all the results showed yield increases from inoculation with either microorganism. The greater increase from the bacteria-pelleted treatment than from the bacteria-water treatment may result from a difference in inoculum density. The A 13 isolate has also increased plant growth in soil treated at 60° C and 100° C/30 minutes, as discussed below.

Growth of the plant may, however, sometimes be reduced by soil inoculation with a microorganism (Broadbent et al., 1971) (Fig. 4.7; Chapter 7).

The successful bacterizations reported by Broadbent et al. (1971, 1974—see below), Price et al. (1971), and Merriman et al. (1974) all used selected bacteria. Isolations were either made directly from the surface of mycelia of a pathogen (A 13 from *Sclerotium rolfsii*) or from selected soil, and were screened in agar plates for antibiotic activity against six or more distinct plant pathogenic fungi. By contrast, bacterial isolates used by other investigators of bacterization have not been so screened. Apparently bacteria selected for pathogen inhibition have provided greater increase in plant growth than most of those not so screened. The question may, therefore, be raised whether the action of the bacteria may be to (a) inhibit nonparasitic pathogenic rhizosphere microorganisms that decrease plant growth without producing other symptoms or (b) break down the phytotoxins they produce. Such a view would be consistent with the statement of Rovira (1972): ". . . at least 20 percent of the production potential [of plants] is lost simply to satisfy an unfavourable microbiological situation." The increased growth obtained in steamed soil is not in-

FIGURE 4.7.
Retardation of growth and malformation of snapdragon seedlings by *Bacillus* sp. (left), and stimulation by *B. megaterium* (right) added to soil steamed at 100° C/30 minutes. Center: Control, no antagonist added. (From Broadbent et al., 1971.)

consistent with this hypothesis, in view of the demonstrated rapidity of recontamination (Schippers and Schermer, 1966). If rhizosphere microorganisms commonly produce substances that inhibit plant growth without specific symptoms, their selective displacement by chemical fumigants may provide a possible explanation of the widely discussed increased growth response (IGR).

Chang and Kommedahl (1968) similarly showed that inoculating corn seed with *B. subtilis* or *Chaetomium globosum* before planting in a field with a moderate inoculum density of *F. roseum* f. sp. *cerealis* 'Graminearum' gave about as good control of seedling blight as treatment of the seed with captan or thiram. In the field there was an average of 34.7% seedling stand in three varieties after 7 weeks without seed treatment or inoculation, 53.7% when inoculated with *Chaetomium,* 54% when inoculated with *Bacillus,* 55.3% when treated with captan, and 63.7% when treated with thiram. The effect was thought to result from antibiotic production, and the success to the relatively short protection period required and the persistence of the organisms in the rhizosphere, providing longer protection against the foot-rot phase than resulted from seed treatment with chemicals.

Tveit and Moore (1954) found in Minnesota that *C. globosum* and *C. cochlioides,* which occurred naturally on oat seed from Brazil, were the real basis of the supposed resistance of these varieties to *Helminthosporium victoriae* in the field. Hot-water treatment of the seed made the Brazilian varieties fully susceptible. *Helminthosporium victoriae* is unimportant on oats in Brazil, presumably because the seed is protected by the prevalent naturally occurring *Chaetomium* spp. If this is so, the Brazilian situation would be another example of effective biological control, which should be investigated in Brazil.

Wood and Tveit (1955) obtained control of *Fusarium nivale* on oats in England. Seed heavily infested with *Fusarium* was sown in the field along with oat-straw cultures of various *Chaetomium cochlioides* isolates. One isolate gave a 38–40% stand of plants at harvest, as against 62% for seed treated with an organic mercury dust, and 25% for the check. In another test, the figures after 2 months were 35–38, 48, and 0%, respectively. Of 47 isolates of *Chaetomium* spp., only 8 gave any protection, and the majority were ineffective. The effective strains did not inhibit *Fusarium* in culture, an unusual circumstance if the metabolite cochliodinol is the active agent (Meiler and Taylor, 1971).

Inoculation of Vegetative Structures. The most conspicuously successful control of this sort is provided by the work of Rishbeth (1963) on *Fomes annosus* by inoculating freshly cut (and therefore uninfected) pine stumps with spores of *Peniophora gigantea,* a low-grade pathogen (Chapter 10). This control is now widely used in England. The principle here is similar to that of the control of *Fusarium oxysporum* f. sp. *batatas* (McClure, 1951) by inoculation of the cut surfaces of sweet potato cuttings with certain isolates of *F. solani,* a low-grade pathogen that occupies the wounded surfaces and prevents infection by the more injurious pathogen. Aldrich and Baker (1970) used *Bacillus subtilis* to inoculate freshly cut surfaces of carnation cuttings; when these rooted cuttings were transplanted to soil infested with *F. roseum* f. sp. *cerealis* (which produces the major uncontrolled disease of commercial carnations), a fair degree of protection was provided to the growing plants. Michael and Nelson (1972) obtained good control of this pathogen by dipping unrooted carnation cuttings in a suspension of bacteria (a pseudomonad isolated from soil), and rooting in a propagation bed inoculated with the pathogen. Rooting was unimpaired, and plants grew well for 4 months, although the fungus could be isolated from the symptomless stems. Perhaps even better results would have been obtained if the inoculated cuttings had been rooted in treated media before being planted in infested soil.

In these examples, the antagonist was provided with a favorable environment for growth and with relative freedom from competitors. Also it was presumably applied in generous amounts, and except for *F. annosus,* needed to be effective only for relatively short periods. It became well established before invasion by the pathogen could occur and it was not expected to disseminate from the point of application to a distant infection

FIGURE 4.8.
Chrysanthemum cuttings rooted in propagation beds inoculated with 1700 sporangia of *Pythium ultimum* per g. Plants at left had almost total root destruction. Plants at right rooted in the same type of medium plus antagonist amendment. The organic amendment was sterilized, exposed to airborne contamination in the glasshouse area, and allowed to "ferment" for a few weeks. It was then added to a propagative bed of near-sterility at 1:7 by volume before the cuttings were stuck. The amount of root decay was minimal. (Photo courtesy of A. G. Watson and California-Florida Plant Corporation, Fremont, California.)

court. Such application of biological control is a type "most likely to succeed" in the present state of our knowledge.

Inoculation of Soil. It is surprising that introduction of antagonists into treated soil has been so seldom attempted, since it is less difficult than introduction into field soil. However, application of antagonists to propagation and seed beds to protect the propagule after it is transplanted has been investigated. Aldrich and Baker (1970) inoculated the carnation propagation medium with *Bacillus subtilis* plus macroconidia of *F. roseum* f. sp. *cerealis* in ratios from 0.5:1 to 500:1. Good disease control was obtained with the higher concentrations; better control probably would have resulted by inoculating the propagation bed with *Bacillus,* and transplanting rooted cuttings to infested soil, a closer simulation of commercial practices.

A commercial test by A. G. Watson (unpublished) showed that chrysan-

FIGURE 4.9.
Differential effect of two *Bacillus* spp. on damping-off and wire-stem of pepper seedlings caused by *Rhizoctonia solani* isolate 15. Left: Disease increased by *Bacillus* sp. over that in the control (center). Right: Disease diminished by *Bacillus subtilis*. Inoculation with *R. solani* at front left in each photo. (From Broadbent et al., 1971.)

themum cuttings rooted in a steamed propagative medium inoculated with random antagonists were well protected against a high inoculum density of *Pythium ultimum* when they were transplanted to infested soil (Fig. 4.8). Moist organic matter of near-sterility exposed to airborne contamination for several days provided the inoculum of antagonists that was placed on the propagative beds. Although this method of obtaining antagonists cannot be recommended, the level of success attained shows that empirical methods may give surprising results. Nair and Fahy (1972) controlled brown blotch of mushroom by inoculating the casing soil (this chapter).

Ferguson (1958) used various saprophytic fungi, and Olsen and Baker (1968) used isolates of *Bacillus subtilis* to protect seedlings from *Rhizoctonia solani*. Broadbent et al. (1971) introduced *Bacillus* or *Streptomyces* spp. into steamed soil and reduced damping-off caused by *Pythium ultimum* and *Rhizoctonia solani*. Some microorganisms inoculated into soil may, however, increase the amount of disease (Fig. 4.9).

Extensive commercial trials were made by Broadbent et al. (1974) in a New South Wales nursery to determine the feasibility of inoculating treated soil with bacteria to increase the growth rate of bedding-plant seedlings. Seed germination was increased for portulaca, delphinium, eggplant, celosia, dahlia, cabbage, and alyssum by *Bacillus* sp. WW 27, and for portulaca, delphinium, and snapdragon by *B. subtilis* A 13. Germination of several species was increased 25%, and that of portulaca 133%. Seedling top weight was increased by WW 27 for pepper, portulaca, and celosia, and by A 13 for delphinium, dahlia, and carnation. The weight increase frequently was 25%, and that of celosia reached 123%. Although marked specificity was shown in plant-bacteria response, and top weight, germination, or both, were sometimes reduced by inoculation with the bacteria, the method was regarded by the nurseryman as having commercial promise. These results, and the data of Merriman et al. (1974) and of Price et al. (1971) on seed inoculation reported above, show that the method may be successfully applied to nontreated field soil as well as soil steamed at 60° or 100° C/30 minutes. The studies are being continued. Seed germination and seedling top weight of portulaca, snapdragon, carnation, and alyssum in the uninoculated checks were greater in soil steamed at 60° than at 100° C for 30 minutes, probably because of phytotoxins formed in the latter.

HOST RESISTANCE

Plants have constantly been selected for resistance to or tolerance of pathogens, as well as for other characteristics. Some of the mechanisms, such as the rhizosphere flora of corn in relation to phymatotrichum root rot, are forms of biological control by any present definition. Others (*Crotalaria* as a nematode trap-plant, *Tagetes* as an inhibitory plant) are also forms of biological control, as is disease control resulting from inhibitory activity of the rhizosphere flora on the pathogen. The dominant factor controlling this rhizosphere flora is, however, the quantity and type of root exudate, controlled in large part by host genes. Exudates from the hypocotyl, stems, leaves, and buds also influence the flora on those structures.

Since phytoalexins may be induced in host tissue by organisms that are not, as well as by those that are pathogenic to the plant, it is probable that some antagonists will eventually be found that incite resistance in this way. This again involves host genes. The host is certainly a crucial part of the biological environment of the pathogen. The continuing selective pressure on microorganisms is involved in biological control in many ways. Why should the selection of the host be treated as an unrelated subject? Any clear distinction between host resistance and biological control is as difficult, in the present state of our understanding, as it is fruitless. When man incorporates resistant genetic material into agronomically or horticulturally desirable varieties, and thus prevents a disease, he is accomplishing one of the most successful types of biological control.

ECOLOGICAL MANIPULATION TO CONTROL WEED MOLDS AND PATHOGENS OF MUSHROOMS

The commercial mushroom (*Agaricus bisporus*) is here viewed as analogous to a pathogen, comparable to *Armillaria mellea, Fomes lignosus,* or *F. annosus,* and its many weed molds and parasites are viewed as antagonists of it. Interpreted in this way, mushroom culture is a remarkable example of commercially successful biological control. The mushroom grower generally suppresses the antagonists, but when he fails, they reduce or prevent production of caps. In effect, he attempts to prevent development of the antagonists, whereas producers of other crops encourage them. Information on conditions that favor these antagonists may, therefore, prove helpful in the usual forms of biological control.

It is not improbable that successful biological control may eventually involve suppression of hyperantagonists of an antagonist to the pathogen, as well as direct stimulation of the antagonist itself. When biological control thus involves second- or third-order interactions, the gap will have been closed between mushroom culture and biological control. The viewpoint here is that the complex second- or third-order microorganism interactions involved in mushroom culture will sooner or later find application in other phases of agriculture, and we therefore discuss them in some detail.

Since mushroom culture began in France after 1600 as an outgrowth of melon production on piles of manure, there has been a remarkable development in culture techniques, empirical at first, but increasingly scientific after 1920. Because of the exceptionally favorable conditions of nutrition, moisture, and warmth provided in modern mushroom houses and caves, many saprophytic weed molds and parasitic fungi, bacteria, and nematodes have to be controlled for successful commercial mushroom production. Environmental manipulation to minimize the occurrence of at least 14 fungus, bacterial, virus, and nematode parasites, and 16 weed molds has provided an unparalleled example of successful biological control of a complex group of microorganisms. As Sinden (1971) said, "Probably more than for any other plant crop the use of chemical control measures can be and is being replaced by ecological control." Although it is true that complete control of the environment has made this success possible, this is merely an advanced example of the improvement in environmental control being developed in all agriculture.

The compost consists of straw-bedded horse manure, hay and straw, hay and rough-ground corn cobs, or various combinations of these, enriched with chicken manure, oilseed meal, brewers' grains, or malt sprouts. The thermal manipulation of this mass provides the necessary chemical composition and biological control for maximal growth of the mushroom. The stages in mushroom culture are as follows:

Composting Phase I

The compost is mixed and arranged in long piles outdoors, or in huge revolving drums, to decompose for 5–10 days (formerly 15–30 days). The temperature and moisture are carefully controlled by repeated turning and by irrigation of the piles. This was formerly done in soil- or cinder-based yards, but changing to large concrete areas eliminated the soil organisms, *Diehliomyces microsporus* and *Chrysosporium luteum,* weed molds that commonly cause the truffle and mat diseases, respectively, and that inhibit mycelial growth of the mushroom.

The piles during this outdoor composting are frequently turned by machinery to maintain aerobic conditions and temperatures of 71–80° C. Because of uneven thermogenesis and cool surfaces of the piles, this stage does not free the compost of weed molds or pathogens, but it does produce a medium unfavorable to several weed molds. If the medium is overcomposted, two organisms, *Papulospora byssina* and *Scopulariopsis fimicola* (brown and white plaster molds), develop and may subsequently inhibit mushroom mycelia. If the medium is kept too wet and partially anaerobic, *S. fimicola* is favored, often giving a black gelatinous medium. Addition of gypsum at the beginning of composting improves porosity and aeration of the medium and reduces the incidence of this fungus. *Trichoderma viride,* strongly inhibitory to mycelial development, becomes troublesome if the compost temperature is permitted to reach 100° C; it also indicates the presence of incompletely decomposed organic matter, as does the inhibitory *Doratomyces* sp. With the present short-composting procedures, these molds are extremely rare.

Sufficient carbohydrates must be left in the compost at the end of this phase to carry the medium through Phase II. The endpoint here is critical, because too little residual carbohydrate will not permit conversion of ammonia and amines to protein, and too much favors development of weed molds such as *Trichoderma lignorum* and *Stysanus* sp. According to L. C. Schisler (unpublished), the presence of *T. lignorum* leads to severe infestations of pigmy mites, which feed on it but not on mushrooms. They cause skin irritation to pickers and also disseminate *Pseudomonas tolaasii* and *Verticillium malthousei. Stysanus* spores cause a severe respiratory trouble similar to "farmers' lung" in workers who clean out the house at the end of the crop.

Composting Phase II

The second phase is usually done in trays in a building under carefully controlled conditions. This usually takes 4–7 days (formerly 10–20 days) at a temperature slowly rising by thermogenesis to 50–60° C under abundant oxygen provided by forced ventilation. This phase is very important as a control measure of weed molds and parasites, but also provides continued decomposition to free the medium of ammonia and amines, and to produce the microbial protein to support mushroom mycelial growth. This phase is continued until the ammonia has been eliminated and the temperature drops, indicating reduced thermogenesis. At the end of decomposition, the medium should have 2.0–2.5% nitrogen, with less than 0.05% ammonia (dry weight). Since *Coprinus fimetarius* (ink caps), *Oedocephalum* sp. (brown mold), and *Thielavia thermophile* are able to use the ammonia and amines, which are toxic to mushroom mycelia, the balance swings to favor the weeds if this process is incomplete. *Coprinus fimetarius* inhibits the mushroom mycelia and is favored by compost with a pH above 8.2 from excessive ammonia.

If adequate forced ventilation over the tray surfaces and through the compost is not provided, anaerobic conditions develop and thermogenesis falls. When this occurs, unidentified compounds toxic to mushroom mycelia are formed; these materials are, however, available to *Chaetomium olivaceum* (olive-green mold), which then flourishes and further inhibits mushroom mycelial growth. Such anaerobiosis may result from excessive compaction of compost, too thick a layer (20 cm is usual), or permitting ambient temperature to rise for several hours above that in the compost. Lambert and Ayers (1952) have suggested that, if the compost is permitted to exceed 65° C in Phase II, the medium also becomes particularly favorable to *Chaetomium*. To some extent this effect may be overcome, for unknown reasons, by continuing Phase II for 5 days more at 43.5–60° C.

Pasteurization

After thermogenesis during Phase II is well established, the peak-heat pasteurization (or in England, methyl bromide fumigation) is carried out. The air temperature of the room and the surface of the compost must be raised to 60–62° C for 1 hour to destroy weed-mold spores, nematodes, mites, and insects. This must be done quickly with copious injections of live steam, and then quickly lowered after 1 hour by full volume ventilation and circulation, to prevent anaerobiosis.

At the end of Phase II, the compost is free of actively growing weed molds, pathogens, and competitors, and is able to support rapid development of mushroom mycelia, even if other fungi are later introduced.

Spawning

Inocula of selected strains of *Agaricus bisporus* are grown in pure culture on rye, wheat, or millet seed, usually by commercial laboratories. This spawn is uniformly mixed throughout the finished compost; the greater the amount of inoculum used, the faster the occupation. The mycelia of *Agaricus bisporus* are antibiotic to bacteria and other fungi, and produce volatile materials that may be inhibitory to some organisms but stimulate some bacteria. However, if spores of *Pythium artotrogus* or an unidentified *Coprinus* sp., for example, get into the compost during spawning, the infestations will spread until checked by the surrounding developed mushroom mycelia. This contamination is prevented by filtering the air and by sanitation (see Spawn Running).

Spawn Running

This stage is performed in special well-ventilated rooms held at 25–28° C. The compost has 2–2.5 parts water per part dry weight. Steam may be injected into the air stream to maintain humidity during mycelial growth. The air-compost temperature differential may be 10° C, as a result of thermogenesis of mushroom mycelial growth. After 10–20 days the temperature falls, indicating that growth is completed, and that the fungus is ready to fruit. During this period the *Agaricus* often is in nearly pure culture, despite the huge quantities (80 tons or more in a house of 8000 square feet of bed) of materials handled! However, some fungi, unaffected by the antibiotics of *Agaricus,* may compete for the available substrate and may even kill the mushroom mycelia with their metabolites. Among these are the weed molds, *Sepedonium* sp., *Geotrichum candidum, Chrysosporium luteum,* and *C. sulfureum* (confetti). These, and *Pythium artotrogus* and *Coprinus* sp., are excluded by filtering the air and maintaining sanitary conditions in the area where spawning, spawn running, casing, and production are carried out. This would also exclude virus-infected spores of *Agaricus,* which otherwise fall on the compost, germinate, and anastomose with the spawn, infecting it. Keeping the floors and halls wet with water will reduce spread by dust. Using large quantities of inoculum in spawning insures a quick start of the mushroom mycelia and prevents the inhibition of it by *Coprinus* sp.

Before the use of concrete slabs in composting yards, the damaging truffle disease was controlled by holding the temperature during spawn growth at 24° C for 1 week, and then finishing this stage at 15.5–18.5° C. Spores of *Diehliomyces microsporus* germinate in 2 weeks at 21–

26.5° C, but below 18.5° C they remain dormant. This enforced dormancy of the spores provided excellent control, even in heavily infested houses.

If the composting has not been completed in Phase II, the thermophilic molds, actinomycetes, and bacteria may sometimes reactivate and raise the compost temperature high enough to kill or injure the mushroom mycelia at this time.

Casing

Placing a casing layer about an inch thick over the spawned compost is necessary for commercial production of mushrooms. The reasons for this effect are still controversial, but the work of Eger (in Baker and Snyder, 1965) in Germany and of Hayes et al. (1969) in England indicates that bacteria (*Pseudomonas putida*) in the casing mixture stimulate cap formation, and are in turn stimulated by volatiles given off by the mycelia (Chapter 2).

In most areas, peat mixed with ground limestone is now used, but pulverized soft marl is used in France. In some localities the spent compost is spread in the field for 2 years to decompose and leach, and then mixed with soil or tuff. If specific bacteria are indeed the sole triggering agent, as suggested, their occurrence in such a range of media is surprising. The moisture-retentive casing material effectively provides abundant water for the rapidly expanding mushrooms.

If the spent compost is to be spread in the field, it must first be pasteurized, since an abundant source of spores of weed molds, parasites, and virus-infected mushroom spores will otherwise contaminate the whole area.

Casing soil is usually treated at 60–70° C for 30 minutes (preferably at 70–75° C with aerated steam) and is best applied to the tray or bed by machinery directly from the container in which it was treated, to minimize contamination. This treatment kills *Verticillium malthousei, Mycogone perniciosa* (bubble disease), *Trichoderma koningi, Pseudomonas tolaasii* (brown blotch), and the nematodes *Ditylenchus myceliophagus* and *Aphelenchoides composticola,* as well as virus-infected mushroom spores and colonies, and other organisms that may have been in it. Casing should be done in a room supplied with filtered air and with strict sanitation (see Spawn Running). Antagonists that survive treatment at 60° C/30 minutes in casing soil apparently are suppressive to *Verticillium* (Chapter 7).

Overheating of casing soil may reduce yield and mushroom size. Nematodes can be killed in it at 50° C for 1 hour, but if other organisms are present, 60–70° C, or even 70–75° C, may be required.

The change from soil to peat for casing helped eliminate *Chrysosporium luteum,* for reasons explained under Phase I. *Peziza ostracoderma* (brown mold) is, on the other hand, especially favored by peat.

Production

Present practice is largely to grow mushrooms in trays, producing 5–7 crops per year, each yielding 3–4 pounds/square foot (14.7–19.5 kg/m²). During production, the rooms are held at 90–95% relative humidity and at 15–17.5°C for high yield. Harvesting begins 15–25 days after casing, and continues for 30–60 days.

These conditions are ideal for *Verticillium malthousei,* the most widespread and probably the most destructive pathogen of mushrooms. It spreads rapidly through the beds. This fungus can be dramatically controlled by reducing relative humidity to 80–85% and the temperature to 14°C, but this extends the harvest season. Spores are carried mainly by pickers and flies, and by water splashing during watering the beds. *Mycogone perniciosa,* on the other hand, spreads slowly in the trays and is epidemic only where multiple contamination of the casing soil by it occurs.

The short picking-cycle helps reduce the spore load of *Mycogone* and *Verticillium.* The best control is exclusion (see Spawn Running, Casing). If the rooms can be pasteurized with live steam between cycles, contamination from that source is minimized.

Pseudomonas tolaasii (possibly a form of *P. fluorescens*), the cause of brown blotch, is omnipresent in casing soils and is spread in watering. An effective biological control is to use *P. multivorans, P. fluorescens,* or *Enterobacter aerogenes* added as a peat culture to the casing soil after it is applied to the compost (Fig. 4.10). The antagonists must be at least 80 times the number of pathogens to be effective. With 100% brown blotch in the check, these antagonists reduced the disease to 7.6, 9.7, and 11.1%, respectively, perhaps by competition for nutrients. Antibiosis to *P. tolaasii* is not shown in culture, and there is no inhibitory effect on growth of the mushroom or the stimulators (*P. putida*) of cap formation (Nair and Fahy,

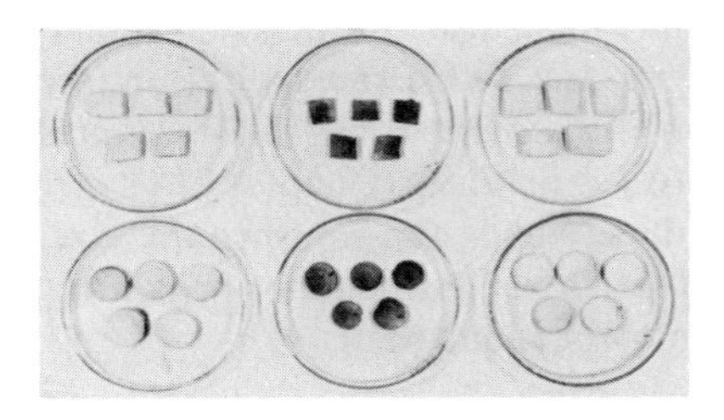

FIGURE 4.10.
Biological control of brown blotch (caused by *Pseudomonas tolaasii*) of mushroom. Transverse blocks of caps (top) and stems (below) inoculated with *P. tolaasii* (center), with *P. multivorans* (left), and *P. tolaasii* plus *P. multivorans* (right). (Photo courtesy of N. G. Nair and P. C. Fahy.)

1972). A commercial trial gave 4135 mushrooms weighing 219 kg, with 1.6% brown blotch in a bed in which the casing soil had been inoculated with antagonists. The check bed yielded 2612 mushrooms weighing 212 kg, with 20.4% brown blotch (N. G. Nair and P. C. Fahy, unpublished).

The disease can also be prevented by environmental control. Chlorinated water has sometimes been used for moistening the beds, but high rate of ventilation and accurate temperature regulation afford better control. Some houses maintain 93±2% relative humidity, and a temperature that does not fluctuate more than 0.1° C. If air movement is adequate and if air temperature does not fluctuate, caps dry within 1 or 2 hours, and infection is prevented, because they absorb water and remain slightly warmer than ambient, because of respiration. If air temperature rises as little as 0.5° C, the cap is cooler than ambient, and moisture condenses on it. Control depends, then, on efficient air-conditioning equipment and well-insulated houses. Control is also possible by lowering the relative humidity to 80–85%, but mushroom quality and quantity may quickly decline (see *Verticillium* above). Another *Pseudomonas* sp. that causes mummy disease is eliminated by proper composting and by treating the casing material. *Hypomyces auranteas* and *H. rosellus* (mildew) also require relative humidity above 90% to infect. They can be checked, even in summer, if the air conditioning is adequate. Sporulation and dissemination can be prevented by covering newly infected caps with calcium hypochlorite (15% powder) or salt.

Filtering the air and practicing sanitation in the production room are desirable to reduce pathogen dissemination (see Spawn Running). Viruses are reduced by picking before the caps open and shed spores (see Spawn Running). This involves reducing the temperature to 15° C prior to weekends so that picking of unopened caps may be accomplished during a 5-day week. It is, of course, important that the spawn used should be virus-free (Hollings and Stone, 1971).

Viruses may also carry over in infected mycelia in trays or beds; the diseased hyphae may later resume growth and anastomose with healthy newly planted mycelia, and infect them. Weed molds and fungus, bacterial, and nematode parasites may also survive between plantings in this way. Since the usual pasteurization will not disinfest parasites in cracks in the wood, it is best to treat the empty trays in the pasteurizing room at 75° C for 1 hour. Since *Agaricus bisporus* is killed in 30 minutes at 43.5° C in aerated steam, it is easier to eradicate than *Geotrichum candidum, Mycogone perniciosa, Trichoderma viride, Verticillium malthousei, Dactylium dendroides,* and *Ostracoderma* sp., which are killed at 54.4° C (Wuest et al., 1970; Wuest and Moore, 1972). *Diehliomyces microsporus* is said to resist 82° C, but this needs confirmation.

All machines for handling compost should be thoroughly cleaned and disinfested with 5% formaldehyde (Hollings and Stone, 1971).

Mushroom culture thus illustrates how a large series of disease agents may be controlled principally by manipulation of temperature and moisture. A correspondingly broad, imaginative, flexible, and ecologically sound approach to disease control has yet to be attempted for other crops.

5

APPROACHES TO BIOLOGICAL CONTROL WITH ANTAGONISTIC MICROORGANISMS

Knowledge proceeds by the development of techniques.
—LORD RUTHERFORD

The most reliable and effective knowing follows from direct and open confrontation with phenomena, no matter how complicated they are. Nature can be trusted to behave reliably . . . and nature's ways are open to direct, intuitive, sensuous knowledge. . . . What is urgently needed is a science that can comprehend complex systems without, or with a minimum of, abstractions.
—T. R. BLACKBURN, 1971

This chapter outlines some of the methods of studying and achieving biological control of plant pathogens through antagonistic microorganisms. The outline does not include all the excellent techniques developed for specific purposes. Many of these have been summarized by Johnson and Curl (1972). Methods of studying control of plant pathogens by host resistance and tolerance, because they are part of an established coordinate science, are not included here, although part of biological control.

What is an antagonist? When the evidence is in, we will probably find that every soil microorganism is antagonistic to some other one under suitable environmental conditions. **Antagonistic potential resides in every soil microorganism.** Any random soil sample should yield antagonists to some microorganism. Metabolites are secreted, and one of these would certainly prove inhibitory to some other microorganism. Finding one that will control the specific pathogen becomes the objective. An investigator seeking a control for *Rhizoctonia solani* might correctly conclude that a given soil had no suppressive effect, whereas another investigator might find the same soil inhibitory to *Phytophthora cinnamomi*. Some broad-spectrum antagonists might prove inhibitory to both.

SELECTING SOIL AS A SOURCE OF ANTAGONISTS

There is an old adage that in preparing rabbit stew, you must first catch your rabbit. The same advice applies to the study of biological control—it has first to be found.

Antagonists are where you find them, but some areas afford better hunting than others. However, no reasonable opportunity should be overlooked. Considering one type of antagonist, the antibiotic producer as an example, one should remember that the discovery of penicillin—and the beginning of the antibiotic age—resulted from a chance contaminant of *Penicillium notatum* in a plate inoculated with staphylococci and held in a cool laboratory (Hare, 1970). On the other hand, since most commercial antibiotics now come from soil microflora, the soil is still the preferred place to seek new antibiotic producers. We should seek antagonists in the most likely places, but not ignore "long shots" from any source.

Antagonists should be sought in areas where the disease caused by a given pathogen does not occur, has declined, or cannot develop, despite the presence of a susceptible host, rather than where the disease occurs.

Some generalizations on locations that should be considered may be helpful.

1. The pathogen is unable to establish. Ten examples were described in Chapter 4 in which an organism intentionally or accidentally introduced into a field soil resulted in no or only brief establishment. When massive quantities of inoculum are added to a soil, the pathogen may for a few months overwhelm the resident antagonists and then disappear, as did the *Rhizoctonia solani* isolate from the Eyre Peninsula introduced into soil in Adelaide, Australia (Chapter 4). With more natural amounts of inoculum, it is doubtful whether the pathogen would gain even a transitory foothold. To determine whether the inhibitory effect is biological, the soil may be steamed at 100° C/30 minutes; if the pathogen can then be established, it is usually considered that the inhibition is biological in origin. Such a soil would be worthy of further study.

2. The pathogen is present, but causes no disease. Three examples were described in Chapter 4. The same test as above will indicate whether this situation results from biological factors; if it does, the soil is

worthy of intensive study. This may be the most fruitful of all categories of pathogen-suppressive soils.

3. The pathogen diminishes with continued monoculture. Three examples were described in Chapter 4. Areas in which the losses caused by the pathogen have essentially ceased would be worthy of intensive study.

4. The host and parasite are native to the area. Since an indigenous host-parasite combination rarely, if ever, produces an epidemic in its native habitat, such an area would be a good source of antagonist populations as well as host resistance, as indicated by *Phytophthora cinnamomi* in Queensland (Chapter 4). **There is a tendency for both host resistance and the presence of antagonists to the pathogen to increase through long association.** If susceptible plants grown in such a soil remain free of the pathogen, the soil is eminently worthy of study.

5. The presence of antagonists in the area is suspected for other reasons. The inhibitory effect of the duff of established pine-forest soil on *Fusarium oxysporum* shown by Toussoun et al. (1969), although probably a type 1 situation (above) for the pine pathogen, has been found in Georgia, and in Victoria and New South Wales, Australia, to be suppressive to a much broader group of pathogens.

Although it is true that an effective antagonist may be found in unexpected places, the chances of success are greatly increased by systematic investigation of sites selected on bases such as the above. Sampling of "any convenient soil" is unlikely, in our experience, to prove productive. It is generally desirable to observe the various types of culture of the given crop in a given area, checking for the absence of the pathogen, or if present, checking on the amount of disease loss produced. The relationship of tillage practices, for instance, to the incidence of disease may supply important clues on biological control.

Special attention should also be given to the method by which the soil sample itself is collected. Although no rigid procedure can be prescribed, there are some suggestions that may increase the chance of obtaining soils that contain effective antagonists.

1. Collect the sample at the most favorable season for antagonists to be actively suppressing the given pathogen. When bacteria and actinomycetes are the antagonists sought, sampling is best done in seasons when the soil has been moist and warm for some time. In a drier season these organisms may be dormant and in low numbers.

This factor is also related to the effect of the abiotic environment on the crop plant and pathogen one is studying (4 below), both of which should be actively growing.

For example, soil samples collected for antagonists to *Fusarium roseum* f. sp. *cerealis* 'Culmorum' should be taken from low moist areas in the field, where bacterial antagonists are apt to be most active (Chapters 4 and 9).

2. Collect the sample at the soil depth where antagonists are most active. This would be in the lower levels of the surface duff in pine forests, and just under the surface leaves in a hardwood forest or certain orchards. When no litter is present, the immediate surface soil, because of fluctuating moisture and temperature, is apt to be rather free of roots and low in antagonists; collection should then be made in the root zone, at 5–15 cm. Populations of all microflora tend to fall off below that level, making for a less favorable collection location.

3. Collect soil whenever possible from the rhizosphere of the host to be protected rather than from the soil mass. The soil around seeds, bulbs, or belowground stems is also preferable to the general soil mass because these are the parts to be protected. If the host is not available at the site selected, roots of other plants are probably still preferable to the soil mass. If possible, it would be better to plant seeds of the given crop in the soil, and to collect samples from its rhizosphere after it has grown for several weeks.

4. If mycelial strands, sclerotia, or host refuse containing propagules of the fungus pathogen are present, these should be collected for study; antagonists are very likely to operate on the surface of these structures or of mycelia themselves (see below and Chapters 6 and 7).

ANTAGONISTIC POPULATIONS OF WHOLE SOILS

It is important to distinguish between antagonistic populations (used to transfer suppressiveness from one soil to another) and individual antagonists (used singly or in mixed groups to inoculate soil of near-sterility), as pointed out in Chapter 3.

The relative potential of the antagonistic population of a soil may be determined by fairly simple testing. Such tests employ the complete microflora as far as possible and help focus the search on the most promising sources of antagonists. A test of this kind should always be tried before launching into a more involved and sometimes less meaningful microbiological analysis.

Inoculated Flat Test

Ferguson (1958) devised a method for determining the suppressive effect of the whole soil microflora on *Rhizoctonia solani* and other pathogens. The nontreated test soil was placed in a shallow flat about 20 cm square, and the surface was smoothed. Seeds of the susceptible test plant were thickly sown over this surface and covered with a thin layer of the same soil. Seeds used in these tests should be surface disinfected with sodium hypochlorite (0.5% available chlorine), but should not be treated with a residual chemical such as a mercurial. A small uniform quantity of the fungus (grown as a mat in liquid culture and washed with sterile distilled water) was buried in one corner of the flat. The flat was maintained in a glasshouse and watered daily for optimum plant growth. When the seedlings emerged, the extent and severity of fungus attack were noted. In

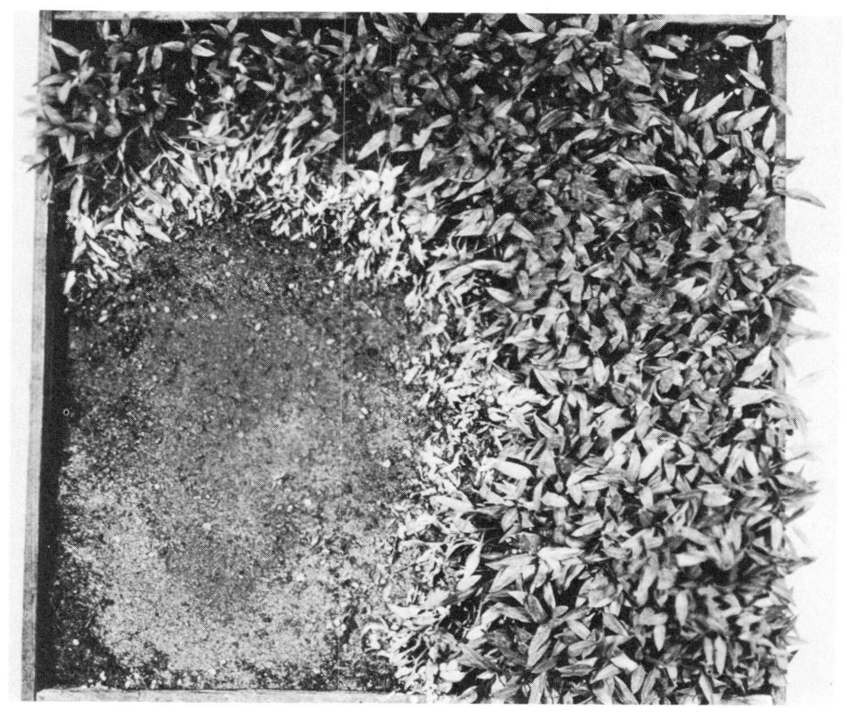

FIGURE 5.1.
Retardant effect of *Myrothecium verrucaria* on damping-off of pepper seedlings by *Rhizoctonia solani* in soil steamed at 100° C/30 minutes. Upper right: Both fungi introduced into corner at time of seeding. Damping-off was slight. Lower left: Same isolate of *R. solani* introduced into corner at same time as above. Damping-off is unchecked. (From Ferguson, 1958.)

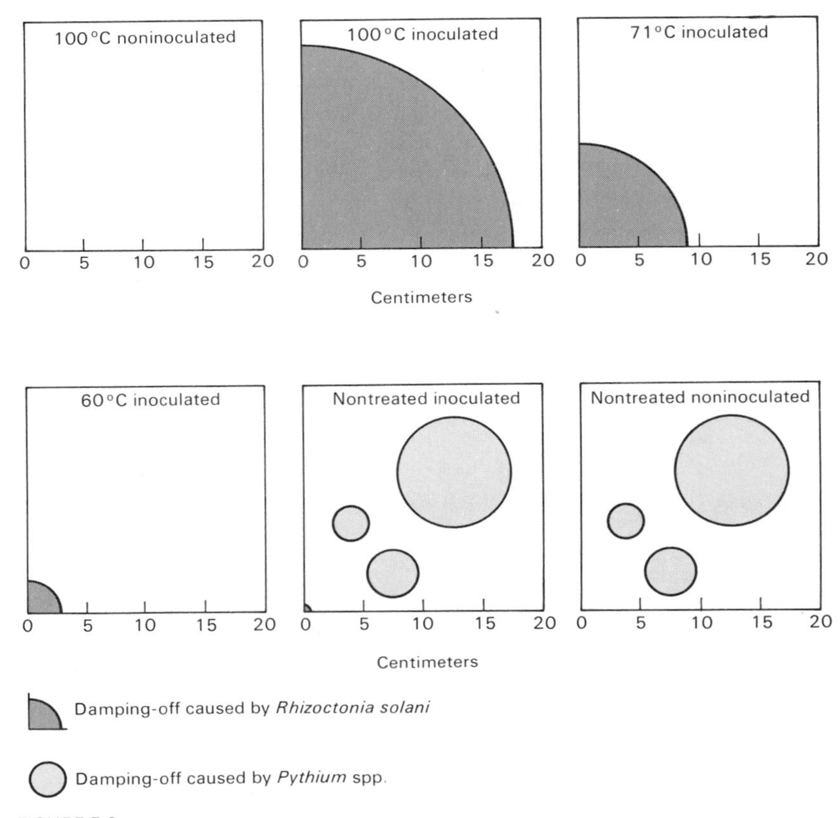

FIGURE 5.2.
Diagram of the relative occurrence of damping-off of pepper seedlings caused by *Rhizoctonia solani*, advancing from inoculum inserted at lower left of each flat. The field soil was treated for 30 minutes at indicated temperatures with aerated steam. It was naturally infested with *Pythium* spp., which caused spots of damping-off in nontreated, but not in treated, soil. (Data from Olsen and Baker, 1968.)

conducive soils, or in suppressive soils rendered nearly sterile (steamed at 100° C/30 minutes), the fungus spread out fanlike from the inoculated corner (Fig. 5.1). Seedlings in the corner showed preemergence damping-off; further out, postemergence damping-off; still further out, attack of older seedlings. With a nontreated suppressive soil there was, on the other hand, loss of perhaps a few seedlings in the corner, but not further out. This could be quantified as the surface area or linear distance of damping-off caused by the pathogen in a given time, or as the percentage of seedlings infected. This method has proved generally useful in evaluating the comparative inhibitory effect of the soil against the given pathogen. Small plastic trays or punnets about 10 × 20 cm are now used.

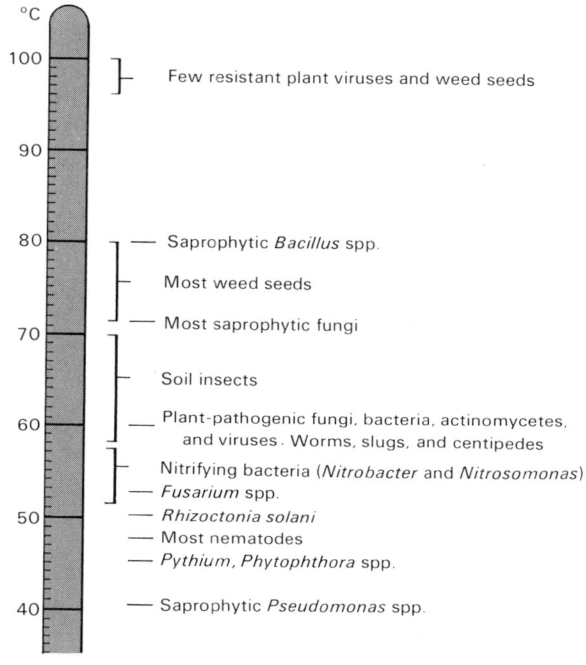

FIGURE 5.3.
Temperatures necessary to kill various groups of soil organisms. Based on 30-minute exposure to moist heat (10 minutes for nitrifying bacteria). (Modified from Baker, 1957.)

Broadbent et al. (1971) showed that, for Australian soils at least, this method gave an erroneous reading for *Rhizoctonia solani.* Isolated single pepper seedlings in the middle of a flat of treated soil inoculated with various single antagonists sometimes showed damping-off, although surrounded by noninfected plants. Isolations from the soil indicated that the mycelia freely grew through it but apparently either made no contact with the seedlings, or were prevented by the antagonist from infecting seedlings. If, however, the soil was steamed to near-sterility, the pathogen moved on an advancing front of infected seedlings in the usual manner. In dense stands of seedlings under humid conditions, the pathogen may spread through the tops, further invalidating the method.

This method can be modified to compare the suppressive effect of a nontreated soil with one treated with aerated steam at 60°, 71°, and 82° C, and with regular steam at 100° C/30 minutes (Fig. 5.2). This will permit assessment of the effect of different fractions of the antagonistic microflora as selectively eliminated by heat (Fig. 5.3).

FIGURE 5.4.
Tests with soil highly suppressive to *Phytophthora cinnamomi* in the field and when additionally infested in the glasshouse. Soil from a Queensland avocado grove untreated (left), steamed at 60° C/30 minutes (center), steamed at 100° C/30 minutes (right). Each lot was then inoculated with a culture of *P. cinnamomi* and planted with jacaranda (*Jacaranda acutifolia*). Photo after growing 4 months at 24° C. The resident antagonists suppressed the fungus and survived treatment at 60° but not 100° C. (From Broadbent et al., 1971.)

Inoculum Blended with Soil

Lester and Shipton (1967) devised a method to assay for antagonism to *Gaeumannomyces graminis* var. *tritici* (*Ophiobolus graminis*) in soils. In this method, 500 g inoculum of the fungus as a sand-wheat meal-water mixture was blended with 2 kg of soil in a plastic bag, then incubated at 20° C for 4 weeks. After this incubation, the sides of the bag were rolled down to form a container, wheat seeds were sown, and after 3 weeks, the number of root lesions or the percentage of roots infected with *G. graminis* was determined. Although this method did not work for eastern Washington soils (Shipton et al., 1973), in Britain, inoculum survival was poor in soils of long-term wheat monoculture, and good in soils from fields of short-term or no recent cereal culture. The technique was based on an earlier one of Zogg (1959), who mixed cultures of *G. graminis* with soils from fields of various rotations, then assayed for inoculum immediately and after 3 and 10 weeks incubation. Inoculum decline was considerably more rapid in some soils than in others, indicating that the soils could be distinguished on the basis of their antagonistic effect. Gerlagh (1968) compared levels of antagonism to *G. graminis* in soils of the polders in Holland, but used sterilized straws infested with *G. graminis* added to the soil (1% w/w) and sown immediately to wheat. Differences in levels of antagonism to *G. graminis* were expressed as differences in disease severity in glasshouse pots.

A similar method by Broadbent et al. (1971) used small plastic trays filled with treated or nontreated soil, in which 1 g (about 1:580 w/w) of cornmeal-sand inoculum of *Phytophthora cinnamomi* was mixed. A suppressive soil in which *P. cinnamomi* was present without causing detectable root rot was tested in this way. Jacaranda seedlings grew well in nontreated and treated (60° C/30 minutes) soil to which *P. cinnamomi* had been added, indicating that the antagonists prevented infection. In soil treated at 100° C/30 minutes and then inoculated, the seedlings were severely diseased (Fig. 5.4). Seedlings grew well in soil after all three treatments when *P. cinnamomi* was not added. The antagonists thus survived 60° but not 100° C treatment, and provided excellent biological control. All antagonists are not this heat tolerant (for example, those of *Gaeumannomyces graminis;* Chapter 4). Results with this *P. cinnamomi*-infested soil gave excellent agreement with field experience and with a soil-tube test (see below). Unless the mixture is incubated for a few weeks prior to planting, for equilibration between pathogen and antagonist, as was done by Lester and Shipton (1967) for *G. graminis,* the inoculum density must be kept within the range encountered in the field.

Growth Through Soil in Tubes and Petri Dishes

Tests for antagonism may be based on saprophytic growth of a pathogen through candidate soils and their associated microbiota (Broadbent et al., 1971) (Fig. 5.5). This method is suitable for laboratory assays, does not require much space or a glasshouse, eliminates to a large extent the environmental variables of a glasshouse, and does not need as much care as flats of seedlings. A thin layer (2 ml) of dilute alfalfa-extract agar was placed in aluminum-capped glass vials (2.5 × 12 cm) and autoclaved. Soil-extract agar was sometimes used but was less satisfactory. The medium acts as a weak food base for the pathogen. Tubes were then inoculated with the pathogens to be tested and held until growth was well started. The candidate soil was moistened with sterile distilled water and placed in the tubes (3 cm deep above *Phytophthora* spp.; 9 cm for fungi of good competitive saprophytic ability). Mucors and other fast-growing fungi overran the tubes filled with some soils. To eliminate them and any resident plant pathogens, the soil generally was treated with aerated steam (60° C/30

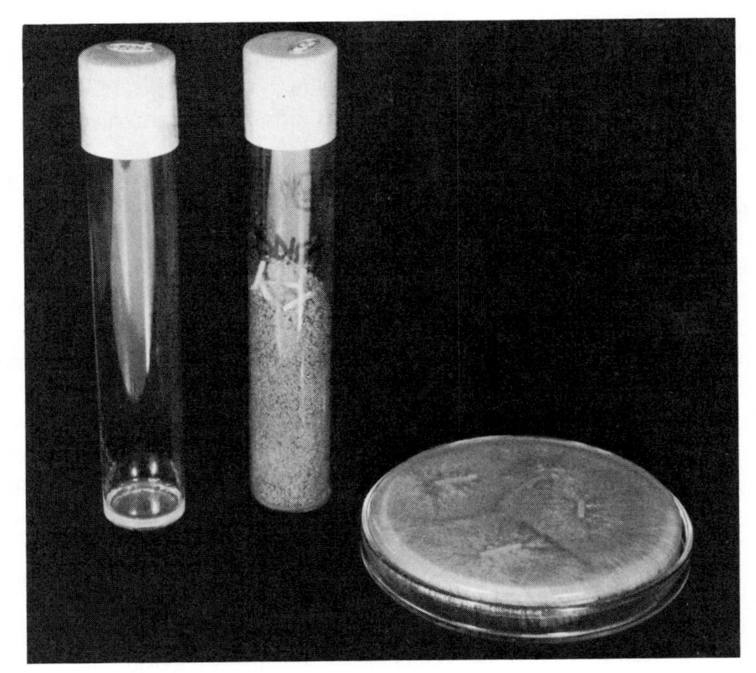

FIGURE 5.5.
Soil-tube method for assaying a soil for antagonism. Dilute alfalfa-extract agar in a tube (left) is inoculated with a pathogen, and when growth is well started, the moistened candidate soil (treated at 60° C/30 minutes) is placed over it (center). After 1–2 months cold-sterilized alfalfa stems are placed as bait on the surface for 1–2 days and then transferred to an agar plate (right) to determine whether the pathogen has grown through the soil. If it has not, the soil has antagonistic potential (Photo courtesy of P. Broadbent.)

minutes), which left many of the antagonists (Baker, 1962; Bollen, 1969) and did not produce phytotoxins (Baker, in Toussoun et al., 1970). The heat treatment unquestionably kills some antagonists, but is a compromise between maximum yield of antagonists and minimum difficulty with extraneous microorganisms. Treatment should be at the lowest temperature found to accomplish both results.

The tubes were held in the laboratory, or preferably in a humidity cabinet to prevent excessive drying. After 1–2 months, fresh alfalfa stems that had been sterilized with propylene oxide were placed on the soil surface for 1–2 days, and then transferred to agar plates to determine whether the pathogen had grown to the soil surface. There was excellent agreement between results with the soil tubes and with the inoculated flats. *Rhizoctonia solani,* for example, grew through all the Australian soils tested, as Garrett (1970) had also found in England. There was also good correlation between the inability of a fungus to grow through the soil in this test and its inability to induce disease in that soil in the field.

Martinson (1963) used perforated soil microbiological sampling tubes to assay the effect of inoculum density, PCNB treatment, and temperature on disease potential of *R. solani;* there was a high positive correlation between growth of the pathogen into the tubes and the amount of damping-off of radish seedlings. The fungus has to grow only a short distance through the soil before penetrating the tube in this method, perhaps providing a situation comparable to that of a root in soil.

Use of petri dishes instead of tubes permitted direct examination of the hyphae. A thin layer of alfalfa-extract agar was poured into a petri dish and inoculated with the test fungus. When the mycelium had formed a sizable colony, the moistened candidate soil (treated at 60° C/30 minutes) was placed in the dish to a depth of 1.5 cm. After 1–2 weeks, sterile alfalfa stems were placed on the surface and then cultured as above. The uncovered dish was then inverted and given a sharp blow to knock the soil from it. This exposed the hyphae on the agar surface so that they could be examined microscopically for lysis. Colonies of bacteria on the surface of lysed hyphae could also be picked off for culturing and more detailed study.

This method is particularly useful for pathogens that produce little or very slow growth in soil (*Verticillium albo-atrum;* Sewell, 1959) and cannot, therefore, be studied in the soil tubes.

Double-Layer and Barrier-Membrane Methods

Williams and Kaufman (1962) used a double-layer agar technique to compare differently cropped soils for relative numbers of microorganisms antagonistic to *Fusarium roseum* f. sp. *cerealis.* A 250-mg soil sample was diluted 1:10,000 in 1.0% carboxymethyl cellulose in water, and then 1 ml of this suspension was mixed in 10 ml melted nutrient agar amended with streptomycin, chloramphenicol, sodium propionate, and ox bile to retard growth of soil microorganisms, particularly fast-spreading types. After 4 days, a second agar layer containing propagules of the *Fusarium* was poured over the soil-dilution layer. Two days later the zones of inhibition were counted.

Neal (1971) used a similar method for determining the approximate number of antibiotic-producing bacteria in rhizosphere soil of wheats susceptible and resistant to *Helminthosporium sativum.* The bottom layer of soil-extract agar + glucose and cycloheximide was inoculated in eight locations with 0.1 ml of the soil slurry and then incubated. The second layer of cooled potato dextrose agar + streptomycin was poured over this and allowed to stand overnight to permit any antibiotics produced to diffuse into the top layer, which was then seeded heavily with spores of *H. sativum*. Areas over inoculated spots in the bottom layer would be inhibited if antibiotic producers were present. By using serial dilutions of 10^{-1} to 10^{-6} in separate plates, it is possible, referring to Table 1 of Harris and Sommers (1968), to transform the counts to numbers per g. This method does not indicate the number of species of antibiotic producers present or their relative efficiency, nor help to obtain them in culture, but it is useful in studies (Neal et al., 1970; Larson and Atkinson, 1970) where approximate numbers of antibiotic producers are involved (Chapter 8). The antibiotics used in the two layers will necessarily vary for different pathogens and antagonists.

Fox (in Baker and Snyder, 1965) placed a soil sample in a petri dish, smoothed the surface, and poured a thin layer of cooled sterile tapwater agar onto it. Small disks of a malt agar culture of *Fomes lignosus* were placed on the center of the surface, and the plates were incubated. Inhibition of growth of the pathogen occurred with some soils before microorganisms became microscopically evident, probably because of diffusion of fungistatic factors through the agar. Some colonies were later lysed by resident microorganisms.

A further modification of this general method would be to seal the moistened soil in a dialyzing bag and to place this in soil or on an agar plate on which the pathogen is growing. This can be used only for a few days, as cellulolytic microorganisms will digest holes in the bag.

Another promising method is to place cellophane, dialysis tubing, a sintered glass plate, or a Millipore or Diaflo filter on the surface of a moistened candidate soil in a petri dish, to seal around the edges of the membrane to prevent movement of organisms from the soil over the edges to the membrane surface, and to inoculate it with the pathogen. The fungus may grow on the nutrients diffusing through the membrane, providing it is not inhibited by materials formed by the soil microflora. Because *Mucor, Rhizopus,* or *Trichoderma* may overgrow the dish, it may be necessary to treat the soil with aerated steam at 60° C/30 minutes to eliminate them.

Since some pathogens may not grow on the soil extract alone, particularly those that require a host stimulus, host seedlings may be planted in the soil to stimulate the pathogen and the antagonists on opposite sides of the membrane. In either arrangement the pathogen and the antagonists could affect each other without contact.

It is known that molecules of different sizes are produced by antagonists, and it may be necessary to try membranes of different pore sizes to find the appropriate combination.

Transfer of a Total Soil Microflora to a Treated Soil

Antagonistic properties of different soils to *Streptomyces scabies* were compared quite simply (Menzies, 1959) by mixing suppressive soil with conducive or steamed soil and measuring the resultant level of scab suppression. As little as 10% suppressive soil provided some restoration of antagonism to steamed soil, and a 50:50 mixture provided essentially complete restoration of the antagonism.

Tests for relative levels of antagonism to *Gaeumannomyces graminis* in Pacific Northwest soils were made by Shipton et al. (1973), using "sterilized" soil or nutrient-amended sand inoculated with known amounts of candidate soils. Oat-kernel inoculum or mycelia of *G. graminis* grown on potato-dextrose agar was blended with the "sterilized" soil or sand while the soils were being mixed. Wheat was planted immediately, and 3 weeks later the seedlings were removed and the roots examined for lesions. Checks included diluted candidate soil without *G. graminis* and nondiluted candidate soil with *G. graminis*. There was generally a 100% root infection with *G. graminis* in "sterilized" soil, and the plants were stunted and chlorotic after 3 weeks. Some soils added at the rates of 1 and 10% w/w caused no suppression in symptoms, but other soils of different areas or rotations caused significant reductions in disease (Chapter 7; Fig. 5.3). This technique has worked in the field at Puyallup, Washington, on acid (pH 5.6) fine sandy loam (Chapter 4; Figs. 7.11 and 7.12). Indications were that its success there was caused at least partially by soil acidity, which is itself somewhat unfavorable to *G. graminis*, and which may have permitted expression of even weak antagonism. This suggests the possibility that glasshouse tests for antagonism to this fungus may be made more sensitive by adjusting soil pH downward, thereby favoring expression of low levels of antagonism. Alternatively, raising soil pH may make the test more rigorous because the pathogen is favored, and a higher level of antagonism is then needed for its suppression. This makes possible the qualitative and quantitative assessment of the suppressiveness of soil in the field (Shipton et al., 1973). It may thus be possible to use the method to intensify antagonism in an existing field without identifying the organisms involved.

Multiplication of Antagonists in Soil

Vojinović (1973) was able to increase the population of antagonists in a soil sample. Wheat seedlings were grown in sand-culture tubes into which mycelia of *Gaeumannomyces graminis* and a water suspension of selected field soil had been introduced. After a few weeks growth, rhizosphere soil from these seedlings was collected and introduced with more mycelia of *G. graminis* into a second series of sand-culture tubes, and these were also planted with wheat. This entire process was repeated. By the third or fourth serial transfer, antagonists were sufficiently abundant to greatly reduce or prevent infection of seedlings by *G. graminis*. This "enrichment" technique selectively increased specific antagonists that develop in the rhizosphere. Gerlagh (1968) increased antagonism of *G. graminis* in soil in the glasshouse by growing four successive wheat crops in a single year. R. W. Smiley (unpublished) also grew four successive wheat crops in a virgin nonsuppressive soil, which then became highly antagonistic.

P. Broadbent (unpublished), however, was unable to apply this method to multiply antagonists to *Phytophthora cinnamomi.*

The soil-water potential and temperature at which the above methods are conducted are important as additional selective influences. They therefore should be in the range known to occur under field conditions where effective protection is provided.

Some antagonists appear to multiply more rapidly on the surface of pathogen hyphae than in soil, as found by Olsen and Baker (1968) for *Rhizoctonia solani.* The pathogen thus provides the substrate for its own suppression. The fact that some antagonists are able to attack several different pathogens (Chapter 7) suggests the interesting possibility that, in soil freed of pathogens by treatment with aerated steam at 60° C/30 minutes, the antagonist might be induced to multiply on a nonpathogen. Thus, some saprophytic basidiomycete might substitute for *R. solani,* or a *Mucor* or *Rhizopus* for *Phytophthora* or *Pythium.*

TESTS WITH INDIVIDUAL ANTAGONISTS

Sooner or later investigations on the suppressive nature of the soil lead to determining the organisms important in the effect, and the mechanisms involved. It is also desirable to isolate specific antagonists from soil for direct use as agents of biological control in treated soil.

Selective Separation of Components of the Antagonistic Microflora

Perhaps the major impediment to development of a successful program of biological control has been the difficulty of manipulating the microflora, either to obtain the desired control or to determine which components were responsible for the effect.

A scheme was suggested (Baker, 1961) for this, developed from quali-

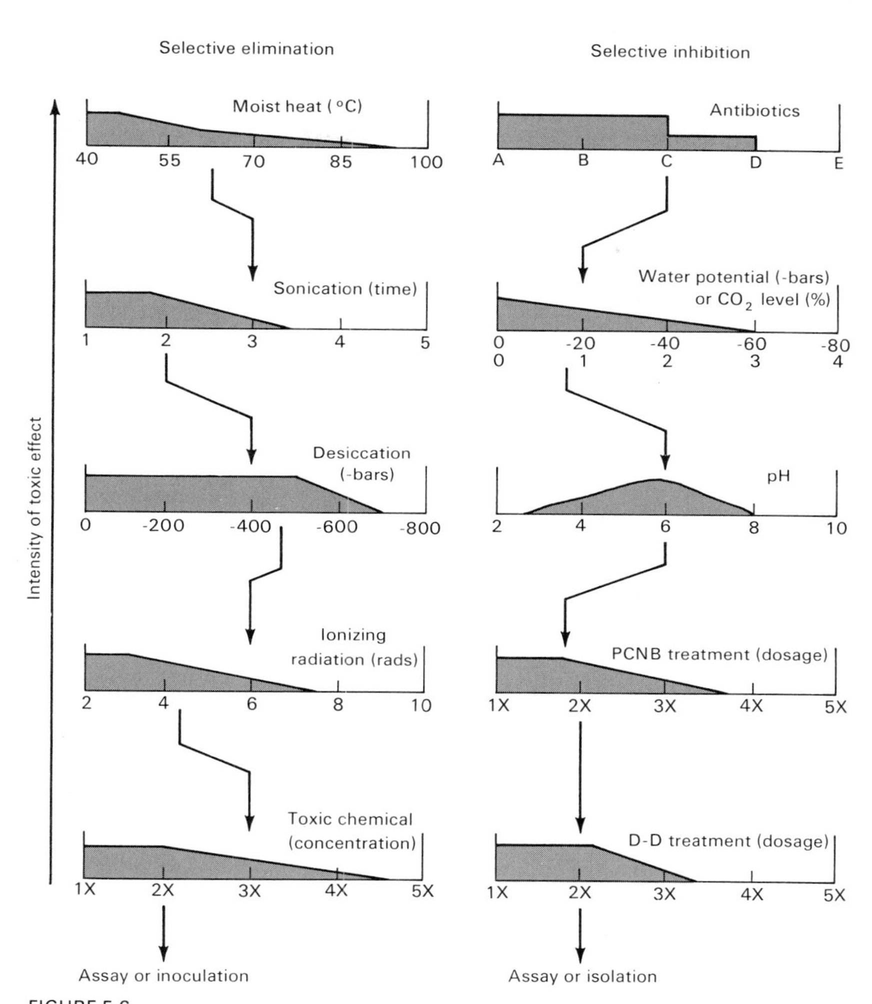

FIGURE 5.6.
Schematic outline of some suggested methods for selectively separating components of the antagonistic microflora. The selective elimination of microorganisms extraneous or neutral to the antagonistic effect provides a means for obtaining a partially purified microbiota for inoculating soil. The selective inhibition of extraneous microorganisms is also useful in isolating antagonists.

tative analysis techniques of inorganic chemistry and from A. G. Lochhead's method for separating soil bacteria into nutritional groups. This is modified here (Fig. 5.6). It is based on the fact that microorganisms vary greatly in tolerance of each environmental condition.

A convenient and accurate assay for measuring the degree of suppression of the pathogen is essential in these methods. Some have already been discussed in this chapter.

Selective Elimination of Extraneous Microorganisms. A successive series of fractional-separation techniques can be used to reduce the surviving biota to the principal antagonists. The death of the pathogen in the candidate soil is unimportant, as it can be reintroduced after the treatments. The intensity or dosage required to reduce or eliminate the suppressive effect of the soil on the pathogen is first determined for each of several physical treatments. These are then successively applied, each at an intensity or dosage that leaves a strongly antagonistic effect, as shown in Figure 5.6. Each treatment presumably will eliminate extraneous microorganisms, and the end result will be a relatively purified group of antagonists. The final stage may be a toxic chemical as shown, or one of the treatments shown in the Selective Inhibition series. This is a method for purifying the antagonists for inoculation of a soil mass for giant cultures.

Details of some of the techniques that may be used are given below:

1. Aerated steam at various temperatures. Pathogens are killed at 60° C/30 minutes, and some (Phycomycetes) at even lower temperatures. *Bacillus* spp. survive heat treatment reasonably well up to 80° C/10 minutes, *Streptomyces* spp. up to 60° C/10 minutes, and *Pseudomonas* spp. not much above 40° C/10 minutes (Fig. 5.7) (Baker, 1962; Bollen, 1969; Broadbent et al., 1971). Heat treatment of a suspension in capillary glass tubes (as subsequently outlined), or by application of aerated steam to the soil, provides a flexible method *par excellence* for stripping off various segments of the microflora on the basis of their thermal tolerance. Treatment at 60° C has the additional benefit of breaking dormancy of bacterial spores (mostly *Bacillus*) and thus increasing their relative numbers in the dilution plates (Fig. 5.2).

   Some antagonists (those inhibitory to *Gaeumannomyces graminis;* Chapter 4) are destroyed in soil at 40–60° C/30 minutes, and others (those antagonistic to *Phytophthora cinnamomi;* Chapter 4) survive 60–80° C/30 minutes. A range of heat treatments is therefore necessary in each new study to determine the temperature-group of antagonists involved.

2. Drying the soil reduces some types of microorganisms (nematodes, bacteria) much more than others, particularly if done rapidly. Aging may also have a selective effect.

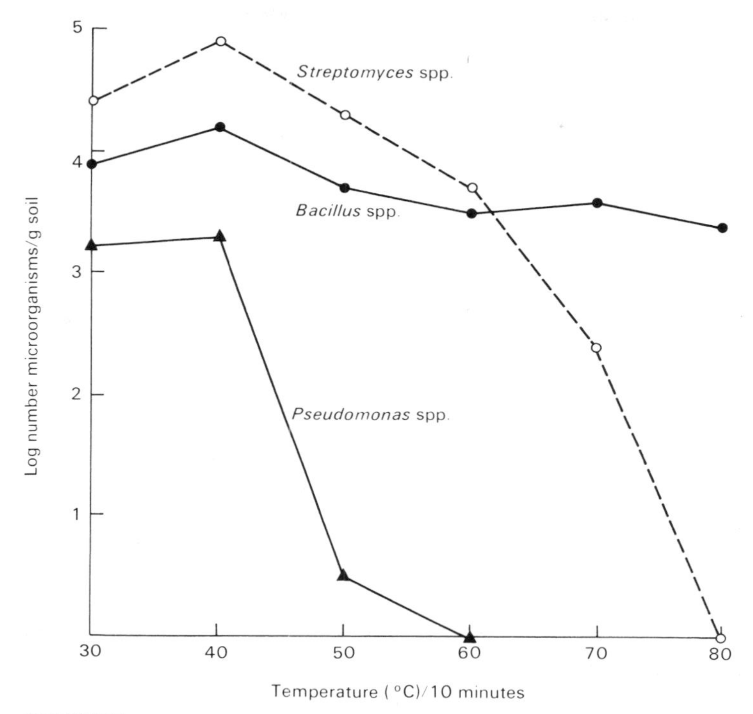

FIGURE 5.7.
Comparative survival of *Bacillus, Streptomyces,* and *Pseudomonas* spp. in soil treated with moist heat for 10 minutes. Results are an average of data from five Australian soils. (Data from Broadbent et al., 1971.)

3. Exposing soil to various carbon dioxide or oxygen levels, or to changed pH, will selectively favor some microorganisms.
4. Ionizing radiation, ultraviolet light, sonication, and similar physical factors may prove selective in effect.

Selective Inhibition of Extraneous Microorganisms. A series of inhibitory treatments applied either successively or collectively can be used to suppress different fractions of the microorganisms selectively and permit determination of the principal antagonists involved. Because some of the materials used will be residual, they must not be toxic to the pathogen, or an assay of effectiveness cannot be made. A possible series is shown in Figure 5.6. This is essentially the principle of selective media, discussed subsequently in this chapter, applied to soil. The method is useful for analysis of the microbiota and for selectively isolating the antagonists involved. It cannot be used in inoculating soil for giant cultures, as most microorganisms are only inhibited and may resume growth when sufficiently diluted. Audus (1970) has reviewed some aspects of this subject.

Some chemicals that show specificity are as follows: PCNB apparently inhibits *Rhizoctonia solani, Sclerotinia sclerotiorum, Sclerotium rolfsii, Streptomyces scabies, Rhizopus stolonifer,* and *Penicillium oxalicum,* but not *Pythium* or *Fusarium* spp.; captan may stimulate saprophytic *Penicillium, Trichoderma,* and *Fusarium* spp., and actinomycetes and bacteria; Dexon inhibits *Pythium* and *Phytophthora* spp.; carbon disulfide and methyl bromide appear to favor *Trichoderma* over *Armillaria mellea;* benomyl apparently inhibits *Trichoderma viride* but not *Armillaria mellea* (L. O. Lawyer, unpublished) or certain strains of *Sclerotinia homoeocarpa;* D-D and Nemagon inhibit nematodes but have less effect on fungi; ethylene dibromide (EDB) is not toxic to nematode-trapping fungi at dosages lethal to nematodes; penicillin and vancomycin antibiotics inhibit actinomycetes and gram-positive bacteria; polymyxin and streptomycin inhibit gram-negative bacteria. Some of the many available antibiotics, which generally are very specific in effect, should be tested in such a series. Because ethylene has a differential inhibitory effect on soil microorganisms (for example, actinomycetes are relatively insensitive) (Chapter 6), it can be used in a series of the type considered here. Domsch (1963) reviewed the selective toxicity of chemicals to soil microflora.

A. D. Rovira and E. H. Ridge (unpublished) demonstrated a striking effect on soil microflora by soil fumigation in 1971–72 in a South Australian field. Chloropicrin (200 pounds/acre; 224 kg/ha, untarped) gave a large rapid rise in population of fluorescent rhizosphere pseudomonads over that in nontreated soil, and this effect persisted for 4 months. There was also a three- to tenfold increase in numbers of *Bacillus* spp. over the check 28 days after treatment, and a twofold decrease in actinomycetes. A combined chloropicrin-methyl bromide treatment (each 200 pounds/acre; 224 kg/ha, untarped) gave similar results.

Methyl bromide treatment (400 pounds/acre; 449 kg/ha, untarped) gave only 10–30% as many flourescent pseudomonads as nontreated soil, and only 1–2% of those in chloropicrin-treated soil. Actinomycetes were increased somewhat over those in nontreated soil, and were more than twice those in chloropicrin-treated soil. This population shift also seemed to persist for 4 months. No clear relationship of fluorescent pseudomonads to incidence of take-all "whiteheads" was demonstrated in these trials. However, the total number of fluorescent pseudomonads was about tenfold greater in chloropicrin than in methyl bromide or check plots, and wheat developed 2% of take-all "whiteheads" in chloropicrin plots, as against 35% in methyl bromide plots. The fluorescent pseudomonads tended to lose their antibiotic ability more rapidly in culture than did *Bacillus* or actinomycetes.

S. Wilhelm (unpublished) found that application of 400 pounds of methyl bromide, or 200 pounds of chloropicrin/acre untarped may increase the severity of infection by *Verticillium albo-atrum*, presumably because the antagonists are more sensitive than the pathogen. However, application of 300 pounds of chloropicrin/acre tarped gives control of the disease caused by this pathogen.

The compound, 2-chloro-6-(trichloromethyl) pyridine (N-Serve) is highly toxic to *Nitrosomonas* spp. in soil and thus prevents conversion of ammonia to nitrites; low toxicity is shown to *Nitrobacter* spp., to the general fungus and bacterial soil microflora, and to many seedlings (Goring, 1962). Dinitro-o-secondary butylphenol (Dinoseb), on the other hand, selectively inhibits *Nitrobacter* (Chase et al., in Gray and Parkinson, 1968) and permits accumulation of nitrites. Other materials are known to have similar selective effects on nitrifying bacteria.

Various dyes, such as rose bengal and congo red, should also be tried. These and many other chemicals should be tested at various dosages, temperatures, and times. Addition of selective amendments or nutrients as a final step will greatly stimulate certain microorganisms.

Methods of Isolation

Isolation from Soil by Dilution Plates. A 1–g sample is shaken for 15 minutes in 10 ml sterile water. In some instances a considerable increase in the number of living cells (as shown by colony count) is obtained if a dilute soil extract (1% w/v, centrifuged and autoclaved) is used instead of water for all dilutions, apparently from avoiding a sudden large change in osmotic water potential.

The efficiency of isolating microorganisms from suspensions of soil or from mycelium, can sometimes be vastly improved by selective heat treatment, with little loss of antagonists. This treatment is especially useful to eliminate fungi such as *Mucor, Rhizopus,* or *Trichoderma,* which overgrow the petri dishes. Draw the suspension into a micropipette or microhaematocrit centrifuge tube, using a hypodermic syringe (with care, about an inch of suspension can be drawn in); remove the tube from the suspension, and draw the liquid back about an inch from the tip; heat the tip of the tube in a tiny oxyacetylene flame for an instantaneous seal without heating the suspension; remove the tube from the syringe, and seal the opposite end in the same way. The tubes are then immersed in a water bath at 50°, 60°, or 80° C for 10 minutes and then plunged in cold water, and the contents are plated in serial dilution (0.1 ml per plate) on potato-dextrose agar plates that have been dried for several days to free them of surface water (Broadbent et al., 1971). Yeast-mannitol agar containing congo red (inhibits spreading bacteria and fungi), soil-extract agar, or potato-dextrose agar are suitable for *Bacillus* and *Streptomyces* spp.; plates for the latter must be held longer, and must therefore have fewer colonies, that is, the dilution must be greater. If fluorescent pseudomonads are being sought, selective King's medium B—as modified by Sands and Rovira (1970) (penicillin G 75,000 units, novobiocin 45 mg, and cycloheximide 75 mg/l of media), or the later version of it (Sands et al., 1972)—should be used.

Some detergents used in washing glassware may be highly toxic to bacteria but not to fungi, even after repeated proper washings. Dishes should be carefully rinsed in tap and distilled water. Glassware may be checked for this effect by making a comparative series, using new glassware or plastic equipment.

Use of the Pathogen as Bait. Fragments of hyphae picked from soil or from the surface of roots may have colonies of bacteria or other antagonistic microorganisms on their surfaces. Such fragments may be placed in water blanks, agitated for a time, and the water used for making dilution plates. A better method is to use mycelia on the surface of agar that has been covered with soil in petri dishes (see above). The fungus colonies can be examined under the microscope, and mycelia that show endolysis may be selected. Bits of these hyphae may be cut out and used for making dilution plates, as above. It is also possible, but more difficult, to touch a sterile needle to one of the bacterial colonies and either to streak it on agar or make a dilution plate. These methods give a reasonable possibility of obtaining colonies of suppressive bacteria adapted to development on hyphae.

A further modification might be developed from the work of Smith (1972) with sclerotia of *Sclerotium rolfsii*. Dried sclerotia leaked nutrients profusely when placed in moist soil. Perhaps placing dried sclerotia in the candidate soil (nontreated or treated at 60° C/30 minutes) for a time would stimulate antagonistic bacteria, actinomycetes, and fungi, which could then be cultured by making dilution plates of sterile distilled water in which the sclerotia have been washed. If the sclerotia have started to decay, dilution plates could be made from the invaded tissue.

Selection of Colonies to be Studied. There is no known way to distinguish potential antagonists from nonantagonists on a dilution plate. Most plates will have 100 or more colonies, mostly bacteria and actinomycetes. Most of the bacteria will look alike when young. The task is one of dealing in probabilities and selecting random colonies. However, the odds can be improved (Broadbent et al., 1971) for finding antagonists on dilution plates by avoiding spreading bacteria and those having brightly colored colonies, which have not been found to have useful antagonistic qualities.

By far the greatest number of colonies will be *Bacillus, Streptomyces,* or *Pseudomonas* spp.—all grist for the mill. Because the young colonies 1–2 mm in diameter may not have developed characteristic features, it is desirable to transfer 25 or more from a given soil sample to agar slants. After a time the characteristic features of the isolate will be evident, and the cultures can be arranged in groups. However, experience will soon teach that although two isolates may appear identical, one will be antagonistic to a given pathogen and one will not.

Use of Selective Media. An additional screening effect in isolation can often be obtained by the use of selective media, produced by addition of antibiotics, adjustment of pH, and other manipulations of the basic medium. Such media will be especially valuable for isolation of antagonistic fungi. Selective media have been reviewed by Durbin (1961), Goldberg (1959), and Tsao (1970). Synergism between antimicrobial agents should also be considered (Mircetich, 1970).

Baits have long been used to isolate plant pathogens from soil (Durbin, 1961). Thus, use of carrot disks to isolate *Thielaviopsis basicola;* peeled rose stems to isolate *Chalaropsis thielavioides;* rice grains to isolate *Chromobacterium violaceum,* seedlings of *Eucalyptus sieberi* or *Lupinus angustifolius,* and apple or avocado fruit to isolate *Phytophthora cinnamomi,* is now standard procedure. It may be equally feasible to isolate antagonistic microorganisms from soil with the proper baits, as suggested earlier for hyphae of the pathogen in question. It may also be possible to isolate antagonists selectively while baiting for the pathogen. Such antagonists may, however, prevent infection by the pathogen being sought.

PRESUMPTIVE TESTS OF ANTAGONISTS IN AGAR CULTURE

Roberts (1874) is said to have first reported an inhibitory effect in culture between "*Penicillium glaucum*" and bacteria. He stated that "the growth of fungi has appeared to me to be antagonistic to that of bacteria, and *vice versá*. . . . there was an antagonism between the growth of certain races of *Bacteria* and certain other races of *Bacteria*. . . ."

The literature since that time includes many studies showing the inhibitory effect of one organism on another in culture. Many of these studies have assumed or implied that the same thing occurred in soil, and that culture studies explained what happened there. When investigators tried to duplicate the results in soil, disillusionment set in; few microorganisms were found to be effective in soil. Furthermore, soils that were conducive to disease produced by a given pathogen in the field might have numerous antagonists inhibitory to that organism on agar media. Why is this?

The reasons include obvious differences between growth in agar culture and in soil, such as the following:

1. Agar tests are useful for evaluating individual antagonists to be applied to treated soil, but cannot be used for antagonist populations, which are usually needed in field soils, as explained in this chapter.
2. Nutrients necessary for the production of an antibiotic by the particular antagonist may be present in agar but not in soil. Since microorganisms may produce adaptive enzymes when a specific substrate is available, they may produce a given antibiotic on agar but not in soil.
3. Agar media tend to favor the antagonist more than the pathogen. Parasitism requires specific enzyme systems and metabolic pathways, and the organism usually has a reduced or nonexistent saprophytic ability. The saprophytic antagonist is better able to use the readily available nutrients of agar media than is the pathogen.
4. The antagonist is free of competition on agar but may be suppressed or forced into dormancy by other microorganisms in soil. The other microorganisms may suppress the formation by the antagonist of the antibiotic, destroy it, or counteract its effect on the pathogen.
5. The pathogen may produce metabolites on agar that inhibit antagonists or prevent formation of antibiotics by them.
6. The environmental conditions (temperature, pH, water potential, nutrients, gas exchange) in agar frequently are quite unrelated to those in field soil; they have generally been designed to give optimal growth of certain microorganisms—a condition that rarely occurs in soil.
7. Tests in agar media tend to select mainly for antibiosis, giving a distorted emphasis of the importance of this type of antagonism in soil. Competition for nutrients and space in agar is usually minimal, since the organisms are planted some distance apart. Hyperparasitism is generally rare. Endolysis caused by bacteria is more common than hyperparasitism on agar, but also infrequent.
8. Spore-forming pathogens in soil may be in a dormant condition, es-

caping the effect of antagonists until stimulated by a susceptible host. In agar tests, the pathogens show the effect of the antagonist.

Because of the obvious disadvantages in the use of agar media, it has become fashionable to dismiss the method as an exercise in futility. This judgment overlooks the useful aspects of the technique. Granting that the method has been improperly used and that wrong conclusions have been drawn from it, it is still useful in an integrated approach to obtaining effective individual antagonists, much as microorganisms are screened for antibiotic production in medicine. The useful aspects of the method will be considered here.

The chances of obtaining an effective antagonist are greatly enhanced by increasing the number of candidates screened (Chapter 3). For this reason, Broadbent et al. (1971) screened more than 3500 isolates from 60 soils. Of these, only about 40% inhibited one to nine pathogens on agar, and of this 40% only about 4% were effective in soil. Undoubtedly an even lower percentage will prove effective under commercial conditions. Obviously, to screen such numbers of isolates in soil becomes excessively expensive and laborious. Moreover, it is rare for a microorganism to prove antibiotic in soil, but not on agar. One such exception apparently is *Chaetomium cochlioides,* antagonistic to *Fusarium nivale* on oat straw, but not on agar (Wood and Tveit, 1955).

Four bacterial or actinomycete isolates may be planted on each plate (Fig. 5.8). The inoculum age and amount should be standardized in comparisons, as should the times between inoculations with the antagonists and the test pathogen, and before taking readings. If the pathogen is slow growing, it may be inoculated in the center of the dish at the same time as the antagonist, but if it grows as rapidly as *Rhizoctonia solani,* it should perhaps be inoculated a day or two later to give the antagonist colonies time to develop and leak antibiotics into the medium. A series of pathogens may thus be tested, each in a different plate, with a common group of antagonists. The replica-plating method (Lederberg and Lederberg, 1952; Stotzky, 1965) may prove useful in this part of the operation. However, the "pitfalls and imperfections that should be recognized by anyone using the technique" were outlined by Palleroni and Doudoroff (1972).

Accuracy of these presumptive tests in agar media could undoubtedly be improved by modifying the type and composition of nutrients supplied, altering salt concentration to raise or lower the water potential, changing the type of agar to alter the diffusion of antibiotics, or using different temperatures. This, however, would need special study for each pathogen-antagonist-soil combination, and no general recommendation can be made. Bacteria growing on potato-dextrose agar showed greater inhibitory effect

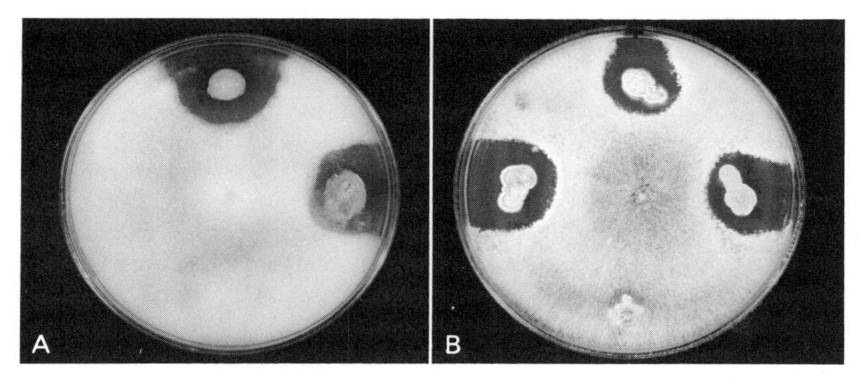

FIGURE 5.8.
Agar plates with four colonies of antagonists tested for antibiotic effect. A. Test with the pathogen *Pythium ultimum* (center); *Bacillus megaterium* (top) and *B. pumilis* (right) were antibiotic, two others were not. (From Broadbent et al., 1971.) B. Test in which *Rhizoctonia solani* was inhibited by three actinomycete colonies, but overgrew a fourth. (Photo courtesy of P. Broadbent.)

against plant-pathogenic fungi than those growing on Czapek-Dox or soil-extract agar, but this affected only the intensity, not the relative ratings (Broadbent et al., 1971). Tests were therefore standardized on the first medium. The width of the inhibitory zone on agar media was not necessarily related to the size of the bacterial or actinomycete colony produced. Tests of the effect of temperature on the inhibition of *Rhizoctonia solani* showed that at 10–37° C there was little difference in effect; the bacteria grew better at the higher temperatures and the fungus more poorly, but antagonistic effect was not appreciably altered. The use of the soil itself as the medium might reduce some of these difficulties but would be more expensive and laborious.

Some pathogens (smuts and Taphrinales) that produce slow-growing yeastlike colonies may be tested for susceptibility to antagonists by a technique now under study. A suspension of spores of *Tilletia caries,* for example, may be uniformly applied over the dried surface of sterile soil-extract agar in a petri dish. Sterilized wheat seeds may be coated with candidate antagonists and placed on the agar surface. Zones of inhibition in the lawn of yeastlike growth can be determined by inspection. The wheat seed provides nutrients for the antagonist, as in field practice. The pathogen may get nutrients from the seed and the soil-extract agar, simulating natural conditions. Candidate antagonists could be obtained by the method of Neal et al. (1970), by placing dilute soil (nontreated or treated with aerated steam to eliminate weed molds) in a bottom layer of wheat-extract agar in a petri dish. Over this is poured soil-extract agar, seeded with smut spores.

In short, the judicious use of agar media has a useful function in a preliminary selection of individual antagonists for subsequent testing in soil.

TESTS IN SOIL

Isolates selected on agar should be subjected to tests in soil as soon as possible. If delay is unavoidable, the cultures should be lyophilized to prevent alteration by mutation.

The final choice of test method here should be made on the type of crop—whether direct-sown, planted in seedbed and transplanted, or grown from cuttings in a propagation bed. The basic strategy for each type is to have an antagonist occupy the infection site before the arrival of the pathogen.

In the soil-tube method, the soil is reduced to near-sterility by steaming at 100°C/30 minutes, and inoculated with a suspension of the given bacterium or actinomycete. Introduction of nutrients from the culture should be held to a minimum to avoid stimulating other microorganisms. The soil should be held for 1–2 weeks to permit establishment of the antagonist. A modification of this, using soil treated at 60°C/30 minutes with aerated steam, indicates the ability of the antagonist to inhibit the pathogen in the soil to which it would be added commercially, and in company with a reduced normal flora.

The soil may be added to soil tubes in which the pathogen is growing, described earlier in this chapter. The ability of the pathogen to grow through the inoculated soil and be recovered from bait sticks placed on the soil surface indicates the effectiveness of the antagonist.

Alternatively, the inoculated soil is placed in a flat or tray, seed of a suitable host is sown, and the pathogen is introduced in one corner or mixed thoroughly with the antagonist-infested soil before being placed in the flat. Seed or seedling transplants of the host are then set in the soil. The effectiveness of the antagonist is measured by the distance and the rate of spread of the pathogen, as evidenced by infected seedlings.

Seeds of the host plant may be inoculated with the antagonist and sown in soil naturally infested with the pathogen. To this may be added nontoxic gum arabic or carboxymethyl cellulose to increase adherence to the seed. Suppressive soil, selectively treated to free it of pathogens, has also been used in pelleting seeds to introduce an effective broad antagonistic flora into another soil in intimate association with the host, with a maximal chance of success (Broadbent et al., 1974; Merriman et al., 1974; Price et al., 1971). Effectiveness of protection is measured by seedling survival, plant vigor, or in the case of smuts, by plant infection.

A hazard of pelleting seeds with nontreated soil was revealed in a field test against take-all of wheat at Puyallup, Washington (R. J. Cook, unpublished). Seeds coated with gum arabic and soil of known suppressiveness to *Gaeumannomyces graminis* were sown in an area where take-all had been severe the previous year. The treatment resulted in an even poorer stand than in nontreated checks. The soil used to coat the seeds was subsequently found to contain a high population of *Fusarium roseum* f. sp. *cerealis* 'Graminearum'. This fungus, introduced as seedborne inoculum, causes severe seedling blight in this locality, but an equal population in the soil causes little injury (Chapter 9). Apparently application of this nontreated soil to the seeds provided a source of antagonists against *G. graminis,* but also introduced another pathogen not suppressed by these antagonists.

Cuttings or transplants may also be inoculated with the antagonist before being planted in infested soil (Chapter 4). The cut ends of cuttings may be dipped in a suspension of the antagonist before "sticking" in the propagation bed. The propagation medium should be steamed at 100° C/30 minutes or more just before planting, for near-sterility. Rooted cuttings may be dipped in a suspension of the antagonist before planting, but this is less desirable. The efficacy of the method is measured by the incidence of disease in cuttings transplanted into pathogen-infested soil.

Propagation or seed beds may be directly inoculated with the antagonist slurry as soon as the propagation medium cools after being steamed at 100° C for 30 minutes. This will establish the antagonist as the dominant microorganism in the medium, and cuttings or seeds may be planted 1–2 weeks later. The test of effectiveness of protection is provided by transplanting the host into pathogen-infested soil.

TESTING MIXTURES OF ANTAGONISTS

There is a better chance of attaining successful biological control with a mixture of several antagonists than with a single one (Chapters 3 and 7). Such complex associations are also more stable (Chapter 1). Antagonists should be complementary, not competitive; that is, they should grow best at different temperatures, water potentials, pH, or on different nutrients so that some will be operative under most environmental situations. It is therefore important to consider means by which congenial mixtures may be obtained.

A mixed group of antagonists could be obtained relatively unreduced from the components found in natural soil, increased, and used for inoculating propagative beds or other limited areas. This is worthy of the most serious consideration for quick, simple, and workable biological control with groups of organisms more or less as they are found. The essential feature here is careful selection of the area from which this soil is obtained (see above, and Chapter 4).

However, since man is an inquisitive animal and tends to hone his methods to a fine point, he will also try to synthesize his own groups of antagonists from individual lines of highly effective microorganisms. Because of the newness of this approach, only a few guidelines can be suggested.

It is desirable that associated organisms should not inhibit each other if they simultaneously occupy a certain niche. In medical research this inhibition is customarily studied on agar plates. Two or more microorganisms are streaked at right angles in cross-hatched lines on the plate. If one is inhibited where two lines cross, it may also be inhibited if they should share a common site in soil. Competitive effects are more difficult to evaluate. The important test is the interaction of these antagonists in the soil in the presence of the host and pathogen. The methods for this may be modified from those outlined already for single antagonists.

PLAN OF ACTION

How can these methods be converted into a plan of action in biological control? Positive thinking provides a desirable framework or approach. This means that one searches for instances where biological control is definitely working, studies them critically to determine the mechanism by which they are operating and why they are effective in each case, and then attempts to adapt these findings, antagonists, or both to other unprotected areas.

There is no substitute for careful analytical examination of the field situation at all stages of the study. Sites with promising potential for antagonists should be selected; the antagonist population should be freed of extraneous microorganisms, concentrated if possible, and tested at once in soil. Attempts may then be made to determine the antagonist(s) responsible for any suppressive effects found and the conditions necessary for it to operate. This is purposeful ecological research, both basic and applied.

Man must learn to create a stable new biological balance beneficial to his long-term purposes. He could not restore the primeval balance, even if he wished to, but he does have the capacity to direct the establishment of new ones by subtle slow nudging of the flora in the desired direction. In this approach he is working with and attempting to direct natural forces, rather than flouting them. For this no specific blueprint can be supplied, as each example is at the frontier of knowledge. Through restrained imaginative research, man can stabilize, preserve, and enhance the biological balance, and do in a short time and on a small scale what nature would eventually achieve in millennia.

6

ROLE OF THE PATHOGEN IN BIOLOGICAL CONTROL

Strangers in the land of soil microorganisms
have little chance for survival: conditions and competition
are so harsh that the soil is a true burial ground for disease—
delicate parasites cannot survive in the company of soil microbes.

—H. S. GOLDBERG AND T. D. LUCKEY, 1959

The pathogen may be the aggressor (parasitic on the crop), the co-aggressor (aiding another parasite), the victim of attack (providing nutrient for antagonists), the defender of the fortress (active resident of an infected plant part), the sleeping guest (dormant resident of an infected plant part), the uninvolved bystander (surrounded by active saprophytes decomposing organic matter), or the helpful friend (mycorrhiza in the crop roots). The assumption of each of these roles is determined by the genetics of the organism, the physical environment, and by the accompanying organisms, including man.

Plant pathogens are necessarily adapted to their biotic and abiotic surroundings and may not be easily subdued or extinguished by manipulation of the environment. They exploit every advantage, no matter how small, often in amazingly resourceful ways. This versatility is an obvious strength of the pathogen and a potential difficulty in biological control.

WAYS THE PATHOGEN CAN OVERCOME ANTAGONISM

Parasites must have the ability to assimilate all or parts of living cells, as well as to counteract any defense mechanism of the host. This requires different enzyme systems, often quite specialized, than are required for saprophytic growth. Perhaps because of this, parasites are also less tolerant of such unfavorable conditions as heat (Baker, 1962; Bollen, 1969) or antibiotics (Weinhold and Bowman, 1968), and less adaptable nutritionally (Misaghi and Grogan, 1969) than are saprophytes. This feature is used by man, for example, in the pasteurization of milk to destroy human pathogens while leaving saprophytes, with minimal effect on the milk. It is also used in soil treatment to destroy parasites while leaving desirable antagonists. It is easier to reduce the virulence of a pathogen through altering environment than it is to increase pathogenicity of a low-grade pathogen.

There are three general ways by which plant pathogens can overcome antagonism and thus nullify a potential system of biological control. Higher plants evade disease and pathogens avoid antagonists in the same ways—escape, resistance, and tolerance.

Escape

The surface 15-25-cm layer of soil in the field contains most of the organic substrates, is generally highest in oxygen and lowest in carbon dioxide, has a great variety of microsites, and consequently is the zone of maximal microbial activity and complexity. Presumably, this is also the zone where antagonism of plant pathogens is the most common. Ability to grow either above or below this zone would therefore constitute a simple means of escaping antagonism, and is, in fact, a means used by several plant pathogens.

Of the *Fusarium roseum* f. sp. *cerealis* ‘Culmorum’ population in eastern Washington, 75% is in the surface 5 cm and 90–95% in the upper 10 cm. *Fusarium solani* f. sp. *phaseoli* in California is largely in the upper 15 cm. The roots of crops may thus extend beyond the zone of infection by some pathogens.

Phymatotrichum omnivorum colonizes plant roots and forms sclerotia as deep as 3.5 m in soil. In this way, the fungus persists in fields where its sclerotia, hyphal strands, or mycelia in crop debris near the soil surface may be quickly digested by microbial activity. This ability of *P. omnivorum* to survive deep in soil has undoubtedly foiled many attempts at biological control through crop rotation or organic amendments. Ratliffe (1929) reported that in a field in Texas, spore mats of *P.*

omnivorum on the soil surface could be traced deep into the soil to infested roots of trees killed by the fungus and partially removed 20 years earlier. Deeply buried infested remains serve as a food base from which new strands are initiated for attack on deeply penetrating tap roots of subsequent crops. Lyda and Burnett (1971) showed that in fine-textured black soils of Texas, *P. omnivorum* is not only tolerant of high carbon dioxide but is stimulated by it to form sclerotia, thereby reinforcing the tendency of the fungus to survive at greater depths.

Sclerotium rolfsii exemplifies the opposite extreme among soilborne plant pathogens, being highly aerobic and tending to attack plants at the soil line or above. Buried sclerotia apparently remain dormant. Griffin and Nair (1968) showed that growth rate of the mycelia of this fungus was not inhibited by oxygen concentrations as low as 3%, but was as the carbon dioxide concentration increased above 0.03%. The inability of *S. rolfsii* to attack plants below the soil surface obviously is not caused by an intolerance to slightly lowered oxygen concentrations, but could relate to its sensitivity to carbon dioxide, which in itself is a product of microbial growth and hence a mechanism of antagonism. Perhaps *S. rolfsii* remains dormant in response to a mild antagonism in soil, namely, carbon dioxide concentrations higher than those of air, and thus avoids more lethal conditions that could ensue if hyphal growth were attempted.

Smith (1973) found that sclerotia of *S. rolfsii* remained dormant on soil over which passed an air flow containing 1 ppm ethylene. Spores of *Helminthosporium sativum* on field soil failed to germinate until the ethylene was removed by passing an air stream over the surface. The production of ethylene in soil was found to reach 20–30 ppm in closed containers, and was thought to commonly reach 1 ppm under field conditions. Soils high in organic matter and nitrogen (that is, favorable to growth of microorganisms) produced more ethylene than was produced by infertile soils. There is thus an equilibrium between nutrients (which provide energy for ethylene production and also stimulate germination) and ethylene (which inhibits germination). Biologically active soils therefore usually produce the greater concentration of ethylene necessary to maintain *fungistasis*, but a large application of nutrients will temporarily stimulate germination (annul fungistasis).

Fungistasis may also be annulled for sclerotia by drying and wetting (Chapter 9). Germination of sclerotia that have been dried and rewetted is not inhibited by the usual concentrations of ethylene that occur in soil (that is, the effect of drying overrides the inhibition by ethylene at usual concentrations).

A similar effect of ethylene on seed of witchweed (*Striga lutea*) has been shown by R. E. Eplee (unpublished) and is discussed by Abeles (1973). Injected into field soil at 15–20 cm depth and 3 pounds/acre (3.4 kg/ha), ethylene caused 90% of the seed to germinate. A layer of trifluralin (Treflan) herbicide was injected at the 5–6 cm level; the *Striga* seedlings were unable to grow through this. A. H. Gold (unpublished) found that ethylene also broke the dormancy of broomrape (*Orobanche ramosa*) seed, but germination occurred only after removal from ethylene.

Another explanation for the greater activity of *S. rolfsii* on, rather than in, the soil concerns the influence of drying on sclerotial germination (Smith, 1972; Javed and Coley-Smith, 1973). Sclerotia maintained continuously moist in soil did not germinate, whereas those dried for a few hours, then remoistened, did. Smith concluded that drying may be an important mechanism that stimulates germination. The soil surface exposed to the sun and wind is quickest to dry, and in many climates may be the only site of drying. Buried sclerotia may not pass through the stage of desiccation conducive to their germination. Whatever the mechanism, *S. rolfsii* avoids a certain amount of antagonism by growing on the soil surface, as *P. omnivorum* does by growing at greater depths.

The ability to grow under conditions of extreme cold, heat, drought, moisture, or pH are other important ways by which plant pathogens may escape antagonism. *Fusarium nivale, Typhula idahoensis, T. incarnata,* and *Sclerotinia borealis* can grow at 0.5° C and even lower, and cause greatest decay of wheat leaves at this temperature beneath snow. The activity of the biological community is undoubtedly very low at 0.5° C, and the opportunities for antagonism are thus greatly reduced.

Fusarium roseum f. sp. *cerealis* 'Culmorum' can grow in soil at water potentials down to −75 to −85 bars, another means of escaping antagonism. Soil bacteria in particular, but also certain actinomycetes and fungi, are inactive at these water potentials. Cook and Papendick (1970) showed that endolysis of germlings of this fungus or their conversion back into chlamydospores in nonsterile soil occurred in progressively decreasing amounts as the soil water potential was lowered below −0.5 bars. In soil drier than −10 to −15 bars, germlings survived and hyphae grew through nonsterile soil with little interference from antagonistic bacteria. Other pathogenic fusaria and *Streptomyces scabies* can also grow in very dry soils. This ability is an especially effective escape method for those pathogens confined to or dependent primarily on infection courts in the tillage layer. This layer is the first to dry because of evaporative water loss, and under dryland agriculture, commonly reaches water potentials of −50 to −100 bars or less. Plants under these conditions do not wilt because deeply penetrating roots supply the water.

Helminthosporium sativum causes severe leaf spot and foot rot of Kentucky bluegrass under dry conditions, and sporulates abundantly on turf debris. By contrast, the leaf spot may be rare in well-watered turf. Colbaugh (1973) found that limitation of the pathogen under wet conditions is caused, at least in part, by microbial activity, which is probably inactive at the lower water potentials. Conidia would not germinate on moist crop debris, but did so when washed free of it. The inhibitory effect could be removed by washing, sterilizing, or drying the debris. Colbaugh suggested that *H. sativum,* because of its ability to grow under relatively dry conditons, escapes antagonism of associated organisms incapable of growth at low water potentials.

Aquatic Phycomycetes probably escape antagonism by virtue of their ability to tolerate low oxygen, and to persist and move through percolation and locomotion to considerable depths in water-saturated porous soils. The opportunities for effective antagonism of discharging sporangia or motile zoospores by fungi or actinomycetes under conditions of waterlogged soil would seem slight. Bacteria tolerant of low oxygen may hold more promise than fungi or actinomycetes. Phycomycetes may also grow through aerial parts of seedlings, as well as deep in soil, and thus escape antagonists.

Propagule dormancy is a very effective means by which soilborne plant pathogens escape antagonism. By remaining within the confines of a thick spore wall, rather than making active growth with a thin hyphal wall, the fungus avoids competitive, antibiotic, or other antagonistic effects of soil microorganisms. In fact, the phenomenon of soil fungistasis as described by Dobbs and Hinson (in Baker and Snyder, 1965) and others (Hora and Baker, 1972; Lockwood, 1964; Smith, 1973; Watson and Ford, 1972) would suggest that propagule dormancy is, in part, a product of antagonism, caused by use of necessary nutrients (the "sink" hypothesis), liberation of nonvolatile or volatile toxins, or both. Again, sensitivity to a mild antagonism characterized by soil fungistasis helps the pathogen avoid a more destructive antagonism. The greater the intensity of microbial activity, the greater the proportion of propagules that will not germinate (Chapter 4).

Many fungi (Xerosporae) produce so-called dry spores in contrast to those produced in a matrix or slime (Gloiosporae). Dry spores are hydrophobic when first produced, but with time, may lose this characteristic in soil. The question arises whether these spores are less attractive to soil bacteria, for example, than are wet spores coated with a layer of nutritious slime. *Penicillium, Aspergillus, Helminthosporium, Stemphylium,* and *Alternaria* are among the dry-spored fungi which move with air currents onto the soil. In contrast, *Fusarium, Verticillium,* and others of the family Tuberculariaceae produce slime spores that wash into the soil. The dry types probably escape a certain amount of lysis by their hydrophobic nature.

Rapid germination and growth is a well-recognized means by which soil-borne pathogens escape antagonism. Chlamydospores of *Fusarium solani* f. sp. *phaseoli,* for example, begin to sprout in soil within 4–5 hours after nutrients become available. Another 20–30 hours is apparently sufficient for the hyphal growth or thallus formation needed for penetration of the bean hypocotyl (Cook and Snyder, 1965). Chlamydospores and endoconidia of *Thielaviopsis basicola* germinate near a host root within 20 hours and penetrate root hairs directly within 24 hours (Linderman and Toussoun, 1968b). *Pythium ultimum* sporangia germinate in 1–2 hours and infect seeds within 24 hours after planting (Stanghellini and Hancock, 1971), and thus would be apt to escape antagonism through speed. Unless the antagonists have a head start, as sometimes may happen with certain organic amendments (Chapter 8), pathogens such as these can escape antagonism simply by winning the race.

Some organisms occur on the surface of the host, causing disease only under special environmental conditions. *Taphrina deformans* on peach (Fitzpatrick, 1934–35; Mix, 1935), *Pseudomonas syringae* on many plants (Leben, 1965), *P. mors-prunorum* on cherry (Crosse, 1965), *P. glycinea* on soybean (Leben, 1969), perhaps *Erwinia amylovora* on pear (Miller and Schroth, 1972), and *Agrobacterium tumefaciens* in the rhizosphere of many plants (Schroth, et al., 1971) are examples of this. The presence of the pathogen on the host reduces the amount of inoculum required to compensate for losses in dissemination, and increases the chance of infection (Chapter 10).

Penetration of the host is the most commonly cited means by which specialized plant pathogens escape antagonism. As stated by P. W. Brian, "Root parasites . . . have . . . escaped from the rigours of saprophytic existence in soil by adaptation to . . . living tissues of plant roots—inaccessible to most soil organisms. Because of their unique capacity to colonize this substrate they can frequently exist, during the life of the host plant, as more or less pure cultures, in which problems of interspecific competition and antagonism do not arise" (in Parkinson and Waid, 1960). Pathologists utilize this fact when they obtain a pure culture from the advancing margin of a lesion. Vascular parasites are especially capable in this regard, being able to move within the xylem to the tops of the plant which, for *Verticillium albo-atrum,* may be many meters above the ground. *Fomes annosus* can escape antagonists indefinitely in dense stands of pine, by internal growth from plant to plant through root grafts (Chapter 8). *Fomes noxious* produces airborne spores in aboveground fructifications and infects stumps (as does *F. annosus*), branch stubs, and pruning wounds of rubber trees. The pathogen then spreads downward to roots and from root to root through natural grafts and other contacts to initiate new infection centers of the brown-root disease in the plantation (R. A. Fox, unpublished). The pathogen thus makes little or no contact with soil and thereby escapes antagonism from soil organisms. **A pathogen that is able to remain within a living host may escape antagonists until it emerges.**

In contrast, most isolates of *Rhizoctonia solani, Pythium ultimum,* and *Phytophthora parasitica* remain more localized within or on basal parts of the host, and are thus more exposed to antagonism by secondary invaders.

Some pathogens escape antagonists through their versatility in many different ecological situations. *Rhizoctonia solani,* for example, is one of the commonest root and stem pathogens of a wide range of plants. The species is also one of the common mycorrhizal fungi on orchid roots (Harley, 1969), assisting them to extract nutrients from the soil; when environmental conditions become less favorable for the orchid host than for the fungus, this microorganism may become an aggressive pathogen in the same roots. This is another example of a host-pathogen relationship in which the pathogen causes damage only under conditions unfavorable to the host (Chapter 2). It may also parasitize other fungi such as *Pythium* (Butler, 1957). Finally, some isolates are able to compete saprophytically in soil for extended periods and are quite able to grow through field soil, and in this way make contact with a nutrient base. A single strain of *R. solani* that can perform all these functions is not known, but some can do several of them (Baker, in Parmeter, 1970). *Verticillium albo-atrum* and *V. dahliae* are able to grow as mycoparasites in the soil (Barron and Fletcher, 1970).

Resistance

Cephalosporium gramineum is an outstanding example of a plant pathogen that resists antagonists by producing an antibiotic. As shown by Bruehl and his students in 1957–68 (Bruehl and Lai, 1968a; Bruehl et al., 1969), this vascular parasite of wheat forms no true chlamydospores, microsclerotia, or other specialized thick-walled propagules within the parasitized wheat stems, and thus depends for survival in soil solely on occupancy of the wheat straw. Unlike *F. roseum* f. sp. *cerealis* 'Culmorum', *Cercosporella herpotrichoides,* or *Gaeumannomyces graminis* var. *tritici (Ophiobolus graminis),* which colonize most if not all of the tissues they parasitize, *C. gramineum* occupies only the vascular bundles and not the accompanying stem parenchyma and epidermis (Fig. 6.1). These latter tissues are ordinarily available for fungus soil saprophytes. However, *C. gramineum* produces a wide-spectrum antibiotic (Fig. 6.2) that reduces

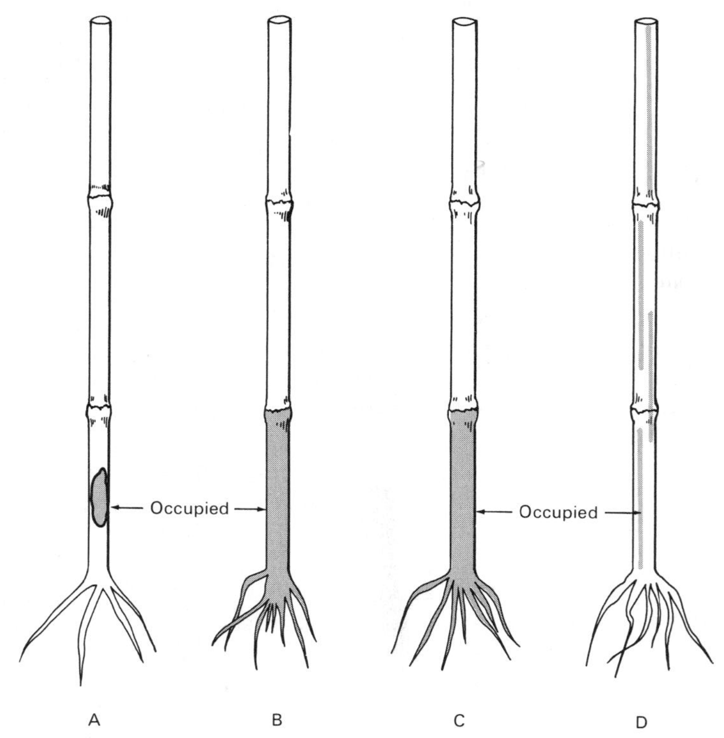

FIGURE 6.1.
Comparative occupancy of wheat stem tissue by four pathogens. *Cercosporella herpotrichoides* (A), *Gaeumannomyces graminis* (B), and *Fusarium roseum* f. sp. *cerealis* 'Culmorum' (C) decay and retain possession of cortical tissue by prior occupancy, Culmorum the most strongly. *Cephalosporium gramineum* (D) invades the xylem, leaving cortical tissue unoccupied; possession is maintained by production of an antibiotic. (Modified from Bruehl, in Cook and Watson, 1969.)

colonization of the straw by would-be colonizers (Bruehl et al., 1969). That antibiotic production is vital to *C. gramineum* in its possession of straw was shown with nonantibiotic mutants; they were overrun by saprophytic soil fungi within 5–6 months after burial of the straw in soil, whereas antibiotic producers retained possession of the straw in almost pure culture for the 2–3 year duration of the tests. Since both types were pathogenic to wheat, antibiotic production apparently is important to survival but not to pathogenicity. All 2717 isolates of *C. gramineum* from nature were producers of the antibiotic, providing even more evidence for the importance of antibiotic to *C. gramineum* in its resistance to antagonists.

Ascochyta chrysanthemi in chrysanthemum petals, and *Sclerotinia*

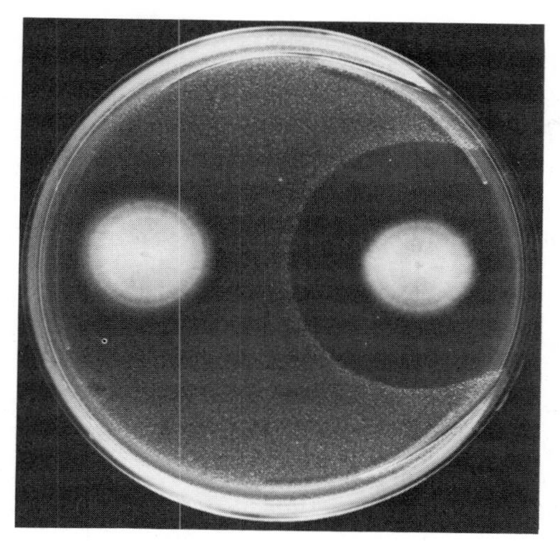

FIGURE 6.2.
Inhibitory effect of antibiotic produced by *Cephalosporium gramineum* on a yeast lawn. Right: Antibiotic isolate from field. Left: Nonantibiotic mutant. (Photo courtesy of G.W. Bruehl and J.W. Sitton.)

camelliae in camellia petals grow out in practically pure culture from infected tissue placed on agar plates. Like *C. gramineum,* these two pathogens produce antibiotics that protect against other would-be saprophytic colonists of the dead petals, and in this way, actively resist being overrun by competitors and other kinds of antagonists.

Rhizomorphs of *Armillariella elegans* (a tropical *Armillaria mellea*) continue to elongate while covered with a water film. When this film dries, more oxygen reaches the structure, pigmentation occurs, and growth slows. Manipulation of soil to maintain well-aerated and well-drained conditions may thus reduce spread of the pathogen by rhizomorphs (Smith and Griffin, 1971).

Pigmentation of the spore or hyphal wall may be another mechanism by which pathogens resist antagonism (Bloomfield and Alexander, 1967; Bull, 1970a,b). Spores and hyphae in soil or on the surface of roots are almost always pigmented, whereas identical structures in host tissues or pure culture are often hyaline. Chlamydospores of *Fusarium,* for example, are hyaline when produced in pure culture, but are brown when removed from soil, presumably because of melanin formation in the thickened walls. Sclerotia are pigmented, often dark brown to black. Hyphae suited to saprophytic survival in soil outside host debris are generally pigmented.

Some of the best examples are brown runner hyphae of *Gaeumannomyces graminis* and the hyphal strands of *Phymatotrichum omnivorum,* both of which must persist on the rhizoplane and thus are in continual close contact with potential antagonists. That the brown pigment somehow results from soil factors, possibly biotic factors, is suggested by the fact that feeder hyphae inside the host roots are hyaline. In addition, *G. graminis* hyphae in contaminated plates become pigmented more rapidly than do hyphae in pure culture.

Pigments appear to be formed usually in response to stimulation by antagonists, and to be directly responsible for the protection afforded (Bull, 1970a,b). Conidia of Dematiaceae such as *Stemphylium, Alternaria,* and *Helminthosporium* occur commonly in soil and apparently have more resistance to lysis than do hyaline conidia of *Glomerella, Fusarium,* and *Cylindrocarpon,* for example. Old and Robertson (1970) found that pigmented conidia of *Helminthosporium sativum* resisted lysis for more than 2 weeks on natural soil, in contrast to conidia of two hyaline isolates, which lysed in a few days on natural soil. Resistance to lysis was associated with an electron-dense layer that contained the pigment.

When the pigment is present before entry of the spore into soil, it is obviously not a response process. On the other hand, it appears to be a response process for those fungi that are hyaline in culture but which become pigmented upon entering soil.

The resistance of the cell walls of some Basidiomycetes to microbial lysis may be caused by the presence of wall-localized heteropolysaccharides, rather than to melanins (Ballesta and Alexander, 1972).

Resistance to antibiotics is common among microorganisms. Certain penicillin-resistant organisms produce penicillinase and thus destroy the antibiotic before damage occurs. There is also evidence for adaptive enzyme formation, whereby an organism exposed to sublethal doses of an antibiotic can produce the proper enzyme and subsequently show resistance to higher concentrations of the antibiotic. Adaptation is also possible by the organism's gradually absorbing less of the antibiotic, because of permeability changes in the membrane. How commonly pathogens adapt to antibiotics of associated microorganisms in nature is unknown, but the opportunity for such adaptation certainly exists. Antibiotic concentrations produced naturally in soil or the rhizosphere probably are initially low at the margins of the localized sites, and this could provide the sublethal concentration needed for developing resistance.

On the other hand, most observations on resistance in organisms to antibiotics suggest that the mechanism is not adaptive but rather permanent and transmissible to progeny, probably because of a genetic mutation. A metabolic process blocked by an antibiotic may, because of mutation, no longer be needed by the organism, or may be bypassed. Antibiotics can also be inactivated or detoxified through special enzyme systems, or be bound to proteins, lipids, or other complex substances by the cell. Finally, there is good evidence that organisms differ in sensitivity to antibiotics because of the cell-wall structure and permeability, or by binding the antibiotic to components of the wall (Goldberg, 1959). Thus, we return to the nature of the cell wall as a means of resistance among plant pathogens to antibiotic-producing antagonists.

As pointed out by Kuo and Alexander (1967), **"for any substance to be effective in nature in protecting . . . a fungal structure from attack by parasitic coinhabitants of the same ecosystem, the protective agent must . . . be long-lived or it must be regenerated as rapidly as it is destroyed."**

Rhizoctonia solani can elude the effects of antagonists, as evidenced by an ability to grow through nonsterile soil several centimeters away from a food base, and possibly even without a food base, using only soil nutrients. The hyphae of this fungus in soil soon become thick-walled and pigmented, which could confer resistance to antagonists. There is also the possibility, however, that this fungus has a physiological resistance to antagonists or their antibiotics that cannot be explained in terms of the cell wall. Evidence for this suggestion comes from work (Olsen and Baker, 1968) with strain differences in *R. solani* in sensitivity to *Bacillus subtilis;* some are dramatically affected by certain isolates of the bacterium and endolyse quickly, whereas others show no apparent reaction to the same isolates.

Although wheat and peas have been grown in the same fields in eastern Washington for decades, *Fusarium solani* f. sp. *pisi,* but not *F. roseum* f. sp. *cerealis* 'Culmorum', has developed populations sufficient to cause yield reductions. Most soils of the area contain at least 500 propagules of the pea *Fusarium* per g, but Culmorum is undetectable. High populations of the wheat *Fusarium* develop in occasional Palouse fields, but disappear within a year or two. Farther west in the low rainfall area, the wheat *Fusarium* persists for years in destructive populations in many fields. *Fusarium* may thus show marked species and possibly even forma specialis (Nash and Alexander, 1965) or race differences in ability to persist in a given soil. Perhaps this reflects a difference in ability to withstand antagonists.

Tolerance

Tolerance, such as that of a host to a pathogen, means that the organism is fully susceptible but succeeds in spite of injury. The production of inoculum in overabundance, or the invasion of the host by multiple infections, are means by which plant pathogens tolerate antagonists, and are extremely important ways biological or any other form of control may be nullified.

The soilborne plant-pathogenic fusaria succeed in spite of antagonism by producing excess inoculum. Populations of 1000–2000 propagules per g in the surface 15–20 cm of soil are common (see below). At soil bulk density of 1.3 g/cc, and assuming 2000 propagules per g, this amounts to 2600 propagules/cc or 2.6 propagules/mm^3, or about 650 propagules in a zone of soil 1 mm thick around each cm length of hypocotyl that is 7 mm in diameter. This inoculum is in addition to the larger pieces of pathogen-infested plant refuse that are not measured by dilution plating, but which are infectious. Fungistasis keeps a portion dormant, but 30–60% germinate in response to exudates from host parts. Of those that germinate, some lyse and some convert back into chlamydospores, but some are successful—perhaps 50 out of the initial 650 propagules. With the numbers involved, even a low success rate could result in a large number of infections; this emphasizes the importance of highly efficient antagonists.

Garrett (1970) pointed out that a certain mass of inoculum is necessary before infection can take place, and that this is essentially a function of the energy required for invasion of the host. **The more specialized the parasite, the fewer the spores required for infection** (Chapter 10); less inoculum is also required to infect leaves than woody parts. A single rust spore requires no external energy source and has a 10–50% chance of infecting a leaf, but a mass of *Armillaria* hyphae must be connected with an external food base to infect a woody root. A dense spore suspension has greater total energy of growth than a single spore; the several thousand hyphae in a rhizomorph have more energy than does a loose unorganized mycelial strand or weft, which in turn has more than a single hypha. The energy available also depends on the energy of the available food base and the distance between the base and the infection site. Although the importance of antagonists in depleting the pathogen's supply of energy is still largely unexplored, it cannot be slight. As pointed out by Brown (1971), Ko and Lockwood (1967), and Lingappa and Lockwood (1964), microorganisms on the surface of the pathogen's hyphae may act as nutrient sinks. Any decrease in energy will reduce the number of successful infections by a pathogen structure, whether spore, mycelium, or rhizomorph (this chapter). Thus, ascospores of *Gaeumannomyces graminis* are able to infect roots of seedlings in sterile but not in natural soil (Weste, 1972). However, Brooks (1965) was able to infect wheat roots from seed planted on the surface of natural soil before they contacted the soil and were contaminated by antagonists.

Hyphal aggregations were thought by Garrett (1970) to have evolved through selection of pathogens that had sufficient mycelial mass to overcome the resistance of woody hosts. Rhizomorphs grow from an apical meristem and attain a critical infection mass near the tip, whereas mycelial strands build up around a leading hypha and attain infective mass back from the tip. Appressoria and infection cushions presumably serve the same purpose as these aggregations. The meristematic tips of all these structures, as well as infection pegs formed by the last two, are probably susceptible to antagonists; this may be the mechanism of control of *Rhizoctonia solani* in the Inoculated Flat Test outlined in Chapter 5.

The above points under escape, resistance, and tolerance are, in effect, Attributes of a Successful Parasite (Chapter 1): production of excess inoculum and of propagules resistant to antagonists; production of antibiotics; possession of different levels of dormancy; rapid germination, growth, infection, and invasion of the host; ability to escape or withstand antagonists; ability to invade debris or possess the substrate. Garrett (1956) listed a similar set of attributes for organisms with high competitive saprophytic ability: high growth rate and rapid germination of spores; good enzyme-producing ability; production of antibiotics; tolerance of antibiotics produced by other microorganisms. No one pathogen need possess all these attributes to survive, but each must have sufficient means to evade potential suppressors. This insures survival of each species and also insures that each ecological niche will be used to the maximum. By having knowledge of the biology of the pathogen, and by using every minor advantage, we can achieve success in biological control.

In addition, most pathogens have certain obvious weaknesses where attempts at biological control should be concentrated. Seeking these weak links is a first principle in other methods of control, and the biological approach is no exception. Of course, a vulnerable stage for biological control in the life cycle may not necessarily be the same point as for chemical or some other control method. Judgment on which stage is vulnerable to biological control will depend on knowledge of the pathogen itself, and will logically be made in terms of potential and useful antagonistic interactions.

VULNERABILITY DURING DORMANCY

A system of biological control that works while the pathogen is dormant or during host-free periods, as during a bare fallow or cultivation of some nonsusceptible crop, would have obvious advantages. **Antagonists that act on a pathogen that survives in soil for extended periods have a long time in**

which to attack it. Those that act only during spore production or infection of the host have a shorter period to function. Moreover, the possibilities for environmental modification by tillage, organic mulches, irrigation, green manures, or other means are greater during host-free periods than when the host is present. Finally, there is a possibility that an inactive pathogen has less defense against exploitation or some other antagonism than would one in a state of high metabolic activity and active growth. The best offense may be to attack while the enemy is asleep. For some microorganisms, however, the intermediate period of spore germination or zoospore release may be the most vulnerable to antagonists.

A plant parasite in soil may maintain its population through periodically infecting its host. A period free of susceptible hosts (fallow or rotation) may thus reduce the amount of inoculum through starvation or senescence. Parasites that produce resting structures may be able to survive for extended periods without a host, but they eventually weaken and die; this is accelerated by the surrounding microorganisms, which ultimately destroy them. The energy reserves of a parasite must be expended in resisting these antagonists; spores thus live longer in sterile than in raw soil. Leakage of nutrients from the resting structures may be accelerated by direct stimulation by the antagonists or perhaps by the osmotic effect of rapid removal of the leachates from the surface of the parasite (Ko and Lockwood, 1967; Griffin, 1972). **As microbial activity increases around a propagule of the parasite, the energy expended during its dormancy increases, weakening the propagule until it is no longer able to germinate or infect its host and thus elude its attackers.** The visible result of this attrition is endolysis (Chapter 2). It resembles the accelerated decline of a person suffering from malnutrition who also has a tapeworm.

The many different methods by which plant pathogens persist in the absence of a host are generally well-known and have been reviewed by Garrett (1970), Menzies (1963), Snyder (1969), and others. Consider the extremes: from complete dormancy of resting sporangia of *Plasmodiophora brassicae* to near-perpetual saprophytic growth of *Rhizoctonia solani* on the remains of many different plants; from ability to form thick-walled, highly resistant resting spores, such as oospores of *Aphanomyces euteiches,* to complete dependence on host tissues as a refuge for mycelia, as with *Cercosporella herpotrichoides;* from fairly specific chemical requirements for germination, as with cysts of *Heterodera rostochiensis,* to those that germinate with water alone, as with teliospores of *Tilletia caries.* Since each species is unique in some way, some should be more prone than others to biological control during the host-free period.

Mycelia in Host Refuse

Gaeumannomyces graminis, Cercosporella herpotrichoides, Cephalosporium gramineum, Fusarium nivale, and *F. roseum* f. sp. *cerealis* 'Graminearum', mentioned in Chapter 4 as generally controllable by crop rotation, exemplify a category of soilborne pathogens unable to form resistant propagules, and thus dependent for survival between crops on occupancy of host tissues. Others in this category include species of *Pleospora, Cochliobolus, Mycosphaerella, Ceratocystis,* and *Gibberella,* in the Sphaeriales or Hypocreales. Some, such as *Cochliobolus (Helminthosporium) sativus,* may have an added ability to survive as thick-walled, dry, pigmented conidia. Others, such as *Mycosphaerella,* may form sclerotia or sclerotium-like structures that in nature are suited primarily for short-term survival.

A significant feature of these organisms is that they are actually above-ground pathogens, damaging to roots in addition to their parasitic activities on aerial parts. Some may be known in a particular region only as root pathogens, but in a region with a different environment, the same pathogen is important on aerial parts. For example, *F. roseum* f. sp. *cerealis* 'Graminearum' is known in the dryland wheat areas of Australia, California, and the northwestern United States as a cause of dryland foot rot, whereas in more humid cereal-growing areas, this same fungus is far better known as *Gibberella zeae,* cause of scab, head-blight, and stalk-rot of corn, wheat, and barley. Pathogens of this type are poorly equipped for persistence in soil other than within the confines of remnant host tissues. Yet soil is the primary and probably the only means by which they can carry over between susceptible crops. This dependence on soil, coupled with a general inability to cope with the biotic environment of soil, is obviously a major weakness. Host tissues quickly decompose, and if the pathogens are given no biennial, or at least triennial, recourse to fresh host tissues, they are readily controlled.

Short-term crop rotations are a practical means for biological control of those pathogens that depend on host tissue for survival and cannot form resistant survival structures of their own. **Pathogens that do not form resting structures or persist in the rhizosphere of nonhosts are more readily controlled by crop rotation than those that do.** The normal flora and fauna involved in residue decomposition are the agents we credit with such control. Specific antagonists introduced or encouraged by selective enrichment from the natural microbial population may be helpful but are not essential. Of greater importance may be treatments of soil or refuse to hasten decay. Rototilling of wheat straw into soil after harvest is sometimes practiced in England and in irrigation districts of the Pacific Northwest to increase the surface area of straw in soil and thereby increase the rate of straw decomposition. This practice may hasten the disappearance of *Gaeumannomyces graminis*. The treatment of apple leaves with urea to hasten their decomposition and provide control of *Venturia inaequalis* (Chapter 10) has a similar purpose. The objective through management, therefore, should be one of reducing the short-term rotation to an even shorter term. Straw infected by *Cercosporella herpotrichoides* decomposes more slowly in soil than does uninfected straw (Macer, 1961), offsetting the low saprophytic competitive colonizing ability of the pathogen (Chapter 8). Soil treatments or management practices must be found that weaken the hold of this pathogen on its straw refuge and permit accelerated decomposition—and biological control.

Sclerotia

Pathogens with sclerotia are also important primarily on the aboveground portions of plants, with few notable exceptions. They use soil mainly as a springboard for infection. Again, we might infer that the aboveground parasitic habit reflects an unsuitability of the pathogen to the soil environment, and thus a vulnerability to control during host-free periods. Among the best known of this category are members of the Sclerotiniaceae, including the genera *Sclerotinia, Botryotinia,* and *Stromatinia.* All cause various forms of decay or watery soft rot of vegetables, fruits, ornamentals, or field crops that may begin by direct infection from sclerotia but which appear first at the soil line or above. Infection may then spread downward to the roots. Sclerotia may also give rise to soil-based apothecia or perithecia, and thus provide airborne ascospores, as in *Claviceps purpurea,* cause of ergot. *Colletotricum coccodes (C. atramentarium)* may form acervuli directly on sclerotia, from which inoculum apparently is disseminated. This fungus attacks belowground parts of solanaceous plants, but is most important on stems near the soil surface. Although certain species of *Mycosphaerella* may survive in soil as sclerotia, the fungus is important primarily on stems and leaves above the soil line. Although infection by *Macrophomina phaseoli* takes place directly from sclerotia, the disease initiates primarily at the soil line or above on stems or fruits in contact with soil.

Subterreanean pathogens with sclerotia also include certain primitive Basidiomycetes or probable Basidiomycetes of the genera *Typhula, Rhizoctonia, Sclerotium, Phymatotrichum,* and *Helicobasidium.* Most are well-known pathogens of plant roots. On the other hand, even within this group the tendency may be toward aboveground rather than belowground activities. *Typhula idahoensis* infects wheat leaves from sclerotia no deeper than 1–2 cm beneath the soil surface. Deeper sclerotia remain dormant awaiting some future tillage operation, when they may be returned nearer to the soil surface. Basidiocarps then form from sclerotia on the soil surface. The first symptoms of infection by *S. rolfsii* are lesions at the soil line; buried sclerotia apparently are harmless. The strain of *R. solani* that causes sharp eyespot on wheat stems forms sclerotia in abundance (Pitt, 1964), whereas the strain pathogenic to wheat roots and responsible for the bare-patch disease in South Australia apparently survives as thick-walled pigmented hyphae. Durbin (1959a) observed that leaf-attacking strains of *R. solani* were the sclerotium-producers, and root-attacking strains generally were not (Chapter 2).

Possibly the sclerotium itself represents the next level of improvement in survival over that of strands of mycelia in crop remains. The important point in terms of potential vulnerability to biological control is that, like seeds of higher plants, sclerotia present a sizable surface target area for attack by microorganisms, are known to leak nutritive substances, and support a significant surface flora. Thus, sclerotia may not be very different from mycelia within crop remains, being dependent for longevity on production of antibiotics or some other physiological process to prevent being overrun by decay microorganisms (Ferguson, 1953; Smith, 1972). Sclerotium-forming pathogens should therefore be controllable by biological means during the host-free period.

Microsclerotia of *Verticillium albo-atrum* may decline in numbers quite rapidly within a 2-year period in cotton-field soils of southern California if cotton is removed from the rotation. Microsclerotia embedded in host refuse survived longer than naked sclerotia. Wilhelm (1955) reported that this fungus can survive for at least 13 years as microsclerotia in dried agar culture; thus septic culture obviously shortens their longevity. Evans et al. (1967) suggested that the rapid decline in numbers in field soil could reflect activities of predacious or antagonistic soil organisms. They suggested further that records of longer survival by this fungus in the fields, and the belief that crop rotation is ineffective as a control, may be caused by growth of the fungus on roots of weeds and other nonhost plants.

The observation that 3 years away from winter wheat will greatly reduce losses from speckled snow mold in northeastern Washington (Chapter 4) suggests that sclerotia of *Typhula idahoensis* similarly can be eliminated through reasonably short-term crop rotations. Although the longevity of sclerotia of *Rhizoctonia solani* in field soil was placed at less than a year by Sanford (1952), others found survival as sclerotia for more than 5 years. Sclerotia of *Mycosphaerella ligulicola* are definitely not long-lived in moist natural soil, being killed in 30 weeks or less, but are capable of longer survival if the soil is sterile (Blakeman and Hornby, 1966). Sclerotia of *Sclerotinia urnula,* cause of mummy berry of blueberry, are dead within 2 years after burial in soil, according to observations of F. Johnson (unpublished) in western Washington.

Long rotations of 3–5 years away from susceptible crops are generally needed to reduce losses effectively in crops susceptible to *Phymatotrichum omnivorum, Sclerotium rolfsii,* and most strains of *R. solani. Sclerotium cepivorum* apparently can survive in the absence of *Allium* spp. for 5 years and possibly more. Other sclerotium-forming pathogens are equally capable of long-term survival. Rotations of this length usually are not economically desirable, and some method of hastening degeneration and death of sclerotia during the host-free period is needed to shorten the rotation to a more practical length.

Ferguson (1953) suggested that sclerotia of *Sclerotinia sclerotiorum* and *Sclerotium rolfsii (delphinii)* resist decay only as long as they remain metabolically active. Sclerotia killed by 5 minutes in boiling water decomposed within a few days when buried in soil, whereas those buried undamaged remained firm and viable for the duration of the experiments. Even exposing cut surfaces of the sclerotium did not necessarily hasten decay so long as the remaining portion was still alive. Apparently, metabolism within the sclerotial tissues somehow provides a deterrent to invasion by surrounding microorganisms—a form of resistance to antagonism. Sclerotia with cut surfaces do not decay, since they form new melanized layers over the cut surface (Smith, 1972). A. M. Smith (unpublished) has shown that exposure of sclerotia of *S. rolfsii* in natural soil to so-called sublethal doses of methyl bromide will cause death of the sclerotia indirectly, through colonization by soil microorganisms. He attributed the effect to fumigant damage or injury of outer cells of the sclerotia, which increased leakage of nutrients and hence invasion of the sclerotia by soil organisms. The microorganisms that decayed the sclerotia were killed by steaming soil at 60° C/30 minutes. Gilbert and Linderman (1971) found that bacteria increased on the surface of sclerotia of *S. rolfsii* in soil because of sclerotial exudation. Volatiles from decomposing alfalfa also increased exudation and numbers of bacteria. They thought that the nutritional status of this bacterial flora determined whether germination was inhibited or stimulated.

Sclerotia apparently support a natural flora, which for the most part is superficial, but which in some instances is known to exist as an embedded and integral part of the entire outer layer. Huber and McKay (1968) have shown that sclerotia of *Typhula idahoensis* will not germinate on agar media unless they are first surface-sterilized. However, it should be noted that mature sclerotia of *Sclerotinia sclerotiorum* from a pure culture also germinated in response to treatment with sodium hypochlorite (A. M. Smith, unpublished). Nevertheless, there is a flora on sclerotia in field soil, a portion of which may be parasitic, but most of which probably is saprophytic. Any process or condition that causes physiological or physical damage would be expected to lessen the ability of sclerotia to resist decay, and thus to contribute to biological control. That *Sclerotinia sclerotiorum* has not become damaging in Santa Maria, California, fields where lettuce and other leafy crops are grown on a continuous basis (Chapter 4) may result from failure of the sclerotia to mature and become more resistant to decay. After harvest, the green lettuce remains are promptly plowed under before the fungus can mature, and a new crop is promptly sown.

Streets (1969) outlined a method by which cotton can be safely grown on an annual basis, using papago peas as a winter green-manure crop to control *Phymatotrichum omnivorum*. "Early in November, pick out all open cotton and broadcast 200 pounds per acre [224 kg/ha] of ammophos 16–20 before turning under the cotton stalks. Plant papago peas . . . between November 15 and December 15. . . . Before turning under peas in the last part of April, broadcast 200 pounds of ammonium sulfate." According to Streets, this schedule can be continued indefinitely with little loss from phymatotrichum root rot. The practice is probably similar to that of using barnyard manures or other organics, as recommended by earlier workers. The papago-pea residue at average yields of 12–16 tons/acre (27–36 metric tons/ha), together with inorganic nitrogen, represents an overwhelming source of microbial nutrients when added to soil. Apparently, the residue quickly decomposes. Probably any tissue of *P. omnivorum* also decomposes quickly, whether as strands and mycelia in cotton remains or as sclerotia. Oxygen would probably be in short supply during peak decomposition, which could cause reduced respiration by the fungus, a concomitant weakening, and then overpowering by the anaerobes or facultative aerobes. Some sclerotia presumably may germinate in response to the residue, but then die from antibiosis and lysis. According to Clark (1942), the wetter the soil amended with manures, the greater the percentage of sclerotial digestion in a given time. He showed further that destruction of sclerotia was largely independent of the kind of manure added, whether of low or high C:N ratio, so long as some organic material was used. Conditions of optimum aeration, water, and temperature for microbial growth were important to sclerotial decline, which emphasizes the importance of total microbial numbers rather than specific microorganisms in decline of sclerotia (Chapters 3, 8, and 9). This may be less true for inhibition of the active parasitic phase of this fungus, as will be discussed in Chapters 8 and 9.

It would appear, therefore, that plant pathogens that survive by means of sclerotia are amenable to practical biological control during host-free periods. Variations in longevity are large within classes, genera, and even species of pathogens. In most instances, however, the sclerotia are vulnerable to colonization and eventual breakdown by soil microorganisms, as evidenced by the very much slower sclerotial death under aseptic conditions than in field soil. Host residues may protect sclerotia and lengthen their life in soil, but on the other hand, management practices that create an unfavorable physical or biotic environment can weaken sclerotia and hasten their rate of death and decomposition.

Thick-walled Spores or Cysts

Pathogens that survive during adverse periods as thick-walled single-celled propagules, because they are notoriously long-lived in soil, are probably least amenable of all to biological suppression during host-free periods. This includes: *Plasmodiophora brassicae, Spongospora subterranea, Olpidium brassicae,* and *Synchytrium endobioticum,* which survive as resting spores or resting sporangia; species of *Aphanomyces, Pythium,* and *Phytophthora,* which survive as oospores, resting sporangia, or chlamydospores; species of *Fusarium, Cylindrocarpon,* and *Thielaviopsis* that survive as chlamydospores; species of smut fungi that survive as thick-walled teliospores; species of the nematodes *Heterodera* and *Meloidogyne* that survive as cysts or egg masses. This category may also include certain strains of *Rhizoctonia* and *Verticillium* that survive as thick-walled hyphae that function much like chlamydospores. Most are true root-infecting fungi and have the capability of indefinitely repeating their entire life cycle belowground. Some may have the added capability of causing disease aboveground, given the proper environment; others such as *Phytophthora infestans* may be important almost exclusively above ground. Generally, however, they are pathogenic in roots, belowground stems, tubers, or bulbs, and are equipped to persist in the environment where such infections must ultimately take place.

Thus, the common opinion that pathogens of this category cannot, in most cases, be controlled through crop rotation seems reasonably well founded. Even where populations of such pathogens decline during absence of the host, there is good evidence that the effect is caused more by the physical environment than by the influence of soil microbiota. Resting sporangia, as well as certain other spore types, particularly several spores of the smuts and higher Phycomycetes, apparently can germinate in water alone. A certain percentage germinate with each moist period, until the entire population has expired. Other spore types are sensitive to desiccation or high soil temperature, and thus die in the tillage layer during the summer months. Nevertheless, there is the inevitable effect of microorganisms on these propagule types as well, slow and erosive if not rapid and complete. Most of these spores are of a size suitable for ingestion by predaceous fauna, although this kind of antagonism probably has little impact on the longevity of a population of thick-walled spores. The more important biological effect is probably the hastening of expiration through consumption of nutrients and gradual erosion of the spore wall.

Old (1969) and Old and Robertson (1969) demonstrated that conidia of *Helminthosporium sativum* on natural soil develop round perforations in their walls that initially are about the size of the cross-sectional diameter of a bacterial cell (Fig. 6.3: A,B). Such conidia eventually died, and bacteria were observed inside them (Old and Robertson, 1970). Clough and Patrick (1972) found similar perforations in chlamydospores of *Thielaviopsis basicola* (Fig. 6.4). One could infer that each hole was digested by an individual bacterial cell attached end-on to the spore wall (Fig. 6.5) (Marshall and Cruickshank, 1973), or perhaps in the manner of *Bdellovibrio bacteriovorus* penetrating bacterial cells (Stolp, 1973).

Volatiles produced by microorganisms, either directly or through decomposition of organic debris, may stimulate suicidal germination of spores or hasten senescence (Chapters 4 and 8). Pathogen resting spores are resistant to antagonists, but are not without weakness. Further research should be directed at accelerating the process of senescence.

VULNERABILITY DURING SAPROPHYTIC GROWTH

Most, if not all, soilborne plant pathogens make some growth outside the host before they penetrate. *Rhizoctonia solani* forms infection cushions from which infection pegs arise. *Fusarium solani*, and probably other root-infecting fusaria, commonly make considerable growth over the belowground stems and roots of hosts before penetration takes place, although some penetrate with almost no growth after chlamydospore germination. Zoospores of Phycomycetes may swim to the vicinity of root tips where they settle and encyst. Nematodes and some bacteria similarly are motile and eventually make contact with the plant surface or a wound. The prolonged ectotrophic growth habit of *G. graminis* as runner hyphae, and of *P. omnivorum* as strands, has already been mentioned in this chapter. The important question at this point is whether variations in ectotrophic growth habit produce significant differences in vulnerability to antagonism. Much of the research in biological control of root-disease

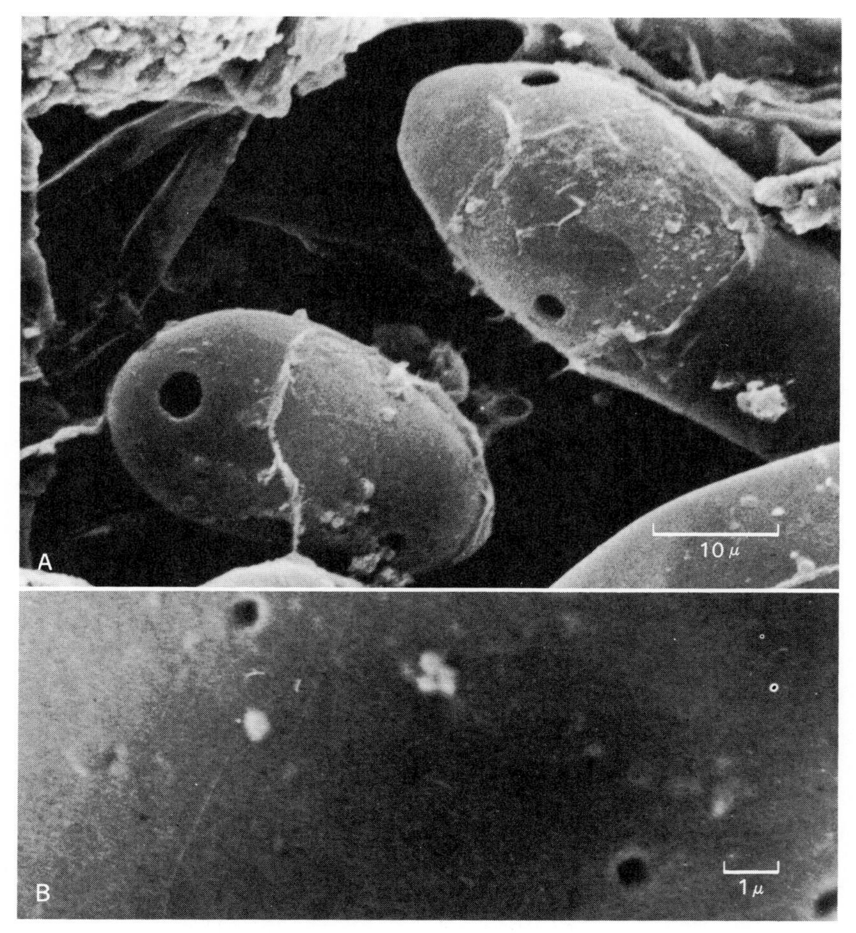

FIGURE 6.3.
Scanning-electron micrographs of fungus spores of *Cochliobolus (Helminthosporium) sativus*, showing perforations of walls induced by microorganisms. A. Conidia placed on nontreated garden soil for 50 days. Note bacteria and hyphae of actinomycetes on spore walls. Conidia held on steamed soil or agar cultures showed no holes. B. Enlarged area of spore surface showing small perforations of wall. (Photos courtesy of K. M. Old.)

organisms has been, and should be, directed at intercepting the growing pathogen in the rhizosphere.

The time spent by the pathogen in the rhizosphere or on the rhizoplane before penetration is particularly important to the success of an antagonist there. For example, the estimated 20–30 hours spent by *F. solani* f. sp. *phaseoli* in forming a thallus on a bean root are sufficient for it to be killed if conditions of soil environment, exudation, and the background

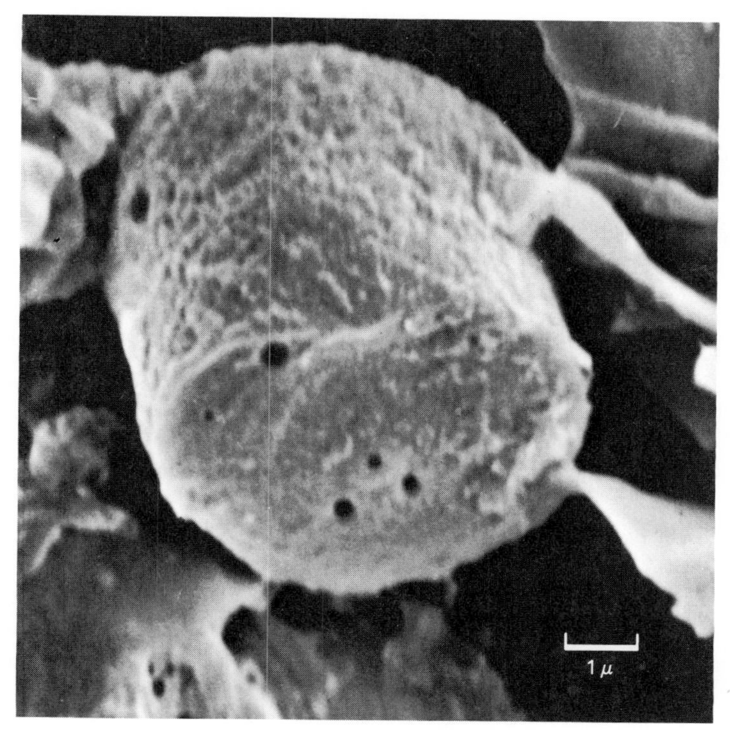

FIGURE 6.4.
Scanning-electron micrograph of chlamydospore of *Thielaviopsis basicola* placed for 8 weeks on nontreated garden soil, showing perforations of walls induced by microorganisms. (Photo from Clough and Patrick, 1972.)

microflora are favorable to the development of large populations of rhizosphere bacteria. As shown by Cook and Snyder (1965), germlings of this fungus lysed within 16–24 hours in an alkaline moist soil supplied with mixtures of amino acids, and within 9–12 hours if yeast extract was added along with the amino acids. This latter combination of nutrients stimulated germination, but subsequently favored such rapid lysis that penetration of the bean hypocotyl was prevented; one such treatment reduced the population of the fungus by nearly 50%. Hypocotyls of bean seedlings exuded some sugars, but only a trace of amino acids (Schroth and Snyder, 1961). Accordingly, chlamydospore germination was low, but germlings that did appear remained active and formed thalli (Cook and Snyder, 1965). In contrast, germination near seeds and root tips was high in response to the abundant exudates there, but the lysis rate was also high. This is consistent with the field observation that hypocotyls and older sections of roots, rather than young roots, have the greatest number of infections.

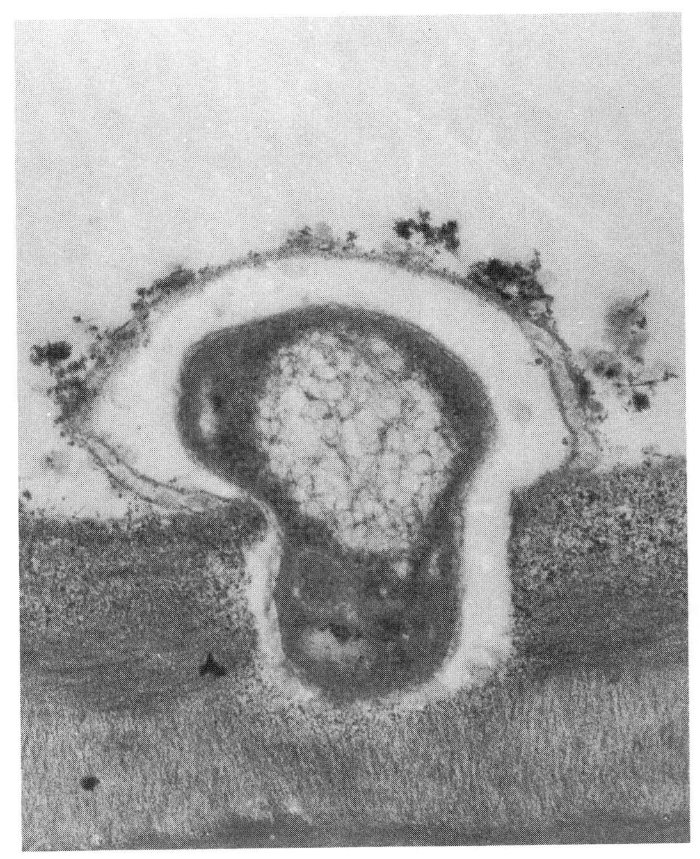

FIGURE 6.5.
Bacterium penetrating dead nodal root of wheat plant 20 weeks old grown in field soil. Ultrathin section, with soil at top and root below. × 34,000. (Photo courtesy of R.C. Foster, CSIRO, Melbourne, Australia.)

On the other hand, Burke (1965b) observed extensive decay of young roots of bean by this fungus in certain Yakima Valley and Columbia Basin soils of Washington. The opportunities for antagonism to *F. solani* f. sp. *phaseoli* near actively exuding bean tissues appear to be less in certain Washington soils than in those of the Salinas Valley. Burke apparently noticed major biological differences among soils of his region in their degree of favorableness to this fungus (Chapter 4).

Noncultivated sandy loam in the Castroville area, California, was suppressive to *F. oxysporum* f. sp. *batatas*, and this suppressiveness increased with cultivation and cropping, even though sweet potatoes were not grown (Smith and Snyder, 1971).

Whereas *F. solani* f. sp. *phaseoli* and possibly the other closely related special forms of *F. solani* may be vulnerable to antagonism in the rhizosphere because they grow too slowly, this may be less true for *Pythium ultimum* on beans and peas. As discussed earlier in this chapter, the germination of sporangia of this fungus within 1–2 hours of the time a root or seed arrives in the vicinity (Stanghellini and Hancock, 1971) and the completion of infection within 24 hours, is more advantageous to escape from antagonists than the 24–30 hours apparently required by *F. solani* for the thallus formation that precedes penetration. Other species of *Pythium* and *Phytophthora* are probably quick to penetrate when direct germination takes place, and Zentmyer (in Baker and Snyder, 1965) has shown that *P. cinnamomi* zoospores can concentrate near host root tips within a few hours after zoospore release from the sporangium. An effective biological control defense against these pathogens may be a preexisting mantle of rhizosphere organisms to block penetration. This may be possible according to Marx (1972, 1973), who found that mycorrhizae on pines protected roots against *P. cinnamomi* (Chapter 7).

Gaeumannomyces graminis exemplifies the opposite extreme for time spent on the rhizoplane. Runner hyphae are confined to this region for the parasitic life of the fungus, and without them, advance of the root lesion will cease. Even the protection afforded by hyphal pigmentation must be inadequate at times. The relatively easy control of this disease with organic and nitrogenous amendments may relate to this vulnerable ectotrophic habit of the fungus. *Phymatotrichum omnivorum* also depends on an extensive ectotrophic growth habit, and is readily controlled by virtually any organic amendment. Shipton (1969) observed that the number of separate root infections on wheat in monoculture soil where take-all is suppressed may be no different from that of wheat in conducive soil, but subsequent progress of the lesions is greatly restricted. The antagonistic factor may be insufficiently rapid to prevent initial establishment of the parasite on the host, but can stop subsequent ectotrophic progress of the fungus. Since *G. graminis* does not exist in soil in populations as great as those of *Fusarium* pathogens, severe disease depends on advance of each individual lesion rather than on multiple infections.

POPULATIONS OF SOILBORNE PATHOGENIC FUNGI THAT PRODUCE DISEASE

The population density of soilborne plant pathogens necessary to produce severe disease has recently been studied intensively with new methods (wet

TABLE 6.1
Population densities for selected soilborne pathogens that produce disease.

Group	Pathogen	Host	Population density[1] (units/g)	Reference
1	*Sclerotium rolfsii*	sugar beet	0.005–0.05/g	Leach and Davey, 1938
	Phymatotrichum omnivorum	cotton	0.01–0.06/g	King and Hope, 1932
	Rhizoctonia solani	sugar beet	0.01–0.09/g	Ko and Hora, 1971
		cotton	0.07–0.13/g	A. R. Weinhold, unpublished
	Gaeumannomyces graminis	wheat	0.01–0.3/g	Hornby, 1969
			0.05–0.11/g	Vojinović, 1972
2	*Verticillium albo-atrum*	cotton	50–400/g	Evans et al., 1967
		cotton	0.03–50/g	Ashworth et al., 1972
		mint	10–100/g	Lacy and Horner, 1965
	V. dahliae	potato	10–130/g	Isaac et al., 1971
	Phytophthora cinnamomi	fir seedling	1–30/g	Hendrix and Kuhlman, 1965
		pineapple	1–3/g	McCain et al., 1967
	Plasmodiophora brassicae	cabbage	>10/g	Colhoun, 1957
3	*Fusarium solani* f. sp. *phaseoli*	bean	1000–3000/g	Nash and Snyder, 1962
	F. roseum f. sp. *cerealis* 'Culmorum'	wheat	100–3000/g	Cook, 1968
	F. solani f. sp. *pisi*	pea	100–6000/g	R. J. Cook, unpublished
	Thielaviopsis basicola	citrus	1000–8000/g	Tsao, 1964
	Pythium ultimum	pea	100–350/g	Kerr, 1963
		pea	100–1000/g	Kraft and Roberts, 1970

[1]Where counts in the original report were given on a volume basis (sclerotia/square foot/4 inch layer), bulk density was assumed to be about 1.3 in calculation.

sieving, flotation, dilution-plate, and particularly, improved selective media). Some interesting relationships are emerging that may apply to pathogens whose population densities have not been determined by previous methods. These relationships are also of fundamental importance in biological control, as discussed below.

The pathogens for which population estimates have been made can be placed into three groups according to whether hazardous propagule counts are recorded at: (1) less than 1/g, often less than 0.1/g; (2) 1 to 100/g; (3) more than 100/g, often more than 1000/g (Table 6.1). This feature appears to be correlated with method of infection by the pathogens.

The first group is represented by *Sclerotium rolfsii, Phymatotrichum omnivorum, Rhizoctonia solani,* and possibly *Gaeumannomyces graminis.* Survival is by sclerotia (mycelia in fragments of host tissue in the case of *G. graminis,* which functionally are analogous to sclerotia). A single infection is likely to cause severe plant damage, because these pathogens have extensive ectotrophic mycelial growth; a single thallus may cover a considerable surface area of the host, and in some instances may spread through or over the soil, or aerially through the tops, and contact other plants. Therefore, relatively low population densities will produce severe disease.

The second group is represented by *Verticillium albo-atrum, V. dahliae, Phytophthora cinnamomi,* and *Plasmodiophora brassicae.* A significant common feature in this group is the capacity to produce secondary inoculum from primary survival structures. Several workers have observed microconidial production on microsclerotia of *Verticillium* in soil. Menzies and Griebel (1967) demonstrated that water added to nontreated virgin soil particularly conducive to sporulation was sufficient to induce sporulation. Microsclerotia may be restricted to direct germination in some soils, depending on the type and biological makeup of the soil. Perhaps this difference accounts for variability in the amounts of inoculum needed to produce wilt in different situations (Table 6.1). *Phytophthora cinnamomi* may produce zoosporangia and infective zoospores from chlamydospores, but may infrequently infect by direct germination, depending on environment. *Plasmodiophora brassicae* depends for infection on zoospore release from resting spores. In these examples, a single survival or resting structure may produce many propagules, greatly increasing inoculum density and making these fungi comparable to the third group.

The third group is represented by *Fusarium* spp., *Thielaviopsis basicola,* and *Pythium ultimum.* All survive as thick-walled resistant spores that apparently do not give rise to secondary inoculum before infection. Moreover, all depend on multiple infections to cause severe damage, although *P. ultimum* probably requires fewer infections to kill its host than *Fusarium* or *T. basicola.* These multiple infections normally coalesce to produce severe disease. High population densities are therefore essential in most instances to allow for germling mortality and to produce the many lesions of severe disease.

There are many pathogens for which population estimates are still unavailable. One is *Aphanomyces euteiches,* cause of pea root rot; infection is by zoospores produced from resting oospores. Based on the above scheme, an oospore count of less than 100/g, possibly less than 10/g, may produce severe disease. There is good indirect evidence that very low populations are adequate to produce disease (Boosalis and Scharen, 1959; Mitchell et al., 1969). Perhaps *Pythium* spp. that infect by zoospores similarly are able to produce disease with low oospore densities. These species would be unlike *P. ultimum,* which germinates only by production of a germ tube. *Macrophomina phaseoli* survives as microsclerotia, but apparently produces no secondary inoculum, and hence may require relatively high densities (Watanabe et al., 1970). Future research will reveal whether these values are accurate, and whether the above scheme has prediction value.

Knowledge of the threshold population densities for soilborne plant pathogens is important in biological control in at least two ways. Where elimination of the propagule by predation, endolysis, or digestion is the objective, population counts are necessary as a test for effectiveness. Perhaps of greater importance, population estimates can be used to determine when a suppressive flora is operative, by revealing whether inoculum is adequate, even though disease does not occur. Indeed, most, if not all populations summarized in Table 6.1 produce disease under favorable conditions, but commonly do not if some component of the biotic or abiotic environment is limiting. Abnormally high populations are then needed to compensate for abiotic adversity or host resistance, or to swamp the antagonists. Thus, Colhoun (1957) reported that less than 10 resting spores of *P. brassicae* per g gave maximum club root when soil conditions were favorable, but when unfavorable, 10^3 to 10^5 spores/g were sometimes inadequate. In eastern Washington wheat fields, apparently healthy crops have been produced during wet years in fields containing 3000 propagules of *F. roseum* f. sp. *cerealis* 'Culmorum' per g, whereas in dry soils where bacterial activity is restricted and the crop is stressed for water, 100 propagules/g may cause severe foot rot. The population density of the pathogen does not necessarily decline in these instances, but their capacity to produce disease is reduced. Many cases of effective biological control can be so characterized because there is little or no influence on population of the pathogen, but the threshold population required to produce disease is thereby greatly increased (Chapters 1 and 4; Fig. 1.1).

STIMULATION OF ANTAGONISTS BY THE PATHOGEN

Higher plants support a microbiota in their rhizosphere or remains that is unique and characteristic for each species (Chapter 8). The question, therefore, arises whether the same holds true for microorganisms, so that each favors the presence of other specific microorganisms. Available evidence suggests that this is the case, that the microorganisms exist in a parasitic or pathogenic relationship, and that this helps to subdue the particular pathogen of the crop. The specific flora that they support contributes to a natural form of biological control. The evidence thus suggests that pathogens may stimulate their own antagonists and perhaps must be present before the antagonism will develop.

TABLE 6.2
Take-all severity (0–5 scale) in wheat in four different soils that had been infested with living, dead, or no *Gaeumannomyces graminis* for about 6 months to increase antagonist populations, and then inoculated with the virulent pathogen at time of sowing. Disease readings 8 weeks later. (Data from Gerlagh, 1968.)

Polder soils	Living *Gaeumannomyces* inoculum	Dead *Gaeumannomyces* inoculum	No *Gaeumannomyces* inoculum
1	0	3.0	3.5
2	0	3.0	4.0
3	0	3.5	4.5
4	1.0	4.0	4.5

Gerlagh (1968) found that antagonism to virulent *Gaeumannomyces graminis* developed experimentally in polder soils in The Netherlands in response to live inoculum of the pathogen, but not with dead inoculum or with no pathogen (Table 6.2). He further concluded that "The *Ophiobolus* strain had to be virulent to initiate antagonism." Vojinović (1972) confirmed this relationship (Chapter 4).

The epidemic pattern shown by phymatotrichum root rot in the southwestern United States, by take-all of wheat, and by certain other diseases that first increase and then decline to an economically tolerable steady state, indicates that antagonists are increasing at the expense of the pathogen. In this pattern, only when the plant pathogen is present in sufficiently large populations do antagonists flourish and cause significant damage to it (Fig. 6.6). This pattern is well known for pathogens of higher plants, insects, and animals, and presumably holds for microorganisms and their antagonists as well. The stimulated antagonists could be hyperparasites (Chapter 7), or perhaps they function in a more primitive manner and destroy by secretion of antibiotics or by digestion of cell walls. In any case, the antagonist is nourished by the plant pathogen and even develops a preference for the specific pathogen, and its populations thus grow as those of the plant pathogen grow.

Hyperparasites are also stimulated by other microorganisms, including plant pathogens. Certain nematode-trapping fungi are fairly specific for nematodes and develop trapping mechanisms partly in response to them (Chapter 7). Hyperparasites likewise can show specificity for certain fungi, and respond accordingly. Even when specificity does not exist, the presence of food in the form of living or dead hyphae or nematodes would be sufficient stimulation for multiplication of organisms capable of utilizing the protoplasmic substrate.

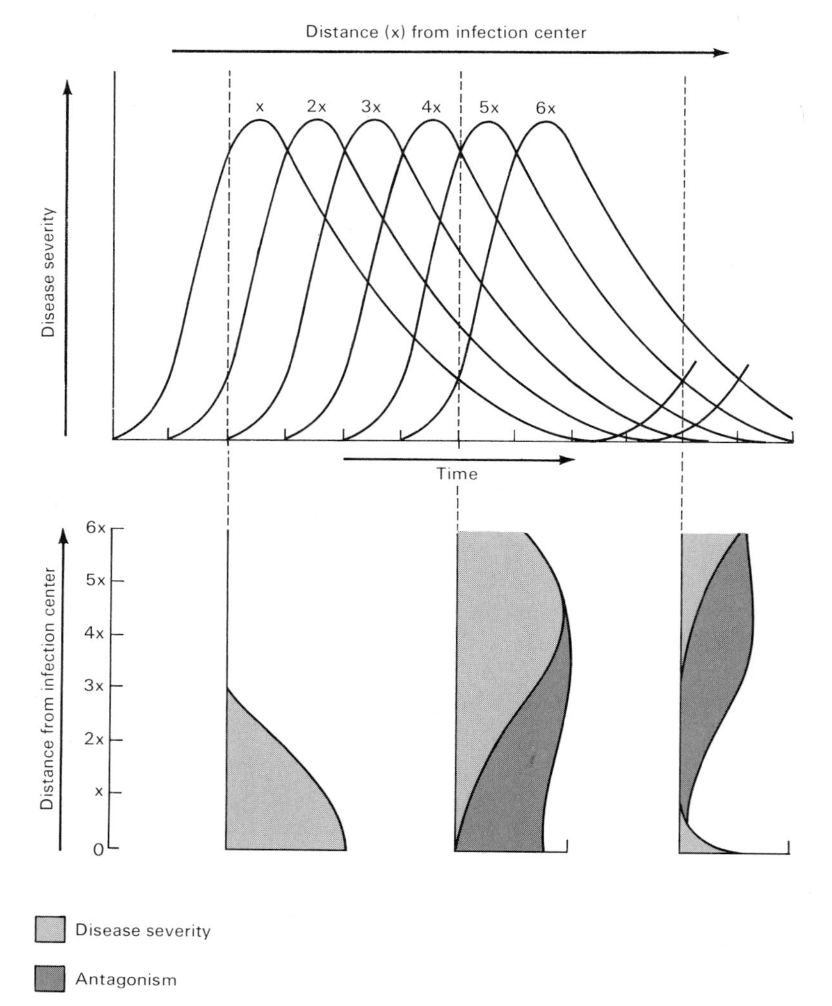

FIGURE 6.6.
Schematic diagram of the theoretical outward progression of a soilborne pathogen, such as *Phymatotrichum omnivorum,* in fairy-ring fashion from an initial infection center. In an established ring, there are advancing young mycelia, maximum inoculum density, senescent mycelia, and dead mycelia progressively from the margin to the center.

Some organisms apparently respond to fungi because of the chitin in the hyphal walls (Mitchell and Alexander, 1962; Vruggink, 1970), but results may vary (Okafor, 1970). Kerr and Bumbieris (1969) found that addition of 1% chitin to soil delayed the increase of *Fusarium* but stimulated *Pythium,* which used chitin as a nutrient. Other fungus pathogens liberate nutritive substances through hyphal walls and thereby support a "hyphosphere" flora, as designated by A. G. Winter.

Old mycelia leak badly, and their senescence is hastened by microorganisms that develop in the nutrients around them; the situation at the growing mycelial tips is less clear. The selective pressure in evolution should have been away from leakage there if it led to lysis from hyphosphere organisms. On the other hand, the high metabolism at the hyphal tips might mean that more metabolites (antibiotics) are leaked there, making the tip less favorable for microorganisms than further back. Also, the mycelial tips may grow away before harmful microorganisms have had time to build up to damaging numbers, since the tip continually pushes into areas of low microorganism populations, as does the root tip (Chapter 8). Whatever the reason, vigorous young hyphae do have increasing numbers of hyphosphere microorganisms back from the tip.

There are characteristic microorganisms associated with hyphae just as there are in the rhizosphere flora of roots. Some may exist in an ectocommensal relationship with the pathogen, but others may be pathogenic to the host plant by their external presence, as *Bacillus cereus* or *Aspergillus wentii* in the rhizosphere of tobacco may be involved in the frenching disease (Steinberg, 1952; Woltz and Littrell, 1968) or *Periconia circinata* in the rhizosphere of milo maize (Leukel, 1948; Gardner et al., 1972) where it forms root-injurious toxins. This kind of fungus action would make food available, and the organisms in closest proximity would derive the greatest benefit. One organism thus becomes "addicted" to another, which may be killed.

CONTROL OF NEMATODES BY ALTERING THE SEX RATIO

After emergence from the eggs, the second-stage larvae of root-knot nematodes (*Meloidogyne* spp.) feed for a time on young epidermal cells of roots before penetrating just behind the root tip. There are numerous records showing that under conditions of deficient nutrition, the population may become largely male (Triantaphyllou, 1973). For example, J. Tyler found in 1933 that infections by single larvae resulted in 0.7% males, those from multiple infections gave 16.4% males, and the crowded nematodes in the second-generation infections were 56.5% males. Declining vitality of the host, larval crowding in galls, removal of the galled roots from the plant, removal of plant tops, infection in unfavorable sites (cowpea leaves), conditions unfavorable to the host, host resistance, age of host, infection of the host by fungi (*Rhizoctonia solani, Verticillium albo-atrum,* or the gray sterile fungus of tomato), and spraying plants with growth inhibitors or regulators have been found by other workers to increase the ratio of males. Crowding or starvation may produce some hormonal disturbance here, as in other animals. Sexual differentiation occurs during the third larval stage. True sex reversal apparently occurs before or during the third larval stage. The host responds to females, but not males, forming giant cells that produce root knots. The males apparently are unnecessary in the life cycle.

Although the evidence for such sex reversal is less convincing for *Heterodera rostochiensis, H. schachtii,* and *H. avenae,* they show a similar increase in the proportion of males, perhaps from differential mortality, since males require less food than the females.

A relatively unexplored opportunity is presented for biological control of these gall-producing nematodes through altered nutrition, and possibly hormonal or enzymatic control. In view of the recent successes in insect control through use of hormones, there should be more studies of this fascinating and potentially important subject.

PATHOGEN INTERACTIONS

It is now well established that two or possibly more pathogens, rather than only one, are required for some diseases. In some of the disease complexes the interaction is probably one where two pathogens together are worse than either one alone. This may be true even where the effect of the two seems more than additive, as measured by symptom severity or rate of host mortality. The plant can withstand a certain mild or even somewhat virulent parasite singly, but cannot withstand simultaneous attack by two or more such parasites. This may be the case with *Pythium ultimum* and *Fusarium oxysporum* f. sp. *pisi* race 2 (Kerr, 1963), where either pathogen alone on the wilt-susceptible Greenfeast pea caused only slight reductions in plant growth or date of maturity, but together they caused severe stunting and early death. Plants of a wilt-resistant cultivar were only slightly stunted and did not die prematurely when exposed to both pathogens simultaneously. If *P. ultimum* increased host susceptibility to *F. oxysporum,* the effect probably should have occurred in the resistant as well as susceptible variety, as has been shown to be true for root-knot nematode-*Fusarium* interactions (see below). Instead, *P. ultimum* made the wilt-susceptible variety die more rapidly than did *Fusarium* alone, possibly because root rot together with vascular infection was more lethal than vascular infection alone. Nevertheless, where two pathogens are more destructive than either one alone, control of one can be considered a form of biological control of the other, operating through the host and the single pathogen.

Nematodes can predispose plants to infection by fungi by opening avenues of entry (Powell, 1971). The ectoparasitic types may be particularly important in this role because of the wounds they make in feeding; wireworms and other soil fauna likewise can open infection sites for soil fungi. The fungi aided may be wound parasites like *Cephalosporium gramineum* (Slope and Bardner, 1965), or they may be the very weak facultative parasites like *Fusarium roseum* 'Gibbosum' that cause decay only in weakened tissues.

Plant-pathogenic fungi may also affect feeding habits and populations of nematodes (Powell, 1971). Root infections by fungi generally depress populations of sedentary nematodes (*Meloidogyne* and *Heterodera* species) otherwise parasitic on that same root, and increase populations of migratory plant-parasitic nematodes. The sedentary nematodes maintain a close relationship with their host during feeding and apparently are adversely affected if the host is damaged by another root disease. The stimulation often observed for ectoparasites may relate to increased exudation of stimulatory substances, predisposition of the plant to nematode feeding, or both. The crowding caused by increased population may account for the alteration of the sex ratio discussed above.

There is particularly good evidence for the importance of nematodes in certain vascular wilts. *Meloidogyne javanica* may cause an increase in propagule count of *Fusarium oxysporum* f. sp. *lycopersici* in the rhizosphere of tomato, and a depression in number of actinomycetes, possibly by affecting exudation from tomato roots (Bergeson et al., 1970). Galls in tobacco caused by root-knot nematode may be readily invaded and decayed by *F. oxysporum* f. sp. *lycopersici.* These and other effects of nematodes on fungus diseases have been reviewed by Powell (1963, 1971). The most important interactions, however, and the ones to which the remainder of this section will be devoted, involve an effect of root-knot nematodes on resistance of plants to fusarium wilts.

Resistance to fusarium wilt in cotton in Texas (Smith and Dick, 1960) showed no consistent inheritance pattern in field studies until the soil was fumigated with ethylene dibromide for control of *M. incognita,* which then permitted identification of a single dominant factor for wilt resistance. Thus, whereas wilt resistance in cotton was previously believed to be controlled by several factors, the inheritance tests in fact were measuring levels of nematode resistance, in addition to monogenic resistance to fusarium wilt. In another study in steamed soil infested with *F. oxysporum* f. sp. *vasinfectum* alone or in combination with pure populations of *M. incognita, M. incognita* var. *acrita, Trichodorus* sp., *Tylenchorhynchus* sp., or *Helicotylenchus* sp., only *M. incognita* and *M. incognita* var. *acrita* together with the *Fusarium* caused wilt in the otherwise wilt-resistant variety, Coker 100 (Martin et al., 1956). A similar effect has been observed with other hosts. *Meloidogyne javanica* combined with *F. oxysporum* f. sp. *tracheiphilum* resulted in more wilt in the otherwise wilt-tolerant cowpea variety Grant than occurred in susceptible Chino, yet when plants of the two varieties were inoculated by dipping cut roots in a spore suspension, only Chino developed wilt and Grant showed almost no vascular necrosis (Thomason et al., 1959). Severe wilt developed in a wilt-resistant tobacco variety inoculated with *M. incognita* together with *F. oxysporum* f. sp. *nicotianae,* and the fungus spread relatively unchecked in vascular tissues of both roots and stems of the resistant variety (Melendêz and Powell, 1967).

The fusarium wilt-root knot interaction has been especially well studied in tomato, and recently reviewed by Walker (1971). In a study by Jenkins and Corsen (1957), wilt-susceptible Red Beefsteak wilted when inoculated with *F. oxysporum* f. sp. *lycopersici* alone; intermediate-resistant Rutgers wilted more rapidly with *Fusarium* plus *M. incognita* var. *acrita* than with *Fusarium* alone. Monogenic-resistant Chesapeake plants all wilted when *M. incognita* var. *acrita* was present together with *Fusarium,* but was fully resistant to the *Fusarium* alone. Sixty percent of the Chesapeake plants wilted when *M. hapla* was added with the *Fusarium.* Wilt developed in high incidence in the resistant tomato Bradley in another study with soil containing *M. incognita* and *F. oxysporum* f. sp. *lycopersici* race 1 or race 2, but was resistant to both races in the absence of the nematode. Culture tests of the plants and the soil after two crops revealed that races 1 and 2 were unchanged in pathogenicity, and no new race had originated (Goode and McGuire, 1967).

A split-root technique was used by Bowman and Bloom (1966) to study loss of wilt resistance of tomato varieties. Severe wilt developed in the otherwise highly resistant Homestead variety when root-knot nematode and *Fusarium* were inoculated either together on the same root system or on separated halves of a given root system. Faulkner et al. (1970) used a similar double-root technique to demonstrate that susceptibility of mint to *Verticillium dahliae* was increased by *Pratylenchus minyus* when the two pathogens were together or on separate root systems of the same plant. However, Conroy et al. (1972) found that there was an increase in susceptibility of tomato to *V. albo-atrum* when *P. penetrans* was present at the same infection site, but not when both were on the same root, yet isolated from each other.

The influence of root-knot nematode on fusarium-wilt resistance, and possibly of *P. minyus* on verticillium wilt of peppermint, obviously involves changes in host physiology. Melendéz and Powell (1967) showed that in tobacco, hyphae of the *Fusarium* were large and fully systemic in the otherwise resistant variety infected with root-knot nematode, but were limited in resistant plants of the same variety free of root knot. As concluded by them, the changes were in the vascular tissues far removed from the root-knot galls. It was also necessary for the nematode to be present in the host for some time to bring about the change.

Beckman et al. (1962, 1967) demonstrated that fusarium-wilt resistance in some plants involves formation of occlusions in the vascular elements by the host, which prevents systemic progress by the parasite (Chapter 8). They have further shown that occlusion is a fairly general means by which plants, including cotton and tomato, exclude microorganisms from the vascular system. The mechanism by which gel and tylose formation occurs in response to would-be parasites is undoubtedly very complex, but involves factors from both microorganisms and host (Beckman, 1966). Perhaps root-knot nematode, which upsets the hormonal balance of the host in promoting gall formation, also upsets the mechanism by which the host forms occlusion bodies to inhibit vascular parasites. Control of one pathogen in these cases would constitute biological control of another pathogen, either operating through the nematode directly, or by allowing normal host resistance to function.

ADDENDUM

Ethylene (page 137) is produced by anaerobic microorganisms resistant to moist heat (80° C/30 minutes), probably ubiquitous *Clostridium* spp., in anaerobic microsites (page 191) that form with vigorous, aerobic microbial consumption of oxygen (A. M. Smith and R. J. Cook, unpublished). All soils produced significantly more ethylene under nitrogen than under air. One soil low in organic matter produced 0.2 ppm ethylene under air, 0.7 ppm under nitrogen, 0.8 ppm under air with 0.5% wheat straw added to increase oxygen consumption, and 0.8 ppm under nitrogen plus straw. Substrate, producers, or both, were limiting. A Queensland soil (page 67) naturally high in organic matter produced 9 ppm ethylene under air, 14 ppm under nitrogen, 15 ppm under air plus 0.5% straw, and 19 ppm under nitrogen plus straw. Only anaerobic sites were limiting. As aerobic growth slows from deficient oxygen and high ethylene, oxygen diffuses back into the system and limits the anaerobes. The system is thus balanced: anaerobes $\underset{O_2}{\overset{C_2H_4}{\rightleftharpoons}}$ aerobes. Whether microorganisms other than *Clostridium,* or volatiles other than ethylene are involved is still unknown.

Nutrients override soil fungistasis. Without this balancing mechanism, organic matter would decompose too slowly. Ethylene must be biostatic at some level above nil-nutrition, or organic matter probably would not persist in soil. For some implications of this, see page 342.

7

ROLE OF THE ANTAGONIST IN BIOLOGICAL CONTROL

Antagonism is the balance wheel of nature. Where there is life, there is antagonism. . . . The soil abounds with powerful antagonists which compete with, parasitize, or poison plant pathogens. . . . The opportunities for playing one soil organism against another to man's advantage are there, and only await man's cleverness in dealing with antagonists.

—W. C. SNYDER, 1960

The vast majority of antagonists are saprophytes. Antagonists exert their influence through competition, parasitism, or antibiosis. Mycorrhizae, which may protect plant roots against root pathogens, are not saprophytes, nor perhaps are the hyperparasites, which can attack the living pathogens. Except for these examples, nearly all are obligate saprophytes. In a sense, biological control is pitting saprophytes against parasites. The first section of this chapter, therefore, concerns the relative biological efficiency of saprophytes and parasites.

BIOLOGICAL EFFICIENCY OF SAPROPHYTIC ORGANISMS

Saprophytes are more broadly adapted to the environment and are able to use a greater range of nutrients than plant parasites, a situation favorable for biological control. Parasites, on the other hand, are more competitive for living host tissue than are saprophytes. The latter are more tolerant of temperature extremes, a fact used to advantage in the selective treatment of soil with heat to eliminate pathogens, while leaving many of the antagonistic saprophytes relatively unharmed (Fig. 5.3; Chapters 5 and 6). The same generalization apparently holds for extremes of desiccation, acidity, alkalinity, anaerobiosis, and high carbon dioxide. Exceptions occur, but in general the survival rate under these various environmental extremes is higher among saprophytes than among parasites.

The general observation that saprophytes are more tolerant than parasites seems to hold (Domsch, 1959) for antibiotics (Weinhold and Bowman, 1968), fungicides (Ferguson, 1958), bactericides, nematocides, and other toxic chemicals.

Lethal dosages of methyl bromide, chloropicrin, heavy metals, and other biocidal materials are generally higher for saprophytes than for parasites. The first report of a widespread soil fungistasis was based on studies of saprophytic fungi, but there now seems no doubt that parasites are more sensitive to this property of soils than are saprophytes.

Saprophytes generally grow more rapidly on nonliving substrates than do parasites. Pathologists, in isolating microorganisms from diseased tissue, take precautions to insure that the slower-growing parasite is not overrun by saprophytes. Phytopathogenic bacteria in mixed culture with saprophytic bacteria can often be distinguished and separated into pure culture by their slower colony growth in bacteriological dilutions on agar. Garrett listed rapid germination and growth as a characteristic of organisms with high competitive saprophytic ability; organisms that grow rapidly stand the best chance of obtaining the limited food supply.

Fusarium offers some interesting comparisons of saprophytes and parasites, since both occur within a given species. *Fusarium solani, F. oxysporum,* and *F. roseum* consist mainly of saprophytes, but each contains a few parasitic forms. Nash and Snyder (1965) reported that in wilt-suppressive soils of the Salinas Valley, California, "saprophytic *F. oxysporum* establishes abundantly and in a variety of clonal types" whereas "neither the pathogenic formae of *F. oxysporum* nor the wilt diseases are found in these soils." Certain saprophytes, but not parasites, occurred in noncultivated soil, and with cultivation the saprophytes did not disappear but held their own and often increased severalfold. They have subsequently observed (Smith and Snyder, 1972) that chlamydospore germination of three saprophytic clones of *F. oxysporum,* each of which was matched with a parasitic clone of the species (respective pairing made on the basis of similar appearance in cultures), was superior in every instance to that of the comparable parasite in soil. Length of germ tubes was also greater for the saprophytic clone in each instance. Meyers and Cook (1972) observed that saprophytic clones of *F. solani* in shake culture may convert into chlamydospores within 8–12 hours after the carbon (glucose) is removed, whereas parasites take as long as 50–60 hours to form them under identical conditions. We have long known that the saprophytic fusaria are almost invariably more aggressive colonizers than parasitic forms.

Saprophytes have broader distribution than parasites, presumably because of their more aggressive nature and their greater tolerance of environmental extremes. The Gibbosum cultivar of *Fusarium roseum,* and *F. oxysporum* 'Redolens,' for example, are extremely vigorous saprophytes and occur very commonly in cultivated and noncultivated soil throughout the world. These two types are probably more common than any other *Fusarium* as secondary invaders of parasitized plants. *Trichoderma viride, Gliocladium roseum,* and *Bacillus subtilis* are likewise broadly adapted and probably more widely distributed than any soilborne parasite of plants.

On the other hand, saprophytes are subject to the same physical limitations of transportation as are parasites, and may not have means for introduction into all sites to which they are adapted. Like parasites, saprophytes are introduced into new areas as contaminant spores and mycelia on propagative stock, with machinery, or by insects, birds, or animals. Many are wind-borne on dust or plant fragments, or as ascospores or dry conidia. Nevertheless, no one species or strain can be successfully and naturally introduced into all possible geographic areas.

New genetic plant material introduced from foreign wild habitats is grown first in one of several U.S. Department of Agriculture Plant Introduction Centers. There it is studied, increased, and finally distributed to breeders and commercial companies for further increase or use as parent material. Parasites may, and apparently do, successfully follow the host through these many steps from origin to farmer, but antagonists of the parasite may not. Parasites have a direct relation with the host plant that antagonists may lack. **The more casual the relationship between a microorganism and the host plant, the less the chance of dissemination of the organism with that plant.** This emphasizes the necessity for examining the populations of many different areas for the antagonists to protect crops. We should not anticipate that soil of a given field already contains all necessary antagonists—they may be unable to get there by themselves.

The relatively vigorous and competitive growth habit of saprophytic organisms is evidence of their physiologically advanced nature. Organisms unable to invade living tissue and also unable to compete effectively as saprophytes were replaced during evolution by those that had one or more of these abilities. Parasites may be less competitive because of their specialized ability to derive food from living tissue. It seems generally true that the *organisms most apt to succeed in mixed populations competing for the same food supply will be those that are no more specialized than is required in the particular situation.* An organism with a specialized metabolic system needed for parasitism eventually will be displaced in competition for a given dead substrate or niche by other organisms lacking this enzyme system. A race of a parasite with genes for virulence for a host but with unnecessary virulence genes will be displaced in competition with a race that has no more than the necessary number of virulence genes (Watson, 1970). Because of this, parasites outside their hosts and in competition with saprophytes for food may find their parasitic capacity a liability, like a traveler with excess baggage.

Some efficient saprophytes also have parasitic ability; *Rhizoctonia solani* and *Pythium ultimum* are well-known examples. In fact, parasitism is largely incidental to saprophytism in the lives of these organisms, and they have thus come to be known as primitive or facultative parasites. Both fungi can make exceptionally rapid growth, are broadly tolerant of environment, are widely distributed, and can live indefinitely as soil-inhabiting saprophytes. Consequently, they are relatively tolerant of the presence of other microorganisms and have proven more refractory to biological control than most root parasites. **The more specialized the parasite, the more restricted the conditions under which it can cause disease, and the more susceptible it is apt to be to antagonistic microorganisms.**

KINDS OF ANTAGONISTS

The antagonists in biological control of plant pathogens include bacteria, actinomycetes, fungi, viruses, higher plants, and predatory microfauna such as protozoa, nematodes, rotifers, collembola, and mites. Reviews on the ecology of these various kinds of organisms are available (Baker and Snyder, 1965; Barron, 1968; Burges and Raw, 1967; Domsch and Gams, 1972; Gray and Parkinson, 1968; Wallwork, 1970). This section considers only the potential of these organisms in biological control.

Bacteria

Bacteria are extremely important in biological control of plant pathogens. They may exceed the number and weight of any other group of microorganisms in soil, and their rapidity of growth and ability to utilize different forms of nutrients under widely different conditions is surpassed by no other group. Clark (in Burges and Raw, 1967) reported from 1×10^6 to 9×10^9 living bacterial cells per g of soil, with a fair representative value of 2×10^9/g, which would give about 2 tons (1.8 metric tons) of bacteria per acre–15 cm of soil weighing 1000 tons (907 metric tons). The mass of actinomycetes is of this same order. Fungi, with about 100 m of mycelia per g of soil, have a somewhat smaller mass than bacteria (Clark and Paul, 1970). These features, and the rapidity with which bacteria begin to multiply in response to nutrients, probably make them of greater importance than any other group as competitors for nutrients, particularly in the rhizosphere and around seeds, but also in soils amended with organic materials. Given a proper physical environment, bacteria can pass through the lag and log phases of growth and attain a climax state in less than 48 hours from the time simple nutrients become available in soil. During logarithmic growth, estimates of the generation time have been placed at about 20 minutes. The addition of water to air-dry soil may result in more than a tenfold increase in bacterial numbers in 12–24 hours, as these organisms use nutrients made available by drying and wetting. The nutritional requirements of soil bacteria have been reviewed by Katznelson (in Baker and Snyder, 1965) and Gray and Parkinson (1968). No organic substrate in soil can escape erosion and digestion by bacteria, whether complex lignins, celluloses and hemicelluloses, proteins, simple sugars, amino acids, pesticides, or even cutin.

Bacteria are especially important as antagonists of pathogens such as *Fusarium* and certain others that produce a germ tube and attack roots rapidly by multiple infections. It is doubtful whether actinomycetes, fungi, nematodes, or other antagonists can match bacteria in intercepting sufficient numbers of pathogen germlings to reduce disease. This may explain why growth and survival of *Fusarium* in different soils, near different plant parts, or with different nutrients seems to hinge so critically on whether the soil environment is favorable to bacteria. Subtle differences in water potential, pH, or temperature can result in striking differences in the success of germlings of *Fusarium* in the presence of soil bacteria (Chapters 6 and 9). *Pythium,* on the other hand, may grow so rapidly that no environmental conditions could provide sufficient advantage to soil bacteria except in special head-start situations. Alternatively, research may reveal a tolerance or resistance of *Pythium* and certain other related aquatic fungi to the presence of soil bacteria and their metabolites, resulting from a long evolutionary association; *Pythium* does well and is even favored by wet alkaline conditions, which are also ideal for most soil bacteria. It would be surprising if escape by rapid growth could account entirely for the success of this fungus in wet alkaline soils (Chapter 9). Resistance to, or even ability to use metabolites produced by bacteria is also a possibility. By contrast, pathogenic fusaria grow best in slightly acid dry soils, and may not have had a long evolutionary association with soil bacteria. This could explain the lack of tolerance of *Fusarium* to bacteria (Chapter 9).

Some bacteria are anaerobic and, along with a few actinomycetes, make biological control possible under anaerobic conditions, as with crop residues that have been turned under and flooded (this chapter). Endospores formed by species of *Bacillus* and *Clostridium* can survive longer or at higher temperatures than most other microbial propagative units. Some bacteria have flagella and possibly may swim to a substrate or target pathogen if the water-filled soil pores are of sufficient size. Bacteria produce antibiotics, although this apparently is not a strong feature of the class. Some pseudomonads produce antibiotics effective against other bacteria, and certain *Bacillus* spp., especially *B. subtilis,* produce antibiotics with activity against all types of soil microflora, including pathogens. Finally, bacteria are small by comparison with other soil organisms and can multiply and form colonies on the surface of hyphae where opportunities for antagonism can be fully realized.

Pseudomonads are common in soil in organic matter, especially in freshly incorporated crop residue and in the rhizosphere of plants. Since these niches are also those favorable to plant pathogens, pseudomonads would seem well adapted for biological control. However, they have not proven very useful as introduced antagonists (Broadbent et al., 1971). Why? As pointed out in Chapter 1, a stable balance of organisms eventually evolves. Because roots, pseudomonads, and root pathogens such as *Pythium, Phytophthora,* and *Thielaviopsis* have probably had a long association, it is reasonable that such a balance would have evolved. Pseudomonads have even been implicated in stimulation of sporangial production by *Phytophthora cinnamomi.* Because pseudomonads that produced potent metabolites injurious to roots would have reduced their own niches, there was selective pressure for mild metabolites. However, to retain these niches they must have produced antibiotics inhibitory to other microorganisms in the rhizosphere; only bacteria and fungi insensitive to these antibiotics shared the sites. Thus, certain root-pathogenic fungi are generally found not to be inhibited by rhizosphere pseudomonads (Broadbent et al., 1971). However, *Gaeumannomyces graminis* is sensitive to certain fluorescent pseudomonads on agar and when introduced into methyl-bromide-treated soil (A. D. Rovira and E. H. Ridge, unpublished; R. J. Cook, unpublished). Rhizosphere bacteria are said (Brown, 1961) to be more tolerant of antibiotics than are bacteria in the soil mass. However, because of the abundance of pseudomonads in the rhizosphere they may well be important in competition with root pathogens. *Bacillus* spp. and actinomycetes, which inhabit different niches in soil and are less common in the rhizosphere, tend to show greater antibiotic activity to root

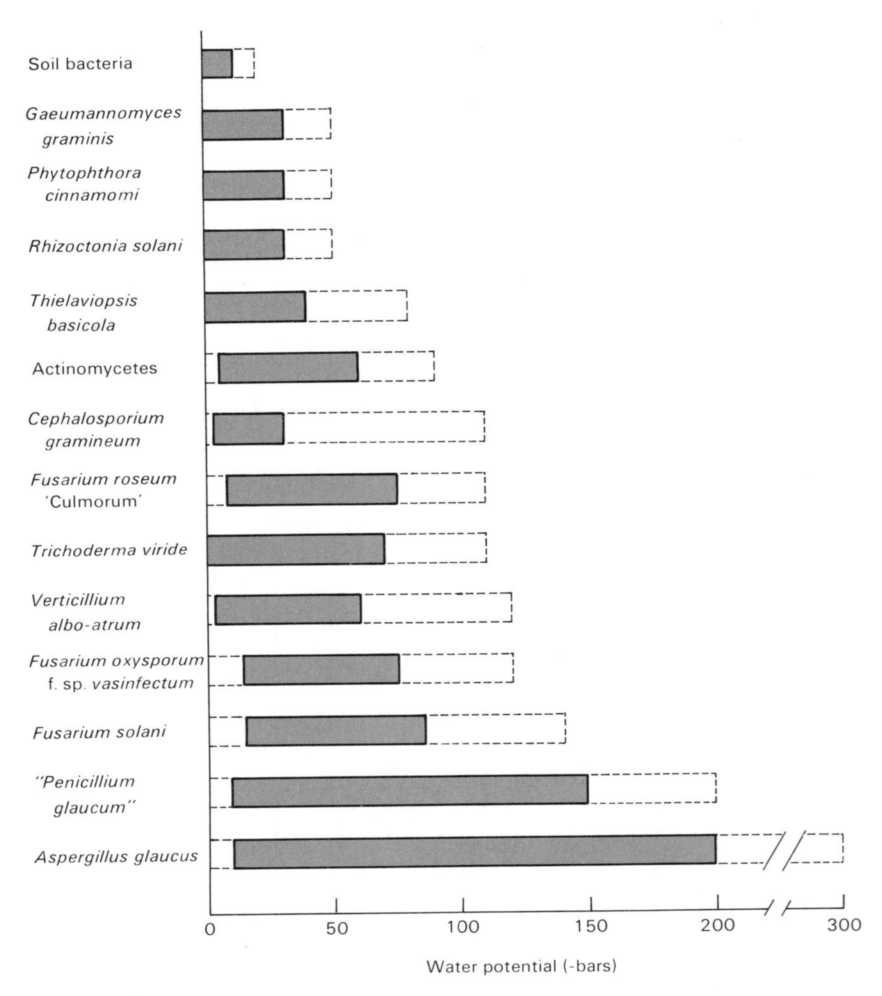

FIGURE 7.1.
Schematic diagram showing the approximate water potentials most favorable for growth of selected genera or classes of soil microorganisms.

pathogens than do pseudomonads. They are also generally compatible for inoculation together. The relationship between bacteria and *Pythium* referred to in this chapter and chapter 9 is another example of the effect of such evolutionary association.

It is now generally considered that bacteria may occur in apparently healthy internal tissues of plants (Matta, 1971). They probably gain entrance through injuries, particularly in roots or cut surfaces, as in potato seed pieces or cuttings, and are carried upward in the xylem. Hollis (1951) found *Aerobacter cloacae, Bacillus megaterium,* and *Bacterium globiformae* in potato stem tissue. Since *B. megaterium* has antagonistic capacity (Chapter 5) and is apparently noninjurious to some plant tissues, it is worth investigating whether dipping the base of cuttings or cut seed pieces in a suspension of this bacterium would provide biological control of vascular pathogens such as *Fusarium* or *Verticillium.*

Bacteria have limitations, such as their relative inactivity in soils not too dry for growth of fungi and actinomycetes and for disease development. Bacteria can grow in culture solutions at osmotic water potentials considerably below −15 bars, yet a soil-water potential of −10 to −15 bars prevents nearly all bacterial activity as measured by cell multiplication or antagonistic effects (Cook and Papendick, in Toussoun et al., 1970). It is not known whether the effect of moderately low water potentials on bacteria is direct, because of an inability of the organism to remove water or to move in the diminished water film (Griffin and Quail, 1968), or indirect, because of a limitation of nutrient diffusion. As the tillage layer dries, soil bacterial activity gradually decreases; growth of fungi and actinomycetes can then increase and invade plant parts with little antagonism from bacteria. Apparently, soil water potentials of −60 to −100 bars and drier would be needed to prevent growth of fungi and actinomycetes (Fig. 7.1). The surface layer of soil in a nonirrigated field could thus become ideal for those organisms. Unfortunately, moist or wet soil is less favorable to actinomycetes and, with rewetting of the soil, their usefulness in biological control is lessened. A combination of bacteria and actinomycetes could be useful in biological control by introduced antagonists. *The higher the potential energy of soil water, the greater the diversity of organisms capable of growth; conversely, the lower the potential energy, the fewer the kinds of organisms capable of growth.*

Actinomycetes

As resident antagonists, actinomycetes are probably second only to bacteria in importance in maintaining a satisfactory biological balance in soil, largely because of their widespread ability to produce antibiotics. Alexander (1961) stated that about three-fourths of the species of *Streptomyces* produce antibiotics. Goldberg and Luckey (in Goldberg, 1959) placed this value for animal pathogens at half of all actinomycetes. Most are effective against bacteria, but many are effective against fungi and actinomycetes as well. In addition, many have the ectoenzyme systems necessary to break down proteins, cellulose, and chitin, which may have potential in the biological control of microorganisms with chitinous walls (Mitchell and Alexander, 1962; Vruggink, 1970), although the evidence is conflicting (Kerr and Bumbieris, 1969; Okafor, 1970). The former workers obtained some control of *F. solani* f. sp. *phaseoli, F. oxysporum* f. sp. *conglutinans, F. oxysporum* f. sp. *cubense,* and *Rhizoctonia solani,* but not of *Pythium ultimum,* with chitin, lobster shells, and to some extent, laminarin. *Fusarium,* but not *Pythium,* has chitinous hyphal walls. Actinomycetes became particularly numerous in response to the chitin, and apparently liberated chitinase, which contributed to a destruction, probably exolysis (Chapter 2), of the *Fusarium* hyphae. Sneh et al. (1971) found the inhibition of *R. solani* in chitin-amended soil "to result from the effect of inhibitory compounds of an unknown nature produced by the soil microflora during chitin decomposition." Mitchell (1963) suggested that this effect might have implications in terms of control of certain diseases with crop residues, and also with the suppressive nature of some soils to pathogen spread.

Actinomycetes, on the other hand, grow slowly and are poor competitors, although their slow growth is partially offset by the abundance of resistant spores formed in soil (Lloyd, 1969), and by their ability to produce antibiotics. Slow growth rate seems to be associated with high antibiotic production. Actinomycetes tend to become active as nutrients are exhausted, which is disadvantageous for control of quick-growing pathogens. They would, then, probably be too slow to control an organism like *Pythium,* which infects rapidly in response to the first available sugars, unless the actinomycetes were already present in the particular niche. *Pythium* could not have had the intimate evolutionary association with actinomycetes that it has had with bacteria in aquatic environments. This could explain why Johnson (1954) and Broadbent et al. (1971) found *Pythium* to be more sensitive to random isolates of actinomycetes than to bacteria in tests on agar and in soil. However, some effective bacterial antagonists of *P. ultimum* have been found (Fig. 7.2). *Cephalosporium gramineum,* pathogenic to wheat in moist soils, is likewise relatively unaffected by bacteria; *Penicillium* seems to be most effective in reducing *Cephalosporium* inoculum (Bruehl and Lai, 1968a). It is clear that some plant pathogens are best controlled by fungi or actinomycetes.

Antagonism by actinomycetes in the rhizosphere may be involved with pathogens that persist as runner hyphae on the roots. Actinomycetes are potentially valuable as antagonists during host-free periods, where time is not critical and where organic amendments (chitin) can be used to encourage their activities (Mitchell and Alexander, 1962; Vruggink, 1970). Actinomycetes do well in dry situations of high organic matter and high temperature, conditions that might be produced with proper organic amendments during a fallow period.

FIGURE 7.2.
Reduction of damping-off of snapdragon seedlings from *Pythium ultimum* by antagonistic *Bacillus* spp. Left: Control, without antagonists. Center: *Bacillus* isolate WA2b67 inoculated in soil gave moderate protection. Right: *Bacillus* isolate WW27 inoculated in soil gave good protection. Containers were inoculated with *P. ultimum* in upper right-hand corner. (From Broadbent et al., 1971.)

Fungi

Fungi have received most attention as antagonists, possibly because they are easier to handle and identify than bacteria and actinomycetes. However, fungi probably rank third as potential antagonists in biological control. This does not imply that they are unimportant; on the contrary, some situations clearly suggest fungi as the important antagonists. The tendency of *Trichoderma, Gliocladium, Penicillium,* and others, to produce potent broad-spectrum antibiotics is well known. Because of their ability to make rapid growth on organic substrates in dry, acid, coarse-textured soils, and thus to reach nutrients available to unicellular microorganisms only by diffusion (Trinci, 1969), they may outperform bacteria and possibly actinomycetes. *Trichoderma, Aspergillus,* and *Penicillium* are able to solubilize insoluble phosphates in soil; *Pythium* and *Rhizoctonia* cannot do this, may grow poorly, and fail to sporulate (Agnihotri, 1970). Saprophytic soil fungi have high competitive ability and undoubtedly play the major role, with help from bacteria and actinomycetes, in the breakdown of pathogen-infested plant refuse. Plant pathogens in straw, wood, leaves, fruit, roots, or other cellulosic material are more apt to be replaced by saprophytic fungi than other organisms. Fungi, therefore, appear to be important in biological control of plant pathogens during the host-free period.

Marx (1972, 1973) showed that shortleaf pine seedlings with naturally occurring ectomycorrhizae were resistant to infection by zoospores of *Phytophthora cinnamomi.* Mycelia of the pathogen were loosely attached to mycorrhizae and firmly attached to nonmycorrhizal roots. Resistance was greatest when mycorrhizae formed complete mantles and a good Hartig-net development. Infection of mycorrhizae occurred if the growing root tip lacked a complete fungus mantle and if the Hartig-net development was incomplete, but *P. cinnamomi* did not progress into the Hartig-net region. This suggests internal as well as external protection conferred by the mycorrhizal fungus to the pine root. Where the cortex was artificially exposed, the mycorrhizae remained resistant to infection. The protection afforded by one mycorrhiza also gave some protection to adjacent root initials, whereas root initials adjacent to nonmycorrhizal roots became severely infected. There was a significant reduction in inoculum density of *P. cinnamomi* in containers with ectomycorrhizal pines during the experiments.

The degree of resistance of mycorrhizae to *P. cinnamomi* varied with the fungus species, and was particularly effective with *Leucopaxillus cerealis* var. *piceina.* This fungus produced diatretyne nitrile when grown in association with shortleaf pine in axenic culture, which totally inhibited zoospore germination of *P. cinnamomi* at 2 ppm, and which was also highly toxic to soil bacteria and many root-infecting fungi. Krupa and Fries (1971) found that *Boletus variegatus* as an ectomycorrhizal fungus on pine roots increased production or accumulation of volatiles (terpenes and sesquiterpenes) two- to eightfold over that of noninoculated controls. They considered that normal volatile and nonvolatile materials produced by roots stimulated fungi in the rhizosphere. By producing antibiotics, mycorrhizal fungi were able to suppress antagonists sufficiently to gain dominance in the rhizosphere and to penetrate the roots. The nonspecific response of the host caused increased production of nonvolatiles in host tissue, restricting mycelial growth, and "resulting finally in the symbiotic state." The root pathogens were then further inhibited by the volatiles produced by the mycorrhizal roots.

Ectomycorrhizal fungi are obviously important in biological control of plant pathogens. Endotrophic vesicular-arbuscular mycorrhizae (*Endogone* spp.) may function in the same manner, although the potential usefulness of this fungus group in biological control is still largely unknown (Baltruschat and Schönbeck, 1972; Mosse, 1973; Ross, 1972). Perhaps mycorrhizal fungi involved could be used to advantage as introduced or managed antagonists. The stimulation of mycorrhizae on Monterey pine in New Zealand by application of superphosphate is one example of management of a resident fungus antagonist (Chapter 2).

Pythium root rot of pecan trees was reduced in Georgia following application of dibromochloropropane (DBCP) and fungicides, which did not, however, significantly reduce either nematode or *Pythium* populations (Marx, in Hacskaylo, 1971). *Scleroderma bovista* became the dominant and abundant ectomycorrhiza on the roots after the treatments, apparently because of elimination of competitors. Plant vigor and root surface increased greatly within a few months, along with marked reduction of root necrosis. *Scleroderma bovista* produced a potent antibiotic to *Pythium* spp. on agar.

Fungi may also be important as antagonists introduced with seed. Fungi used for this purpose probably would have the greatest potential if they were selected on the basis of antibiotic production. The control would be most effective against seed and seedling diseases caused by pathogens carried on the seed. Fungi also have potential as antagonists introduced into treated soil, whether steamed in the glasshouse or fumigated in the field. Fungi are generally more adept at retaining possession of their substrate than are actinomycetes or bacteria. Introduced into treated soil along with organic matter, they might maintain possession of it against establishment of plant pathogens.

Some biological control of plant pathogens by fungi occurs naturally in glasshouse soil mixes steamed or chemically treated to near-sterility. *Trichoderma viride, Peziza ostracoderma, Pyronema confluens, Neurospora* spp., or *Mucor* spp. rapidly colonize and conspicuously sporulate on soil treated at 100° C, and may reduce invasion by plant pathogens. Some of these are known to be effective antagonists. However, they do not noticeably establish in soil treated at 60° C, presumably because of the surviving resident antagonistic actinomycetes and bacteria. The more severe the treatment, the more they luxuriate, but as other microorganisms are returned to the soil by air, tools, or planting material, they become inconspicuous, dormant, or suppressed.

Verticillium malthousei was also found by Moore (1972) to develop on only 27% of the mushrooms if the casing soil was treated with aerated steam at 60° C/30 minutes, but if the treatment was 98° C/30 minutes, a 93% loss of mushrooms resulted (Fig. 7.3). Germination of *Verticillium* phialospores was reduced by surviving thermal-tolerant antagonists.

It is suggested that *soil bacteria are generally effective as scavengers and are thus important in competition;* Bacillus *and* Pseudomonas *spp. and some others are also important in antibiosis. Actinomycetes are poor as scavengers and in competition, but are excellent antibiotic producers. Fungi are effective in competition (possession) and hyperparasitism, and some effectively produce antibiotics (possession).* It is further suggested that *bacteria are effective in the rhizosphere, and that bacteria, fungi, and actinomycetes are effective on the organic debris or crop residues during*

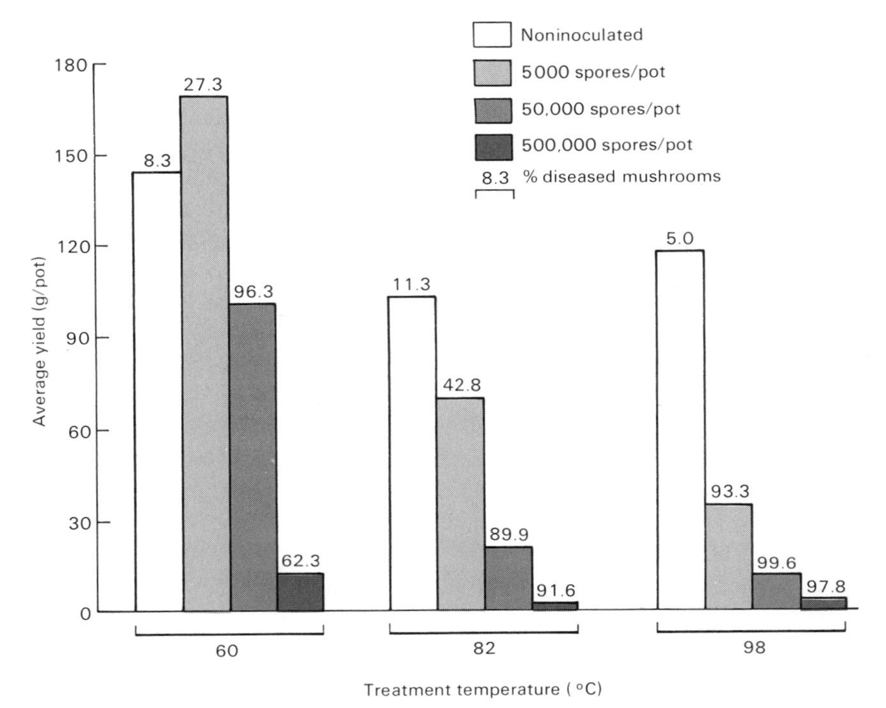

FIGURE 7.3.
Yield of healthy mushrooms and percent of diseased mushrooms 27 days after casing with clay loam (19% moisture, w/w) treated with aerated steam for 30 minutes at indicated temperatures. Spores of *Verticillium malthousei* mixed into treated soil at three different dosages and held for 3 days before use in casing spawned compost. (Data from Moore, 1972.)

the host-free periods. Bacteria and fungi have proven effective as antagonists introduced with seeds to protect against preemergence seed decay or seedling blight.

Viruses

Viruses occur in a number of fungi and cause serious commercial loss in cultivated mushroom crops (Hollings and Stone, 1971). They are transmitted through anastomosis of hyphae, through fusion of spore germ tubes with mycelia, and to a limited extent, by mites. It has not yet been possible to transmit most of them mechanically, which is a limitation to their possible use in biological control. Virus infection of fungi increases hyphal leakage (thus increasing antibiotic production), and decreases growth rate and sometimes the host virulence of the fungus. No commercial use of viruses in biological control of plant pathogens has so far been attempted.

Lapierre et al. (1970) observed virus particles in the hyphae of each of two weakly pathogenic strains of *Gaeumannomyces graminis* var. *tritici*, which also had a diminished production of the sexual stage. Ascospores produced by these two strains were free of the virus, and again gave rise to highly pathogenic isolates. The two infected strains were obtained originally from fields in France cropped to wheat for 6 and 12 consecutive years; the wheat crops had apparently shown decreased take-all. Rawlinson et al. (1973) confirmed these results in England. They rarely found the "virus-like particles" (VLP) in a first-year wheat crop, but the particles became progressively more common up to the ninth year. "The evidence is not yet sufficient to define the role, if any, of VLP in take-all decline." They were able to transmit the virus by anastomosis of hyphae in two instances.

Lemaire and Jouan (in Anonymous, 1973) showed that inoculation of virus-infected strains onto wheat inhibited further attack by an aggressive strain. "Furthermore, it was almost impossible to establish a highly virulent strain in a soil infested by a hypovirulent strain." Lemaire et al. (1971) suggested a combination of (a) amendment of fish, soybean, or alfalfa meal or gelatin with or without antagonistic bacteria, and (b) virus infected "hypovirulent" strains of *Gaeumannomyces*. A virus of *G. graminis* also has been found in Western Australia (A. Sirasithamparam, unpublished).

It is unlikely that a mycovirus can account for the decreased take-all in wheat-monoculture soils in Washington, since isolates recovered from suppressive soils are virulent when introduced into a conducive soil.

Lapierre (in Anonymous, 1973) reported viruses in *Sclerotium cepivorum*, *Verticillium dahliae*, *V. fungicola*, *Fusarium roseum*, *F. oxysporum*, *Mycogone perniciosa*, *Penicillium* sp., and *Chromelosporium* sp., as well as *G. graminis*. Spire (in Anonymous, 1973) added *P. stoloniferum*, *Diplocarpon rosae*, *Ustilago maydis*, *Alternaria tenuis*, *F. moniliforme*, *Helminthosporium maydis*, *H. victoriae*, *H. oryzae*, *Piricularia oryzae*, *Botrytis* sp., and *Verticillium* sp. to the list. Virus infected strains of *P. oryzae* and *H. maydis* were more virulent than noninfected strains, whereas viruses had no detectable effect on *S. cepivorum* or *H. oryzae*.

Animals

Antagonistic animals are predacious and of undetermined importance in biological control. Nematodes, protozoa, and others ingest spores, cells, or mycelia, and may extract cell contents, but the importance of this is unknown. Carriage of antagonistic microorganisms from site to site by these animals within the soil may be important. Like the aquatic fungi, however, nematodes and protozoa require, for mobility, water-filled soil channels of adequate size to accommodate them, and so cannot function in dry situations.

FORMS OF ANTAGONISM

The forms of antagonism generally recognized are competition, antibiosis, and predation or hyperparasitism (Chapter 2). Entomologists have tended to use parasites and predators, whereas plant pathologists usually have used antibiotic organisms and competitors. Where resident antagonists are used, it may be that no one form of antagonism is best, but rather control would depend on the total effect of all forms. An understanding of the strengths and weaknesss of each form of antagonism is basic to any purposeful effort toward biological control.

Predation and Hyperparasitism

Attempts to use predators and parasites of insects have shown that the curve of beneficial populations tends to lag behind that of the pest. Reduction of damage, rather than elimination of the pest, is usually the objective. Plant pathologists similarly have demonstrated that populations of parasites increase in response to high populations of a host. Hyperparasites of plant pathogens are presumably no exception; the population of the plant pathogen is necessarily fairly substantial before the population of a hyperparasite will increase. We might also expect that the pathogen population should be active rather than dormant, because hyperparasites, like parasites, tend to attack hosts in a susceptible active condition. Fungi parasitic on other fungi (mycoparasites) have recently been reviewed (Barnett and Binder, 1973).

There are several means by which mycoparasites attack fungus structures. They may penetrate mycelia directly, the parasitic hypha growing within the host mycelium, as does *Rhizoctonia solani* in various Phycomycetes (Butler, 1957), *Didymella exitialis* in *Gaeumannomyces graminis* (Siegle, 1961), or *Mycena citricolor* in Mucor (Fig. 7.4). The mycoparasite may coil around the mycelium of the host, with or without penetration, as does *Trichoderma viride* (Weindling, 1932). Enzymes may be produced that digest the mycelial walls, or antibiotics may be formed that inhibit growth or cause endolysis (Fig. 2.2), as does *Trichoderma viride* (Weindling, 1934), or which release amino acids that inhibit growth, as in *Didymella exitialis* (Siegle, 1961).

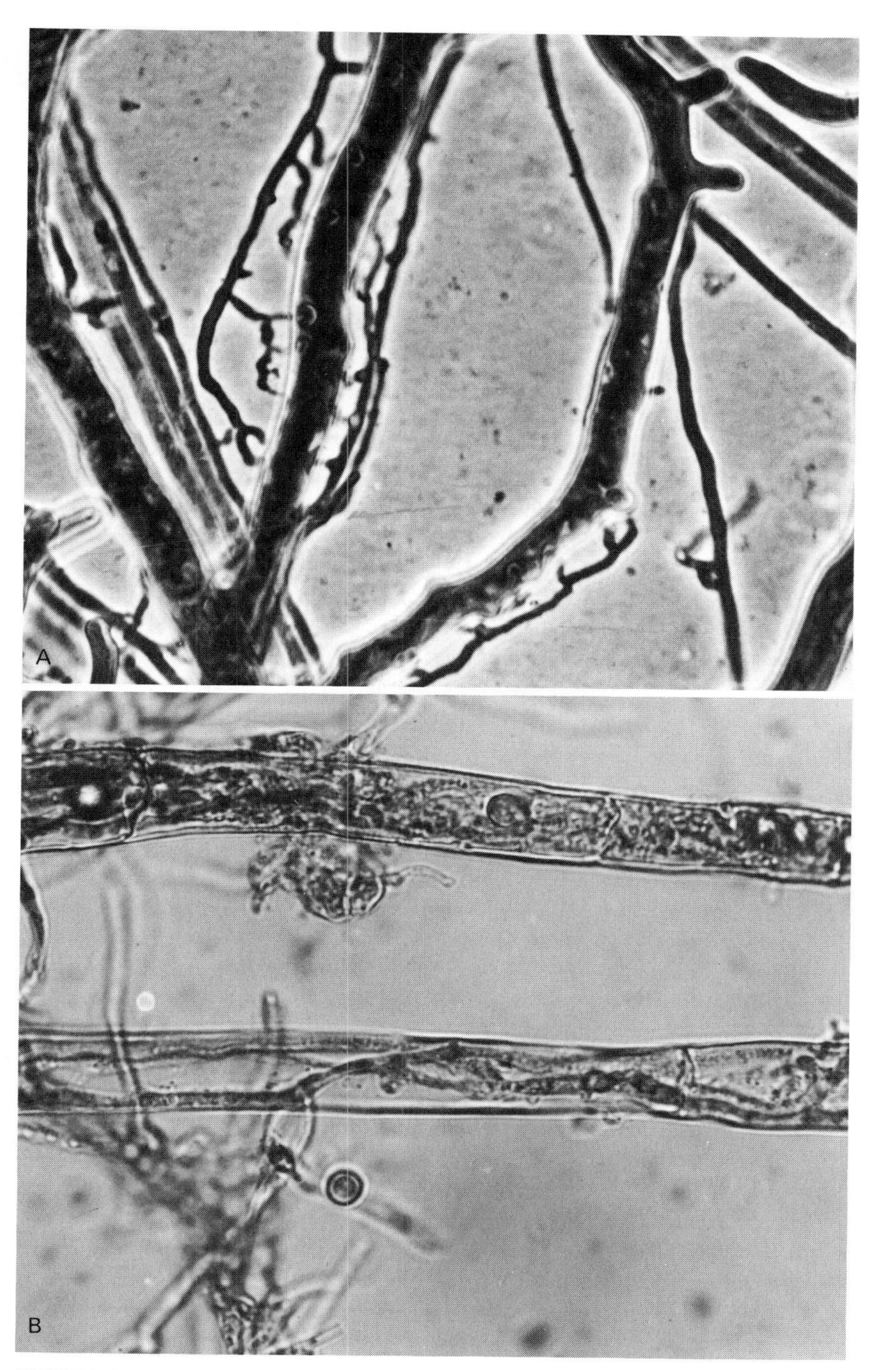

FIGURE 7.4.
Mycoparasitism of *Mucor plumbeus* by *Mycena citricolor,* cause of American leaf spot of coffee. A. Early stage in penetration of *Mucor* mycelium, showing infection pegs formed by *Mycena.* Phase contrast; × 600. B. Hyphae of *Mycena* growing inside *Mucor* mycelium. ×730. (Photos from Salas, 1970.)

The parasitic process in invasion of rust pustules by fungi has not been clarified, but the evidence suggests that the tissue of the host plant is so altered physiologically by the rust (Yarwood et al., 1953) as to become favorable for a fungus such as *Fusarium roseum* or *Tuberculina maxima* (Chapter 10), which kills the tissue and thus destroys the resident obligate rust parasite (Wicker and Woo, 1973).

Hyperparasites would seem of questionable value as antagonists against primary infections by plant pathogens. Actual contact between hyperparasite and host is necessary to kill the pathogen, and little destruction occurs in advance of the point of contact, The same holds for predators. Time is important in the rhizosphere or near any plant part, once inoculation of the pathogen has occurred (Chapter 6). By the time a hyperparasite or predator could respond to and contact the pathogen, or respond to exudates from the plant, infection by the pathogen may have occurred. It may be necessary to provide an alternate fungus host (a *Mucor* for a *Pythium,* a saprophytic for a parasitic *Fusarium*) to increase populations of hyperparasites, and perhaps other types of antagonists as well, in advance of pathogen activity.

Some pathogens, such as soilborne fusaria, usually cause primary infections and rarely depend on secondary hyphal growth or spore production, except at the end of the disease cycle when damage is already complete. Hyperparasites and predators would seem of limited value for biological control of such organisms. Others of this type include species of *Thielaviopsis, Verticillium, Pythium,* and *Phytophthora,* which germinate and infect rapidly by multiple primary infections from high densities of soilborne propagules. Also included are nematodes that invade the host from large populations of soilborne larvae. These pathogens tolerate many forms of antagonism and especially hyperparasitism by producing abundant inoculum (Chapter 6). The value of hyperparasites would seem particularly questionable against primary infections by aerial pathogens, because invasion is by showers of airborne spores that quickly infect foliage of hosts under favorable conditions of moisture and temperature.

Hyperparasites should be most effective against survival or secondary spread of pathogens, because there is maximal time for contact and destruction of the pathogen. Basidiocarps, uredia, aecia, perithecia, pycnidia, acervuli, or other sexual or asexual fruiting structures may be invaded and rendered functionless. Conidia of *Helminthosporium sativum* and *Thielaviopsis basicola* on natural soil may develop perforations in the walls, apparently caused by enzymatic action in sites of intense microbial activity (Old, 1969; Old and Robertson, 1969; Clough and Patrick, 1972; Chapter 6). Sclerotia are commonly invaded by many hyperparasites in soil, and this presumably reduces their survival and inoculum potential.

Wells et al. (1972) found that *Trichoderma harzianum* rotted sclerotia of *Sclerotium rolfsii.* They obtained 99.5% disease-free 77-day-old tomato seedlings, as compared with 21.9% free of *S. rolfsii* in the check, when soil inoculated with *Trichoderma* was applied to the soil surface along the row 14, 23, and 34 days after planting. The objective was "to overwhelm temporarily the infection court with *T. harzianum* and a fresh food base." The antagonist was not injurious to the plants. Moody (in Anonymous, 1973) found that sclerotia of *Phomopsis sclerotioides,* cause of the black root rot of cucumber, were decayed by *Gliocladium* sp.

Established lesions may likewise be invaded by hyperparasites and limited in capacity to produce inoculum. In general, hyperparasites may not prevent severe disease in one year, but by reducing survival or inoculum production they may reduce infection of a subsequent crop (Chapter 10).

Hyperparasites might be useful against pathogens characterized by long periods of exposure in the rhizosphere or on the rhizoplane. Boosalis and Mankau (in Baker and Snyder, 1965) point this out in relation to sedentary plant-parasitic nematodes, which protrude from the roots they parasitize. Other examples might be *Gaeumannomyces graminis* and *Phymatotrichum omnivorum.* Again, a hyperparasite may not prevent initial infection, but could prevent subsequent spread of the runner hyphae or strands along the root and to other roots. Hyperparasitism could account for take-all decline with wheat monoculture (Chapter 4) and explain why the number of lesions on wheat roots generally does not decline although lesion enlargement is checked (Chapter 6). It could also explain why ability of the fungus to cause severe disease declines but the populations of *G. graminis* do not. **An efficient parasite does not eliminate its host.** Furthermore, hyperparasitism could explain why decline in disease severity occurred experimentally when live, but not dead, inoculum of the fungus was added to field soil (Chapter 6).

The specificity indicated in Chapter 6 for antagonists of *Phymatotrichum omnivorum* could perhaps be explained by hyperparasitism. Fungi are likely to be important antagonists of this fungus in its active state (this chapter). As *P. omnivorum* grows, hyperparasites would increase and cause it to decline. The hyperparasite would then also decline, making possible renewed activity of *P. omnivorum* and a repetition of the cycle (Fig. 6.5). An equilibrium between pathogen and hyperparasite is ultimately attained, much like that between plants and parasites in the wild.

Plant-parasitic nematodes are widely parasitized by bacteria, sporozoans, fungi, and probably viruses, and are preyed on by protozoa, nematodes, tardigrades, enchytraeid worms, mites, springtails, and fungi, according to Boosalis and Mankau (in Baker and Snyder, 1965). Of these, the sporozoans, predacious fungi, and predacious nematodes seem to be most effective. Giuma et al. (1973) reported that germinating spores of three *Nematoctonus* spp. produced toxins (polysaccharides) that non-specifically immobilized nematodes before infecting them. It was suggested that these materials might be useful in nematode control.

Phycomycetes of the genera *Catenaria, Myzocytium, Protascus,* and *Haptoglossa* are the fungi most commonly parasitic on nematodes. The predacious fungi (Duddington and Wyborn, 1972; Verona and Lepidi, 1971) have diverse and fascinating trapping mechanisms (Fig. 7.5). They occur most commonly in soil high in organic matter. The more than 50 known species of nematode-trapping fungi are in the Hyphomycetes and Zygomycetes of the genera *Arthrobotrys, Dactylella,* and *Dactylaria.* The types of trapping mechanisms are adhesive (hyphal network, lattice of lateral branches, and stalked knobs) and mechanical (nonconstricting and constricting loops). In all types the trapped nematode is penetrated, and mycelia digest the contents.

From this food base, mycelia and other traps are formed in soil. The fungi presumably grow as saprophytes in organic matter; few form traps in pure culture, but all apparently do in the presence of soil microbiota. Nematodes tend to follow hyphae in moving through soil pores, and are thus trapped. Boosalis and Mankau (in Baker and Snyder, 1965) commented on the use of predacious fungi that "while their effectiveness *in vitro* and in pot tests has been encouraging, their application on a field scale has been singularly disappointing." The presence of a microflora antagonistic to the trapping fungi was thought to be a factor in their ineffectiveness.

Predacious nematodes are either stylet-bearing or carnivorous with a toothed mouth (*Mononchus*). The stylet type may inject a digestive enzyme into its prey and then suck out the contents, as do *Eudorylaimus* and *Thornia.* Some (*Seinura*) inject a fast-acting paralyzing toxin as well. Predacious nematodes are common in soil but of undetermined value in biological control.

Linford et al. (1938) found that when organic matter is added to soil, the populations of saprophytic fungi and free-living nematodes, but not phytophagous nematodes, increased rapidly. The trapping fungi and predacious nematodes then prey on the plant-parasitic forms, lowering their numbers. They found a total of 52 parasites and predators of root-knot nematodes in Hawaiian soils. Linford explained this as a third-order effect: fungi proliferated on the organic matter; free-living nematodes fed on the fungi and increased in numbers; predacious nematodes increased at the expense of the free-living forms and also indiscriminately attacked larvae of root-knot nematodes. Nematode-trapping fungi were also fa-

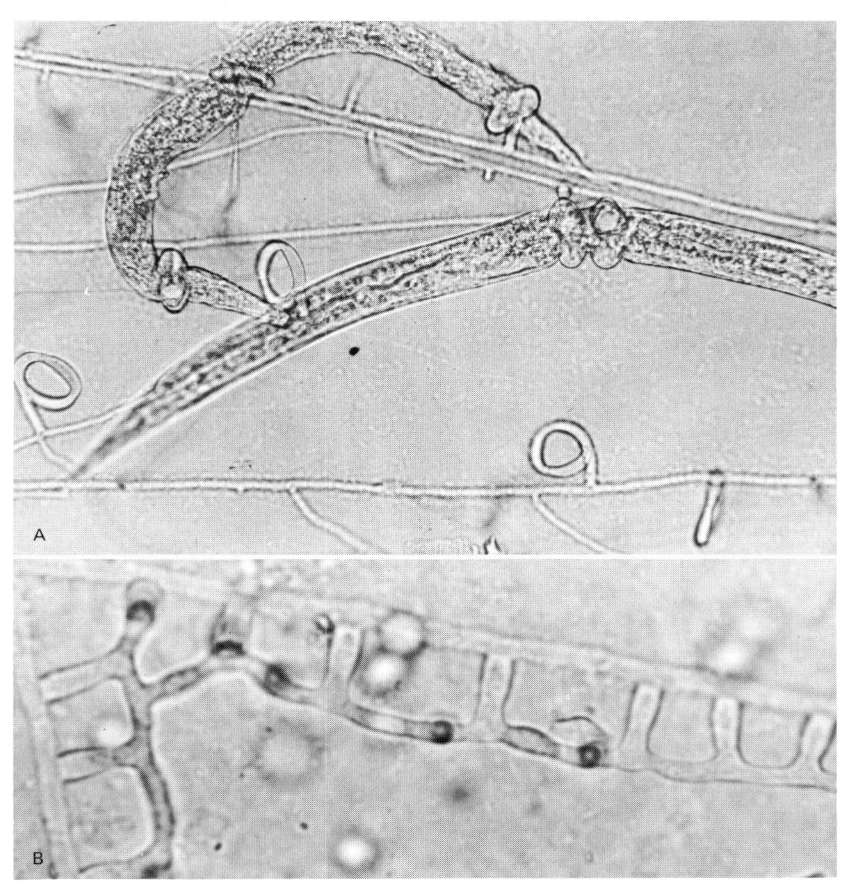

FIGURE 7.5.
Nematode-trapping fungi. A. Citrus nematode *(Tylenchulus semipenetrans)* larvae trapped and parasitized by the constricting-loop fungus *Arthrobotrys dactyloides*. B. Adhesive scalariform trapping fungus, *Monacrosporium gephyropagum*. (Photo A from Boosalis and Mankau, in Baker and Snyder, 1965; B, courtesy of R. Mankau.)

vored by organic amendments. There is much evidence that increasing organic matter reduces crop loss from nematodes in the field (Fig. 7.6) (Boosalis and Mankau, in Baker and Snyder, 1965; Johnson, 1962). There is attrition of phytophagous nematodes from this microbiota, but it has yet to be used for effective biological control. Perhaps there is a suppressive antagonistic effect of soil microorganisms on the parasitic and predacious fungi that restricts their effectiveness. There seems to have been little effort to study soils where plant-parasitic nematodes are not causing serious

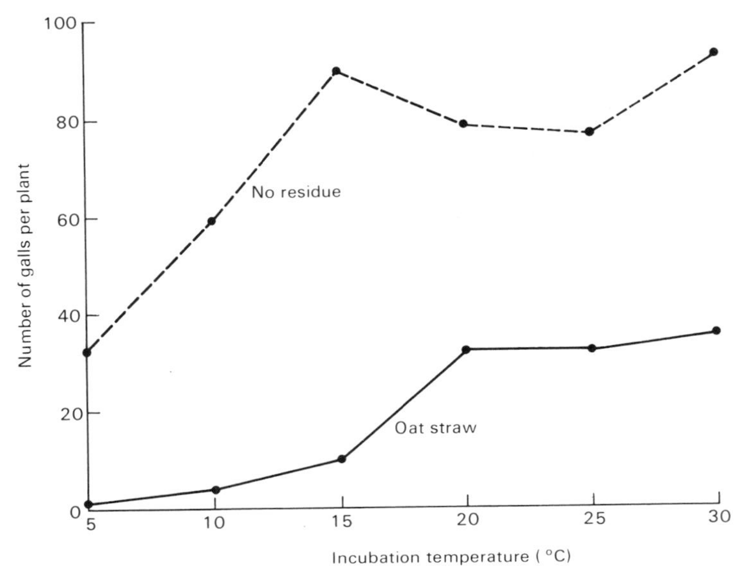

FIGURE 7.6.
Root-knot galls from *Meloidogyne incognita* on tomato plants grown for 10 weeks at different temperatures in unamended soil and in soil with equivalent of 10 tons/acre of oat straw. (Data from Johnson, 1962.)

damage, although environment and hosts are favorable for the disease. Analyses of such examples might provide clues of conditions necessary for effective biological control, as they have for other root pathogens.

Competition

Microorganisms are assumed to compete for oxygen, water, nutrients, and space. Clark (in Baker and Snyder, 1965) pointed out correctly that microorganisms do not compete for water because availability is determined by energy status (water potential), which microorganisms do not significantly affect. Water potential should be categorized with pH and temperature, factors that microorganisms tolerate or respond to but may not affect directly. To compete for water, microorganisms would have to use sufficient water to lower the relative humidity of the soil mass that encloses them. This they do no more than they lower the temperature by absorbing heat. Clark also thought that microorganisms do not compete for space, and rarely for oxygen. He concluded that "basically competition among microorganisms is for a substrate in the specific form and under the specific conditions in which that substrate is presented." In terms of antagonism in biological control, however, it appears that competition for oxygen, nutrients, and possibly even space, are important.

Oxygen. Griffin (1968, and in Toussoun et al., 1970) reviewed work on the relation of concentrations and diffusion of oxygen to the biology of organisms in soil. He suggested that the rate and path of diffusion of oxygen to any given thallus, propagule, or cell embedded in soil or near a root are extremely important, and in wet or microbially very active soils may be limiting to the microorganisms. Diffusion of oxygen is much slower through water than through air. In addition, cells of microorganisms line the path of diffusion, and each withdraws a supply for itself, with the result that at some point away from the source the oxygen concentration reaches zero. Greenwood and Goodman (1964) studied the maximum diameters of moist drained soil crumbs where oxygen drops below the critical concentration in the centers; they arrived at a figure of 3 mm. Their conclusions are in accordance with the general opinion that anaerobiosis is common within microsites in soil, explaining why obligate anaerobes occur commonly in soil. The bulk soil contains more total oxygen near the surface than in deeper horizons, but microsites with limited oxygen are also more prevalent there because of more abundant carbon sources. Clark (in Burges and Raw, 1967) suggested, "In many soils, it is highly probable that microsites of anaerobiosis occur in greater profusion in the plow layer than in the subsoil. . . . the plow layer loses its oxygen much more rapidly than does the deeper profile, simply because the greater number of microorganisms, and thus . . . oxygen demand, occurs in the top soil."

Anaerobiosis is not uncommon in soil where plant pathogens might otherwise be active. Oxygen deficiency could occur near a root if rhizosphere organisms were exceptionally active or if the soil was very wet. It might also occur within a cell where it is ultimately used, if at the surface of the pathogen a film of organisms consumes oxygen that would otherwise diffuse across the cell membrane. This effect could apply where oxygen has already neared the critical concentration at the time of its arrival at the surface of the pathogen. Each microbial, plant, and animal cell along the path of diffusion thus helps reduce the oxygen concentration below the critical level. Clark (in Baker and Snyder, 1965) stated that "if this exhaustion takes place very rapidly, . . . then the period of competition must be correspondingly short, and the importance of such competition should be negligible." In terms of a useful antagonistic effect, however, the more quickly competitors can exhaust the oxygen supply, the more important they are to a system of biological control.

Nutrients. A concept of concentration gradients and diffusion is also useful to depict competition among microorganisms for nutrients. When a fresh organic substrate, a seed, or a root enters soil, outward diffusion of nutrients from them begins about as quickly as does water absorption.

Multiplication of microorganisms does not begin until leakage from the substrate has started. At this stage diffusion of these materials exceeds demand, hence nutrition may be adequate to initiate growth at a distant point. Stanghellini and Hancock (1971) reported that sporangia of *Pythium ultimum* and chlamydospores of *Fusarium solani* f. sp. *phaseoli* may germinate up to 10 mm away from a bean seed 12 hours after planting the seed in moist soil. *As the lag period for growth of the microorganisms passes, and as cell division and nutrient utilization accelerate, the lowest nutrient concentrations disappear, the concentration gradient away from the source steepens, and the most distant cells from the source are left without nutrients.* Within a few hours only those cells in actual contact with the source will receive sufficient nutrients for continued growth. The concentration gradients for simple sugars and nitrogen will probably be steepest and these materials least available; should diffusion suddenly cease, microbial growth would begin to decline abruptly. Unless the propagule some distance from the source can germinate and begin growth with the first contact with nutrients, it will probably not do so until the flush of growth subsides and a new source of nutrients appears and starts the cycle again. Growth of microorganisms is intense at a point just behind the root tip until root elongation moves the nutrient source away, or until toxic metabolites (volatile or nonvolatile) become limiting. The organisms may then become quiescent until an emerging lateral root again supplies nutrients, and the cycle is repeated.

If a high number of microorganisms were for some reason already growing actively when an exuding plant part first appeared, the nutrients liberated might quickly be absorbed and a steep gradient consequently maintained from the outset. The effect would be fewer germinated pathogenic propagules. This effect would be what Baker (1968) has called "shrinking the rhizosphere to a rhizoplane," which was accomplished experimentally on beans with soil amendments of high C:N ratio. Exudates, especially the sparse supply of nitrogenous materials, presumably were immobilized as fast as they were liberated from the bean hypocotyl. Consequently, any chlamydospore germination of *F. solani* f. sp. *phaseoli* which did occur probably was on the rhizoplane. A steepening of the nutrient gradient can explain why Cook (1962) and Adams et al. (1968b) observed greatly reduced chlamydospore germination of *F. solani* f. sp. *phaseoli* near bean seeds planted in barley straw- or cellulose-amended soil, even though the bean seeds exuded an abundance of nutrients stimulatory to the chlamydospores. Adams et al. (1968b) referred to the effect as "raising the degree of fungistasis. . . ." This is in accordance with the evidence of Lockwood (1964) and his students that inability of spores to germinate in soil (*fungistasis*) is generally caused by insufficient nutrients. Fungi that required exogenous nutrients for germination in soil also required them for germination in pure culture, and conversely, those capable of germination on endogenous reserves did so in soil as well as in pure culture, with a few notable exceptions (Ko and Lockwood, 1967). Moreover, fungi slow to germinate were most sensitive to fungistasis (Steiner and Lockwood, 1969). It was suggested that by the time these fungi were ready to germinate, the interim microbiological activity would have created a starvation environment. The role of ethylene in fungistasis (Chapter 6) should also be noted.

A fact not always appreciated is the rapidity with which a limited supply of diffusates or added nutrients can be consumed by a soil microflora. For example, G. S. Campbell and R. J. Cook (unpublished) measured carbon dioxide evolution from flasks containing about 100 g of untreated Ritzville silt loam infested with chlamydospores of *F. roseum* f. sp. *cerealis* 'Culmorum' and amended with 2500 ppm carbon as glucose and 250 ppm nitrogen as ammonium sulfate. Carbon dioxide evolution reached a maximum rate, then tapered off 15 hours after the amendment was made to the soil, at about −1 bar water potential. It was slower with lower water potential, where fewer organisms were able to grow. Addition of fresh substrate after the tapering off had begun caused a renewed burst of carbon dioxide evolution. Chlamydospores of the *Fusarium* began to germinate within 4–5 hours after nutrients were added. Onset of germ-tube endolysis and new chlamydospore formation coincided with the times of diminished carbon dioxide evolution, suggesting that nutrients had been used and that lysis or chlamydospore formation was a response to starvation. Alternatively, ethylene (Chapter 6) could have reached inhibitory levels in 15 hours, but this may have been briefly offset by introduction of more nutrients. In either case, it is doubtful that, 15–20 hours after adding carbon and nitrogen, exudates from any plant part such as a bean seed, would be an effective supplementary source of nutrient.

The saprophyte *Phialophora radicicola* frequently grows on roots of wheat, oats, and barley in a manner similar to that of *Gaeumannomyces graminis,* and apparently competes with it for nutrients, providing "a significant degree of biological control" (Balis, 1970; Deacon, 1973).

Competition for nutrients, oxygen, or both is probably responsible for a significant percentage of the failure and endolysis of germlings of fungi like *Fusarium* in the host rhizosphere. The number that starve, endolyse, or convert into a resting state presumably varies with the environment, such as soil water potential and quantity and quality of nutrients in the exudates. It may also vary with antibiosis in the rhizosphere, which could raise the level of nutrients required, or reduce growth rate of the pathogen and thus permit the greater share of nutrients to be used by less sensitive competitors. Conversely, *the use of nutrients by competitors could increase sensitivity of a pathogen to antibiosis.* It has been shown repeatedly for antibiotic tests on agar that levels of nutrition greatly affect tolerance of an organism to an antibiotic. Thus, the combined effects of a pathogen weakened by low nutrition and then subjected to antibiosis, or the reverse sequence, is probably an important combination in the biological control of plant pathogens with antagonists.

Although antagonism by competition for oxygen or nutrients is obviously common and important in soil and in the rhizosphere, introduction of organisms into soil for biological control probably would be of limited value. The first barrier is the time factor, particularly with those pathogens that are quick to germinate and penetrate and have multiple infection. Assuming that a nutritional or oxygen deficiency could be created in the rhizosphere quickly enough to stop the pathogen before infection, it would probably require every available rhizosphere organism to do the job, and not simply one or more specific introduced antagonists. There already is maximum activity of every rhizosphere organism capable of utilizing oxygen and components of the exudates under the prevailing environment. Multiplication is probably logarithmic once the lag periods have passed, as long as nutrients are available and metabolic products do not accumulate. In spite of the speed with which oxygen and particularly nutrients are consumed, a sufficient number of infections for severe root damage still occurs. The additional contribution of an introduced organism, even a fast-growing saprophyte, could hardly be significant.

Preventing pathogen prepenetration growth near plants through preempting the nutrients would probably require very rapid absorption as such nutrients became available. As already implied, addition of organic amendments is possibly one means to this end. Leguminous refuse and other rich amendments should be especially effective in stimulating an active microbiota that would not discriminate between nutrients from the amendment or the exudate. This might explain why leguminous refuse plowed under has been so effective in biological control of plant pathogens. Alternatively, amendments high in C:N ratio could be used to create deficiencies for specific nutrients, as barley straw to immobilize nitrogen, or lignin to selectively create a deficiency for carbon (Baker, in Cook and Watson, 1969).

An equally effective role of amendments in causing antagonism through competition might be between crops, as with a cover crop plowed under before fallow. Under ideal conditions of temperature and moisture, as during summer rainfall or irrigation, rapid decomposition of the amendment could cause anaerobiosis. This condition, in turn, would be unfavorable to the continued low-level respiration of dormant propagules, and as a result they might die, either directly from lack of oxygen, or indirectly from destruction by anaerobes. The reduction in population of *Verticillium albo-atrum* in Washington and Idaho potato fields by plowing under alfalfa, followed by flooding, probably is an effect of this kind. This was observed by Menzies in 1962 and Watson in 1964 (Cook and Watson, 1969). Control of *Phymatotrichum omnivorum* by green manures might result from intensified microbial growth, which robs the sclerotia of oxygen, nutrients, or both, causing weakening and death.

Space. Competition for space could be a form of antagonism to ectomycorrhizae of roots since, in essence, the fungus covers the root, leaving no openings for penetration by a pathogen. Since the antagonism of mycorrhizal fungi on pine to *Phytophthora cinnamomi* also involves production of an antibiotic (Marx, 1972), the importance of space to this type of antagonism may be minimal.

Competition for space may be involved in the possession of substrata by prior colonists, as described by Bruehl and Lai (1968a) for four wheat pathogens, by Barton (1957) for *Pythium mamillatum,* and by Leach (1939) for *Armillaria mellea.* Indeed, the phenomenon seems fairly general that **one organism already in the substrate, whether by more rapid growth or by good fortune, will retain possession of that substrate even when confronted by vigorous saprophytes** like *Fusarium roseum* 'Culmorum' or *Trichoderma viride. Pythium mamillatum* is a highly effective saprophytic colonist of wood fragments previously soaked in a glucose solution only so long as another organism is not already present in the block (Barton, in Parkinson and Waid, 1960). *Cephalosporium gramineum* produces an antibiotic that wards off would-be invaders (Chapter 6). It is unlikely, however, that possession by virtue of prior colonization is always a result of antibiosis. Use of available nutrients or of available space within the substrate would seem an equally plausible explanation in some instances.

Substrate occupation as a form of antagonism in biological control could be useful against pathogens of high competitive saprophytic ability. *Rhizoctonia solani, Pythium* spp., and certain fusaria are quick to colonize dead substrates not already occupied by another organism. In this way they increase their inoculum density and at the same time acquire a food base. Prior colonization of the substrates by other fungi could prevent invasion by such facultative-type saprophytes. For such prior colonization, an antagonist need not have great competitive saprophytic ability, but rather it need only have first chance for colonization.

In the Pacific Northwest, and probably in most areas where wheat is grown, straw is thoroughly colonized by airborne saprophytes as standing stubble and chopped residue before tillage, and these fungi prevent subsequent colonization by Culmorum after tillage (Cook, 1970; Fig. 6.1). Occupancy of straw by Culmorum is through parasitism while the plant is still alive, not through saprophytism in the fallow year (Cook and Bruehl, 1968). The lack of prior colonists could explain why green residues plowed under are more conducive to inoculum increase by facultative-type pathogens than are mature residues. Green residues may have resident epiphytes (Chapter 10), but the tissues often are not yet occupied and are thus available on a first-come basis. Mature tissue is more likely to be occupied by airborne saprophytes. Plant residues allowed to decay on the soil surface before incorporation in soil would thus be less favorable to *Pythium, Rhizoctonia,* and the highly competitive fusaria than would fresh residue plowed under immediately. This, of course, is what occurs under wild or natural conditions.

Antibiosis

Antibiotics have been defined (Gottlieb and Shaw, 1970) as "organic substances that are produced by microbes and are deleterious at low concentrations to the growth or metabolic activities of other microorganisms." Some definitions set 10 ppm as the upper concentration limit. Other workers would also include materials produced by higher plants or animals. Goldberg and Luckey (in Goldberg, 1959) limit the definition to "special inhibitory products" and exclude lactic acid, ethanol, enzymes, or other similar substances. Antibiosis, on the other hand, could result from production of an alcohol or change in pH of the environment by production of simple substances not commonly considered as antibiotics. Metabolites of all types could thus qualify as mechanisms of antibiosis in soil. *Antibiosis,* therefore, *is the inhibition of one organism by a metabolite of another.* Microbial toxins have recently been treated in detail (Ajl et al., 1970–72); they are usually considered to be poisonous to higher animals. The antibiotic penicillin may be a toxin to a person sensitive to it. The terms thus have a functional rather than a chemical distinction.

Weindling showed in a series of classical papers in 1932–41 that *Trichoderma viride* was parasitic on other fungi and produced the antibiotics gliotoxin and viridin, which were stable only in acid solution (Garrett, in Baker and Snyder, 1965; Dennis and Webster, 1971). He also found that acidification of citrus seedbeds favored inhibition of *Rhizoctonia solani* by *T. viride.* It should be noted, however, that such acidification would also reduce the activity of bacteria and actinomycetes.

Antibiosis has certain advantages over other forms of antagonism. The toxic substances produced may diffuse in water films and water-filled pores through soil or on substrates, or through air-filled pores in the case of a volatile, and thus actual physical contact between antagonist and the subject is not necessary for effect. Distances of inhibition along roots, over surfaces of seeds, or through organic materials are not known, but in agar culture can be several millimeters or even centimeters away from the source. The zone of influence of an antibiotic organism, therefore, is probably a sphere whose radius varies inversely with the molecular weight of the chemical produced, the sorptive nature of the matrix, and the ability of other organisms to utilize the antibiotic, and directly with solubility and the soil-water content. In any event, the zone of influence would be greater than the volume of the organism. The distance of antagonism by antibiotic-producing microorganisms is therefore greater, and commonly more rapid and effective, than that of competitors or hyperparasites. Pathogens may grow for a time on reserves after readily available nutrients are depleted. Whereas hyperparasites are generally too slow to produce effective inhibition or death, antibiosis is fast and generally results in endolysis. Endolysis appears to be more common than exolysis in soil. Furthermore, antibiosis may continue for a while after growth of the antagonist ceases, because antibiotic release continues briefly after colony death. In fact, antibiotic release from a living colony apparently comes largely from senescent cells within that colony. There is always some leakage of antibiotic from an active cell or colony, but as senescence sets in, permeability changes allow it to flow out. Therefore, antibiotic action will provide more of a steady-state form of antagonism than will hyperparasitism or competition.

Antibiosis works best where nutrients are abundant or in excess. The target organism is, however, less affected under these conditions than under deficient nutrition. Since antibiotic production depends on availability of nutrients, little or none will be produced except at specific microsites such as leaking plant parts or organic debris. As stated by Goldberg and Luckey (in Goldberg, 1959), "few organisms produce detectable quantities of antibiotics in 'average soil'. . . . Microbial populations have exhausted the most common limiting nutrient (usually a carbon energy source) in an intense flurry of metabolism, growth and reproduction while a good supply of nutrients was available." They also point out that the production of an antibiotic provided a selective evolutionary advantage in holding an occupied niche, but has not proved useful in greatly extending the territory because of sorption and decomposition of antibiotics in soil. We should expect, therefore, that increase of antagonism by microbial production of antibiotics, and probably other forms of antibiosis as well, will depend on an adequate supply of organic amendments. This has been shown by Weinhold and Bowman (1968), using antibiotic production by *Bacillus subtilis* on soybean refuse for control of common scab of potato (Chapter 4). The bacterium colonizes the soybean substrate, and like *Cephalosporium gramineum* in wheat straw, probably produces a metabolite that enables it to maintain possession of its substrate. By providing an annual supply of the residue to the tillage layer, the number of microsites may be so multiplied that zones of antibiosis begin to coalesce and restrict activities of *Streptomyces* in the same or adjacent microsites, such as lenticels of potato tubers.

If the antibiotic production of *B. subtilis* on soybean refuse only inhibits *Streptomyces,* then an increase in *Streptomyces* numbers would recommence when application of soybean refuse is discontinued. This interpretation is supported by: (a) scab severity increases when soybean refuse is no longer applied; (b) scab severity neither increases nor decreases with application of soybean refuse.

Surfactants may operate in conjunction with antibiotics, and induce leakage from mycelia (Gottlieb and Shaw, 1970), causing them to mat down or "wet" in agar culture. The *B. subtilis* group produces such surfactants (Frobisher, 1968).

A number of arthropods secrete volatile fungistatic and fungicidal materials (Roth, 1961). *Scaptocoris divergens,* a soil-burrowing hemipteron, was prevalent in Central American banana plantings and produced material fungistatic or fungicidal (depending on dosage) to *Fusarium oxysporum* f. sp. *cubense* and f. sp. *lycopersici, F. roseum* f. sp. *cerealis* 'Graminearum,' *Sclerotium rolfsii, Botrytis cinerea,* and 12 soil saprophytes (including *Trichoderma viride*) (Timonin, 1961). It was bactericidal to *Pseudomonas solanacearum,* and nematocidal to *Meloidogyne incognita.* Feeding by the insect on tomato and banana roots was apparently not very damaging to plants in pots (Timonin, 1961). The toxic material is a mixture of seven aldehydes, two furans, and two quinones (Roth, 1961), and is secreted by the insect when disturbed (it possibly leaks small quantities at other times). It is now "doubtful that Scaptocoris is responsible for the absence or low incidence of the [Panama] disease in certain areas." However, this is yet another type of biological control that might lead to a useful application, even as antibiotics have come into use in medicine.

Ants of the tribe Attini maintain extensive underground fungus gardens, which provide their primary and probably sole food source and energy reserve. This is a mutualistic association between the ants and their fungi, "being essential to the survival of both organisms" (Weber, 1972 a,b). Bits of fresh leaves are cut from plants and, along with insect feces and carcasses, are carried to the nest, licked and scraped, cut into smaller pieces, macerated, anointed with liquid feces, placed in the garden, and inoculated with tufts of hyphae. The fungi (*Leucocoprinus, Lepiota, Auricularia,* and *Tyridomyces* spp.) apparently are fairly specific for a given ant species. The fungus grows over the porous mass, producing inflated hyphae on which the ants feed. A dominant single microorganism is selectively maintained, despite gross exposure to contamination. Bacteria and yeasts are present, but kept in check. Gardens not cared for by ants are quickly overrun by bacteria and molds. Pure cultures can even be maintained by ants in agar plates despite surrounding contaminants. There are apparently three methods by which this dominance is maintained:

a. Interspecific microbial competition is controlled by high inoculum density. These fungi are poor competitors outside the nests, and grow poorly on media with protein nitrogen, but well on hydrolyzates of them. The fecal liquid, rich in amino acids, briefly supplies an available nutrient source. The feces are also rich in proteolytic enzymes, which break down the leaf proteins, supplying a sustained nitrogen source. This favorable nutrient source, plus the quantity of initial inoculum, favors competitive dominance of the fungus.

b. The ants may weed out other fungi or abandon contaminated garden chambers. Prolonged licking of all materials brought into the nest, as well as rocks and roots in it, greatly reduces contamination by bacteria, yeasts, and fungi (Weber, 1972 a,b). The use of numerous small chambers instead of a few large ones is also advantageous in this regard.

c. Some fungus-growing ants produce inhibitory chemicals that further assure dominance of the food fungus (Maschwitz et al., 1970; Schildknecht and Koob, 1971). The metathoracic gland of *Atta sexdens* and five other ants produces myrmicacin (strongly inhibitory to weed fungi such as *Alternaria, Botrytis,* and *Penicillium*), phenyl acetic acid (inhibitory to bacteria such as *Escherichia coli* and *Staphylococcus aureus*), and indoleacetic acid (promotes growth of the food fungus). These materials, exuded onto the gardens, assist in keeping the food fungus dominant. One species, *Atta colombica tonsipes,* was thought to produce no inhibitory material (Martin, 1970).

The attine ants have thus evolved a highly successful means of biological control, which assures the success of their food fungi.

Probiosis

The term *probiosis* was independently introduced by Lilly and Stillwell (1965) and Sussman (in Baker and Snyder, 1965) to describe the stimulation of one microorganism by another. There are many examples of this type of effect, only a few of which can be mentioned here; others are given in Chapter 1. They may be important in biological control, since stimulation of a second organism may influence its antagonistic effect on the pathogen or the pathogen's ability to avoid injury—a sort of second-order antagonism or inverse biological control.

Zoosporangial production by most isolates of *Phytophthora cinnamomi* requires the presence of soil bacteria (such as *Pseudomonas* or *Chromobacterium* spp.), and does not occur in soil of near-sterility (Ayers, 1971; Ayers and Zentmyer, 1971; Manning and Crossan, 1966; Zentmyer, 1965). Other bacteria hasten the breakdown of zoosporangia (Chapter 4). Chlamydospore formation by *Fusarium solani* occurs *in vitro* in the presence of *Protaminobacter, Arthrobacter,* and *Bacillus* spp. The bacteria may produce substances that induce such formation (Ford et al., 1970), or the process may reflect starvation of the fungus from sudden depletion of nutrients (Meyers and Cook, 1972; Chapter 9). Salas and Hancock (1972) showed that *Mycena citricolor,* cause of American leaf blight of coffee, is induced to form its basidiocarps by *Penicillium oxalicum* (Fig. 7.7) and four other species, but does this weakly in the presence of other fungi. Some isolates of *Mycena* produce basidiocarps axenically, but production was improved by the inducing fungi. The inducing material appears to be a thermostable weak organic acid. Hawker (1957) found that perithecial formation in *Sordaria (Melanospora) destruens* apparently was stimulated by thiamine and biotin supplied by a number of fungi.

THE IDEAL ANTAGONIST

Most of the attributes of a successful parasite (Chapter 1) are also attributes of the ideal antagonist, which unfortunately does not exist. It should: produce inoculum in excess; resist, escape, or tolerate other antagonists; germinate and grow rapidly; invade and occupy organic substrates. However, these are only some of the desirable features of an antagonist. Others are:

a. It should survive and grow in the rhizosphere or spermosphere (to prevent infection), or in the hyphosphere, near the pathogen's resting structures, or in the soil mass (to reduce survival).

b. It should produce, even when growing slowly on simple available

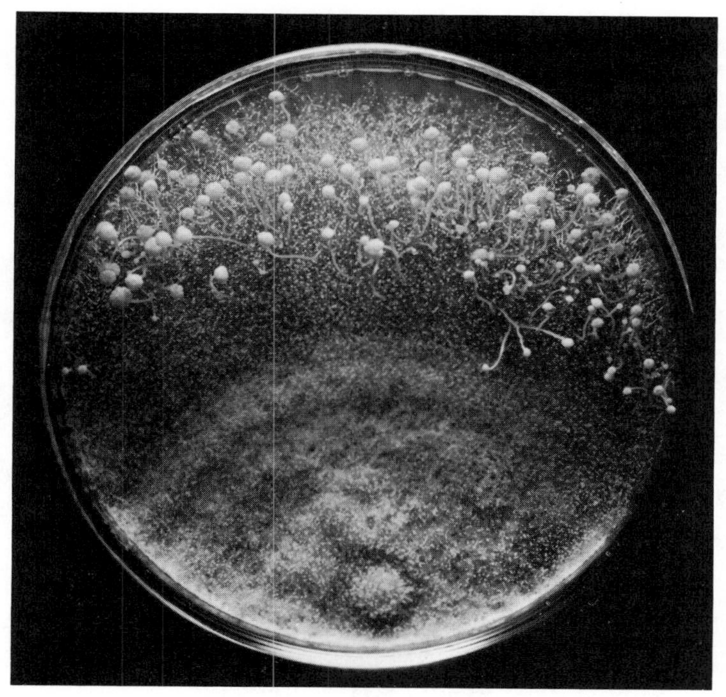

FIGURE 7.7.
Induction of basidiocarps of *Mycena citricolor,* cause of American leaf spot of coffee, when grown in proximity to *Penicillium oxalicum* on agar culture under diurnally fluctuating illumination. *Penicillium* at bottom. (From Salas and Hancock, 1972.)

substrates, a broad-spectrum highly toxic antibiotic that is effective at low concentrations, and that is not readily sorbed or degraded in soil. However, the antibiotic should not persist for long periods in soil (Watson and Ford, 1972). A broad-spectrum antibiotic would provide protection from other soil saprophytes, but would also insure effectiveness against an array of strains of the pathogen or other pathogens. Certain strains of *Bacillus subtilis* were found (Olsen and Baker, 1968) to be effective against only one strain of *R. solani* in California, whereas other strains had broad-spectrum activity. Broadbent et al. (1971) found in Australia that some antagonists from soil inhibited only one, and others all nine test pathogens, with an average of three to four inhibited by each antagonist (Fig. 7.8).

There are nine recognized species and many strains of

FIGURE 7.8.
Differential effectiveness of two antagonistic *Bacillus* spp. (upper and lower rows) against three isolates of *Rhizoctonia solani* in preventing damping-off of pepper seedlings. *Rhizoctonia* isolate 75 at left; isolate 18 at top right; isolate 15 at lower right. (From Broadbent et al., 1971.)

Trichoderma, some of which produce volatile or nonvolatile antibiotics active against fungi, and some do not. There are also variations in the effect on different plant pathogens (Dennis and Webster, 1971).

An organism with a broad-spectrum antibiotic could be useful against many different pathogens of one or several crops, and at the same time could possibly establish and persist in soil with an existing complex biological balance.

A broad-spectrum antibiotic may have some of the inherent disadvantages of broad-spectrum soil fungicides in creating a temporary partial biological vacuum. On the other hand, an antagonist whose antibiotic has a very narrow spectrum of effectiveness may create problems for the investigator, who must then combine many compatible and complementary isolates for satisfactory control. Perhaps one organism that produces several antibiotics of different but complementary spectra would be a satisfactory compromise.

c. The antibiotic produced by one antagonist should not be inhibitory to other associated antagonists. *Bacillus subtilis* isolates generally do not inhibit each other or saprophytic actinomycetes.

d. The antibiotic produced should not cause damage to the host, as may be the case with frenching disease of tobacco and yellow strapleaf of chrysanthemum, perhaps associated with *B. cereus* or *Aspergillus wentii* in the rhizosphere, and root rot of milo from *Periconia circinata* in the rhizosphere (Leukel, 1948; Steinberg, 1952; Woltz and Littrell, 1968). Since meristematic tissues are more sensitive to toxic substances than older ones, tests that measure phytotoxicity should be made with root tips or germinating seeds, especially since it may be desirable for the antagonist to occupy zones of soil near these plant parts. Ferguson (1958) showed that *Myrothecium verrucaria* was greatly favored when cellulose was added to soil; damping-off caused by *Rhizoctonia solani* was suppressed, but the amount of toxin produced was so great that pepper seedlings were stunted (Chapter 5). The cellulose of pepper seed coats acted in much the same manner.

e. The antagonist should be adaptable to large-scale commercial production and handling (Chapter 11). A spore or propagative unit resistant to heat, desiccation, and the antagonistic effects of other organisms, such as produced by members of the genus *Bacillus* or certain Ascomycetes, would facilitate handling and applications of the antagonist, as well as its persistence in soil during unfavorable periods.

f. Spore germination should occur readily, or at least as quickly and prolifically as that of the pathogen. Return to the dormant phase, on the other hand, should be less rapid than that of the pathogen.

g. The antagonist should be more adaptable than the pathogen to environmental extremes. Ideally, it should have broader optima of temperature, water potential, and pH, and be capable of growth at lower and higher values of these factors than the pathogen. Nutritional requirements of the antagonist should not be narrower than those of the pathogen.

Although the ideal antagonist does not exist, some antagonists may approach it and prove useful for inoculation on seed or with organic matter against many different pathogens. It may even prove practical or essential to breed for the ideal antagonist.

INOCULATION WITH AVIRULENT ORGANISMS RELATED TO THE PATHOGEN

A different type of introduced antagonist—a microorganism related to the pathogen but avirulent to the given host—has also been tried (Matta, 1971).

Langton (1969) placed susceptible tomato cuttings in suspensions of different proportions of microconidia of *Fusarium oxysporum* f. sp. *lycospersici* and f. sp. *pisi* before planting. The greater the ratio of *pisi,* the less severe the wilt symptoms produced. Heat-killed microconidia of *pisi* did not produce this effect. Similar results were obtained with *F. oxysporum* f. sp. *melonis* by Meyer and Maraite (1971), inoculating simultaneously with virulent and low-virulence isolates; the protection was explained as resulting from competition for nutrients and space.

A group at Ohio State University has studied the interaction of a *Cephalosporium* from roots of healthy tomato with F. *oxysporum* f. sp. *lycopersici* in tomato seedlings (Phillips et al., 1967). Repeated isolation of inhibitive strains of the antagonist from healthy survivors of inoculation tests with *lycopersici* was said to intensify the protective effect. The longer the interval between inoculation with the *Cephalosporium* and the pathogen, the greater the protection. A similar effect on *F. oxysporum* f. sp. *batatas* by *F. solani* has already been mentioned (Chapter 4).

The reduction of vascular wilt by prior inoculation with an avirulent organism may be explained in terms of the occlusion reaction shown to exclude microorganisms from the vascular system of certain plants (Chapter 8). Beckman and associates (1962–72) showed that in compatible interactions, this occlusion reaction occurs too late to prevent upward spread of the pathogen (Chapter 8). Prior inoculation with a nonpathogen may initiate the occlusion reaction, which can then block the pathogen (Beckman and Halmos, 1962).

New and Kerr (1972) reported that when roots of peach seedlings were dipped in a suspension of selected avirulent *Agrobacterium radiobacter* biotype 2 and wounded before planting in a soil infested by a virulent isolate of *A. tumefaciens,* formation of crown gall was prevented. The avirulent isolate, which was physiologically (except for pathogenicity) and serologically indistinguishable from the pathogen, had to be in a population at least equal to that of the pathogen for protection. Since both isolates establish in the rhizosphere, it was expected that the protection would persist in the field, and this is now under test in South Australia, Victoria, and New South Wales. Peach seeds inoculated with this avirulent bacterium gave 31% galled plants, while checks had 79%, and thiram-treated seed had 71% (Kerr, 1972). The mechanism of control was thought not to be antibiosis, host hypersensitivity, or competition for infection sites, but rather to be a bacteriocin (A. Kerr, unpublished).

The effective biological control of brown blotch of mushroom, caused by *Pseudomonas tolaasii,* through application of related *Pseudomonas* spp. to the casing soil, was discussed in Chapter 4.

Inoculation of plants with a mild strain of a virus to protect them from a more virulent strain is discussed in Chapter 10.

The mechanism of effect of these and numerous other similar instances of partial protection (Chapter 8) are still speculative, and none of these examples has apparently been demonstrated under commercial conditions.

RECONTAMINATION OF SOIL

The rapidity with which treated soil of near-sterility may be recontaminated from the air is often not realized, and it is even less often appreciated what antagonistic effect this contamination can produce. Schippers and Schermer (1966) showed the effect of airborne antagonists on the ability of *Verticillium albo-atrum* to grow from infected seeds of *Senecio vulgaris.* Surface-disinfected seeds sown on autoclaved sterile soil in petri dishes gave 39.2% infected seedlings; in this soil exposed to laboratory air for 3–10 minutes, only 20% developed, those exposed for 1–6 hours only 14.2%, and those exposed for 24–72 hours gave only 7.5%. Seeds placed on raw soil for 8 hours gave 28.8% infection when transferred to sterile soil; if left on raw soil for 48 hours before transfer, only 3.8% infection resulted.

The effect of air-contaminants on a pathogen in fumigated soil is indicated in Figure 7.10. *Gaeumannomyces graminis* caused less disease of wheat in nonamended fumigated soil (methyl bromide) than in the same soil amended with 1% or 10% soil from an arid virgin area. The organisms naturally present in the virgin soil showed little or no antagonism to *G. graminis*, but probably buffered against rapid establishment of airborne contaminants. Air contaminants presumably luxuriate in the nonamended fumigated soil. The biological community established through random contaminants from the air apparently was more suppressive to *G. graminis* than the one established by amendment of the fumigated soil with nonsterile virgin soil (R. J. Cook, unpublished).

This may explain the often-cited results of Sanford (1941), in which a virulent soil culture of *Rhizoctonia solani* introduced 1:1 into autoclaved soil produced only a 5% disease rating on potato stems, as against a 49% rating in nontreated soil. Soils were mixed "under conditions that avoided as far as possible, additional contamination." Since introduction of increasing amounts of steamed soil into a naturally infested soil caused a decreasing percentage rating of disease, contamination by airborne antagonists seems probable. These antagonists apparently were more effective than those found in the nontreated soil.

BIOLOGICAL BUFFERING BY RESIDENT ANTAGONISTS

The resiliency of the biological balance achieved in soil has been demonstrated many times by the fact that an introduced organism will not establish, or it disappears shortly after its addition to soil (Chapter 3). This property of soil, known as *biological buffering,* involves the total effect of the resident soil microorganisms in a complex microbial community. The degree to which the effect can be separated into components of antibiosis, competition, exploitation, or combinations thereof, is still unknown. The significant feature is that the effect can be detected, possibly even measured, and that it varies from soil to soil and with different cropping practices (Chapter 5), suggesting that the effect can be managed.

Attempts to Increase or Decrease Disease Potential of Soil

Soils suppressive to the establishment or function of the wilt fusaria, *Streptomyces scabies, Rhizoctonia solani, Phytophthora cinnamomi,* or others can be described as biologically buffered against these pathogens. If introduced, the pathogen does not cause disease, or causes it temporarily, and may disappear. The existing biological balance of the soil must be upset by some shock (reduction of the native population by fumigants, steaming, or alteration of the food base by addition of some organic substrate), before introduction of the pathogen can be successful. It may be equally difficult to increase the potential of disease in a soil by adding more of a pathogen already there.

Experiments of K. F. Baker and W. C. Snyder in 1942 (unpublished) with the fusarium wilt of China aster illustrate these situations. A dried and ground giant cornmeal culture of a virulent isolate of *Fusarium oxysporum* f. sp. *callistephi* was used in field and glasshouse tests for varietal resistance in asters. A field used for many years by a southern California seedsman to eliminate wilt-susceptible individuals from commercial aster lines had a high disease potential because of abundant residual inoculum and the high soil temperatures in the inland valley where it was located.

In method 1, cornmeal inoculum was placed (1 g/linear 30 cm of row) 2.5 cm deep, and seed of several aster varieties sown 1.2 cm above the inoculum by using two modified Planet Junior drills in tandem. Method 2 used the same seed lots, but dipped in a dense spore suspension of the pathogen and dried; these were sown without cornmeal inoculum. In method 3, the seed was treated with mercuric chloride to eliminate surface inoculum, and then rinsed, dried, and sown with no cornmeal inoculum. Seventy days later in this field there was a very poor survival with method 1, a fair stand with method 2, and a good stand with check method 3. After 133 days there were no survivors in method 1, and practically none in the others.

A similar test was conducted in a nearby field in which asters had never been grown. After 55 days there were almost no survivors with method 1, moderate wilt with method 2, and no wilt with check method 3. After 118 days there were no survivors with method 1, many lethal late infections with method 2, and only a few diseased plants with check method 3. Similar, but less severe losses resulted in a test in a cool coastal valley in soil that had never been planted to asters.

Finally, tests were made in freshly steamed soil in benches heated to 24–25° C, and in unheated benches in the glasshouse and outdoors. The cornmeal inoculum was varied from 0.35 to 2.02 g/linear 30 cm of row, but otherwise applied as in the field. There was little effect from an increase in quantity of cornmeal inoculum, probably because the lowest dosage was adequate against asters in soil freed of antagonists. Plants were killed faster in heated than in cool soil. The addition of spores to the seeds sown gave only slightly less kill than the cornmeal inoculum in heated soil, but caused little disease at lower soil temperatures. Susceptible varieties reacted the same whether in cool or heated cornmeal-inoculated soil.

TABLE 7.1
Relationship of inoculum density of *Rhizoctonia solani* in soil to the control of pre- and postemergence damping-off of pepper seedlings by fungicidal seed dusts.[1]

Dilution, infested/ steamed soil	Percentage of damping-off		
	Seed nontreated	Seed dusted with Thiram	Seed dusted with Captan
Nondiluted	51	48	49
1 to 4	49	39	43
1 to 16	32	28	32
1 to 32	32	11	13
1 to 64	29	6	9
Noninfested	0	0	0

[1] Data by Otniel Tjahjakartana, University of California, Berkeley, 1970. The soil to be diluted had been heavily inoculated, and the infestation increased by planting peppers.

However, resistant selections had few survivors in heated cornmeal-inoculated soil, and few diseased plants in cool cornmeal-inoculated soil.

At least five factors were operative in determining the level of disease in these tests: (a) the quantity of inoculum in the soil or on the seed; (b) the soil temperature; (c) the level of host resistance; (d) time; (e) the level of biological buffering by resident soil microorganisms. The higher the level of host resistance, the higher the soil inoculum density, soil temperature, or both, or the lower the level of biological buffering must be for *F. oxysporum* f. sp. *callistephi* to cause a high incidence of wilt. To some extent, a low inoculum density, such as spores on seeds, can be offset by an elevated soil temperature, and a cool soil by increasing the inoculum density. The longer the plant is exposed to this pathogen, the less the chance of its survival. Low inoculum density, low soil temperature, or both, in soil freed of antagonists, will give disease losses equal to those with high inoculum, high temperature, or both, in field soil. Conversely, sufficiently high inoculum density may override the limiting effects of temperature, resistance, or biological buffering (Fig. 7.3), as it does fungicidal seed treatment (Table 7.1). Thus, the several factors involved may, to a considerable extent, compensate for each other.

It would appear that in an association that has stabilized for several years, there is an upper limit to the incidence of disease, which may be increased with considerable difficulty by large increases in inoculum or a much more favorable environment. A consistently high disease potential, as required in selection for resistance, requires a near-absence of antagonists, as in freshly steamed pathogen-inoculated soil. Such steamed inoculated soil gives decreasing disease losses with each successive planting, probably because of accumulated contamination by antagonists. Furthermore, better elimination of susceptible individuals may be obtained by inoculating soil previously not infested by the pathogen than by inoculating a soil of already high inoculum density. The decreasing order of disease provided by these methods, then, is: freshly steamed soil plus inoculum → soil steamed, inoculated, and planted several times → noninfested field soil plus inoculum → field soil infested for some years plus inoculum. This is also the progressively increasing sequence of antagonists in the soils. It is likely that the decrease of a disease with continued cultivation (Chapter 4) is a manifestation of the same phenomenon of increasing antagonist population, or of the type of biological balance that develops in an undisturbed wild stand.

The observations of Weinhold and Bowman (1968) on the increase and decrease of potato scab with different green manures suggests that biological buffering can also prevent decline in disease severity. Incorporation of green barley or pea residue into soil before each potato crop for 8 years failed to prevent the annual increase in scab over that with potato monoculture without green manures. In contrast, the green soybean residue prevented scab increase, but was ineffective in reducing disease incidence. When green-pea residue was replaced bý soybean in the fifth year, scab severity remained at the same moderately severe level and did not increase or decrease. Thus, whereas Baker and Snyder could not increase disease in a heavily infested field, Weinhold and Bowman were able to establish a ceiling but could not decrease disease. The questions that arise are why did soybean residue buffer at the level it did (see Antibiosis above), and is it possible to buffer at a predetermined level?

Field Inoculation with Suppressive Soil

The microbiological community of desert soils does not prevent invasion by plant pathogens once irrigation and cultivation of pathogen-susceptible crops begin (Chapter 1). By comparison, well-established, fertile, agricultural soils of high organic matter and high populations of many kinds of saprophytic organisms are not nearly so favorable, and are sometimes suppressive to pathogens that normally flourish in irrigated desert soils. These two situations represent extremes in natural biological buffering. The microbial community of desert soils is relatively simple, qualitatively if not quantitatively, until such time as natural introductions of saprophytes fill the biological vacuum, and the organic-matter content can be increased to support a more complex microbiota. Clark et al. (1960) showed that the nitrifying capacity of a virgin desert soil remains low until cultivated for a few years, or amended with a small amount of soil from an established ag-

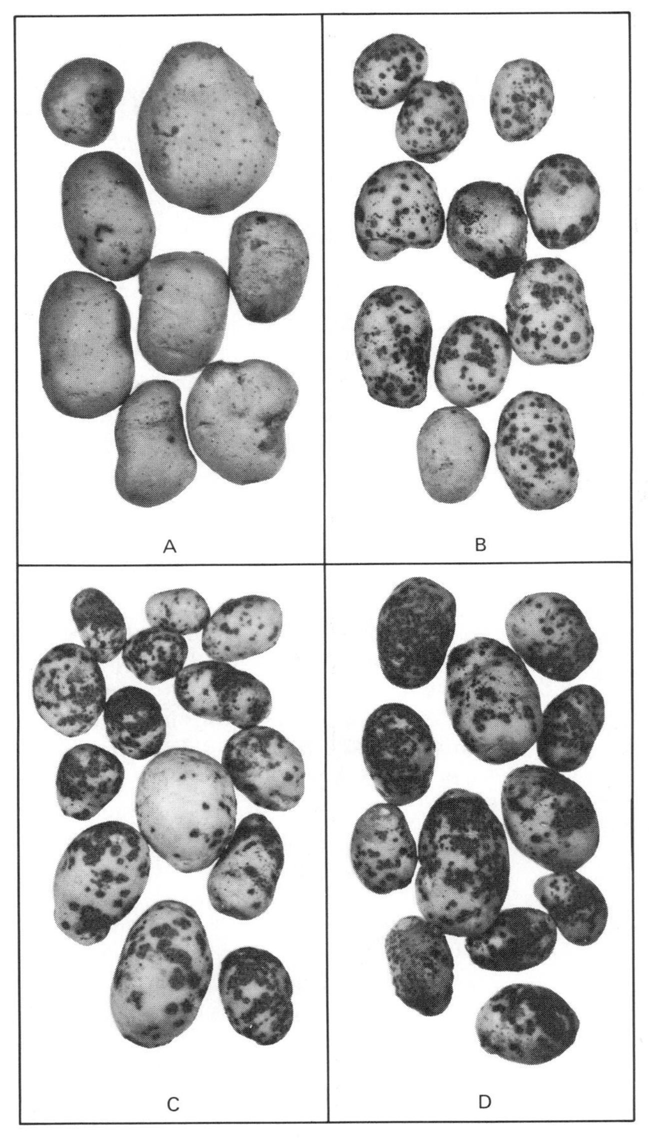

FIGURE 7.9.
Representative potato tubers grown in various soils infested with *Streptomyces scabies*. A. Suppressive soil alone. B. Autoclaved suppressive soil. C. Conducive soil alone. D. Autoclaved conducive soil.

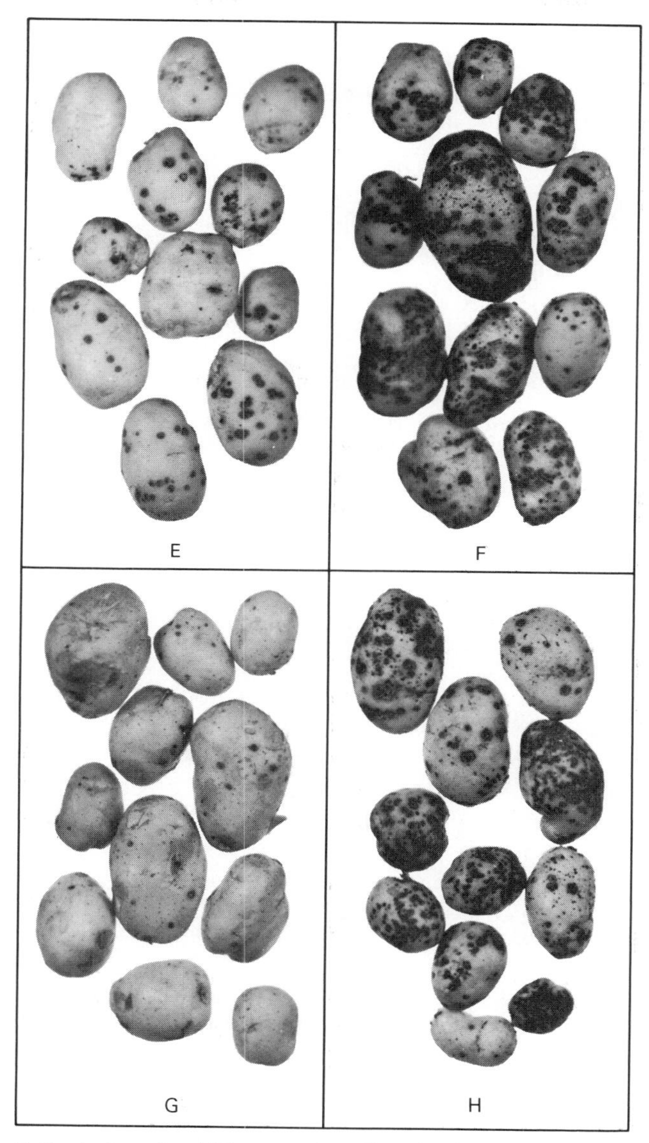

E. Conducive soil + 10% suppressive soil. F. Conducive soil + 10% autoclaved suppressive soil. G. Conducive soil + 10% suppressive soil + 1% alfalfa meal. H. Conducive soil + 1% alfalfa meal. (From Menzies, 1959.)

ricultural field. Since the microflora is not balanced for the new conditions imposed by cultivation, it is subject to dramatic changes, as when a pathogen establishes and quickly multiplies. On the other hand, just as biological instability can be advantageous to the pathogen, so it can be advantageous to the establishment or functioning of antagonists, or more generally to biological buffering at desired levels. **Virgin soils are more reactive to microbiological manipulation by cultivation practices or with organic amendments than older fertile agricultural soils where biological buffering has stabilized.**

Menzies (1959) showed that antagonists in a soil suppressive to *Streptomyces scabies* could be transferred to an untreated conducive (nonsuppressive) soil. Addition of suppressive to conducive soil at ratios of 9:1 or 1:1 gave complete control of scab, as did suppressive soil alone; at a 1:9 ratio, 43.9% of the tubers were scabbed, and conducive soil alone gave 37.1% scab. Since autoclaving the suppressive soil destroyed the effect, it probably is biological in nature. Adding 1% alfalfa meal plus 1% suppressive soil (as a slurry, or applied separately) gave averages of 15.6–16.7% scabby tubers, 1% alfalfa meal plus 10% suppressive soil gave 11.3%, 1% alfalfa meal alone gave 66.9%, no amendments added gave 62.4%, suppressive soil alone gave 0% (Fig. 7.9, pages 210 and 211; Frontispiece). The slurried (inoculated) alfalfa meal was effective at the 1% rate, but not at 0.1 or 0.01%. One wonders whether incubation of the alfalfa meal (perhaps sterilized before inoculation) after adding the suppressive soil would so increase the antagonistic flora that greater dilution would be effective. Menzies commented that this "does not seem to be a temporary effect from organic residues, because it persisted for a 5-year period of continuous potato cropping. . . ."

The factor(s) suppressive to *Gaeumannomyces graminis* in certain long-term wheat-field soils in eastern Washington was successfully transferred by Shipton et al. (1973) to fumigated soil (methyl bromide, 1 pound/100 square feet; 1 kg/20.5 m^2) in both glasshouse and field trials. One field trial was at Lind, in the 1-acre experimental site where take-all did not develop significantly in spite of five consecutive irrigated wheat crops (Chapter 4). Whereas take-all was severe in small blocks in the plot, fumigated then reinoculated with *G. graminis* and sown to wheat, it was again suppressed (although not as much as in nonfumigated areas) by disking back into the fumigated plots a small amount (less than 0.5% w/w to a 12–15 cm depth) of nonfumigated soil from an adjacent area.

In the glasshouse, a 1% (w/w) contamination rate of fumigated soil from Lind with soil from any of the three long-term cereal fields so prevented disease progression that plants looked no different than the noninoculated checks (Fig. 7.10). In contrast, identical contamination

FIGURE 7.10.
A pot test showing the difference between two soils in antagonistic properties to *Gaeumannomyces graminis*. All five pots contain the same methyl bromide-fumigated Ritzville silt loam. From left to right: Added inoculum (1% w/w with oatmeal inoculum) of *G. graminis;* wheat showed severe take-all in 3 weeks. Added inoculum plus 1% w/w virgin soil; showed extremely severe disease. Added inoculum plus 1% w/w soil from a long-term wheat field; showed mild disease. The remaining two pots received the same respective soils (noninoculated checks); no disease developed.

rates using soils from arid or semiarid virgin areas permitted severe disease development, with blackened stem bases, stunted plants, and chlorotic leaves. Virgin soil from a Pullman site was more strongly suppressive than virgin soil from an arid or semiarid area. In addition, soil from a long-term dryland wheat field at Lind not previously cropped to irrigated wheat was included in the glasshouse test, and was considerably more suppressive than arid or semiarid virgin soils, although not as suppressive as the soil from annual cropped fields. Several variations of this glasshouse test have been carried out, with the same results. Thus, the decreasing order of disease in this series is: fumigated soil plus inoculum → fumigated soil, contaminated with soil from arid virgin area, plus inoculum → fumigated soil, contaminated with soil from grassy virgin area, plus inoculum → fumigated soil, contaminated with soil from wheat fields, plus inoculum → nonfumigated soil plus inoculum → nonfumigated soil, no added inoculum → fumigated soil, no added inoculum. The above observations suggest that soils show degrees of biological buffering in relation to a given pathogen that can be measured and transferred, and that naturally vary with the history and ecology of the site. Fumigated or desert soils are the most poorly buffered, virgin soils of the semiarid areas in eastern Washington are next, dryland wheat-fallow soils of the semiarid area near Lind, and the virgin grassland area of the higher-rainfall Palouse show considerable buffering, but do not equal that shown by soil of fields with good fertility, supplemental or total irrigation, and cropped annually to wheat. There is also the possibility that an antagonism specific for *G. graminis* developed in the long-term cereal fields (Chapter 6).

	A	B	C	D	E
Plant height (cm)	24.0	49.7	72.1	50.7	42.4
Disease index (0-4)	2.7	1.8	0.5	1.8	2.2

FIGURE 7.11.
View of adjacent wheat plots (borders indicated by stakes 1.2 m apart) at Puyallup, Washington, at harvest (August, 1972), showing biological control of *Gaeumannomyces graminis* under field conditions. The plots had been fumigated (methyl bromide, 1 pound/100 square feet; 1 kg/20.5 m^2), amended with about 0.5% (w/w) of suspected conducive or suppressive soil from eastern Washington, and then inoculated with *G. graminis*. Although take-all was slightly suppressed by some treatments in the first (1971) wheat crop after amendment, highly significant differences in disease suppression did not show up until the second crop, shown in this photo. The treatments illustrated are: A, native Puyallup soil (check); B, no fumigation, no amendment (check); C, suppressive soil from Quincy; D, suppressive soil from Pullman; E, conducive virgin soil from Quincy. Disease index and plant height were measured April, 1972 and May, 1972, respectively. Values given for the five treatments above are averages for four replications.

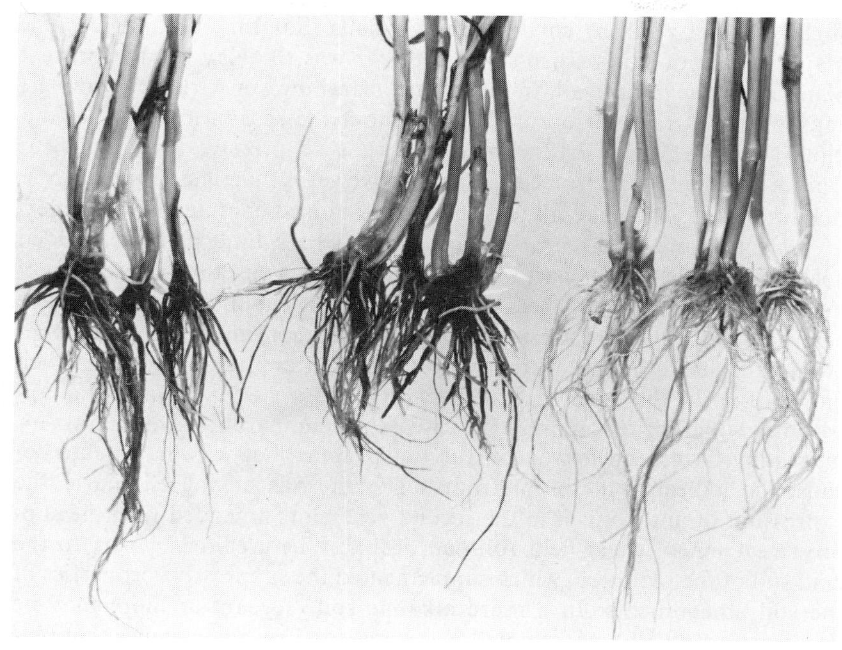

FIGURE 7.12.
Typical plants from plots in Figure 7.11. Left: No soil amendment (plot B). Center: Plot soil amended with conducive soil (plot E). Right: Plot soil amended with suppressive soil (plot C).

A long-term field trial was initiated in September 1970 at Puyallup, Washington (100 cm annual rainfall), where take-all is characteristically severe in recropped wheat, and where *G. graminis* var. *avenae* causes a brown patch disease on turf. The soil is an acid (pH 5.6) fine sandy loam. In one experiment, an area was fumigated with methyl bromide and then amended (about 1% w/w, rotovated to a depth of 12–15 cm) with nontreated eastern Washington wheat-field soil suppressive to *G. graminis,* or virgin conducive soils (Shipton et al., 1973). An adjacent experiment involved the influence of bulk and rhizosphere soil pH on take-all (Chapter 9), and included fumigated and nonfumigated areas with and without lime (CaO), which raised the bulk soil pH to about 7.0–7.1 (Smiley and Cook, 1973). The entire area was then sown to Nugaines winter wheat mixed 1:1 v/v with dead oat kernels infested with *G. graminis.* The area was thereafter cropped to wheat without additional treatment or inoculation.

The first year, disease was most severe in fumigated plots, was somewhat less where suppressive soils had been added, but was reduced slightly or not at all by conducive virgin soils (Shipton et al., 1973). The major benefit of suppressive soils that year was to delay but not prevent plant death due to take-all. In the second year, however, virtually complete suppression of take-all occurred throughout the season in only the four plots (Figs. 7.11 and 7.12) amended with a suppressive soil from near Quincy that had been cropped 12 consecutive years to irrigated wheat.

In the third year, take-all was suppressed almost completely, regardless of prior treatments, except in those areas where lime had been added nearly 3 years earlier as part of the pH study. This onset of suppression of take-all after only three wheat crops is unusually rapid, and related to the acid soil. Possibly there was no more actual antagonism or greater number of antagonists in these plots than in other soils after 3 years, but that which did develop in this acid soil (also unfavorable to *G. graminis*) was apparently sufficient to suppress the disease. A comparable level of antagonism may have been present in the limed areas, but was inadequate because the additional inhibition from soil acidity was lacking. Similarly, the expression of antagonism in the second year plots amended with the suppressive Quincy wheat-field soil can probably be credited in part to the acid soil of the plot area, which supplemented the suppressive properties of the soil amendments. In a more alkaline soil, 3 years or more may be necessary before antagonism will be expressed. This again shows that biological control is not rapid, that it may depend on some treatment 2 or more years earlier, but that it may be expressed more rapidly if given the proper environment in which to operate.

As Menzies commented in 1959, "The experiments . . . suggest that a disease-controlling equilibrium, which may take many years to appear naturally, may be induced in one season by appropriate treatments."

8

ROLE OF THE HOST IN BIOLOGICAL CONTROL

By more or less profound modifications of their root system,
many plants become adapted to different soil environments. . . .
But some species . . . exhibit much earlier tendencies to change than others
and a widely different degree of flexibility is also shown. . . .

—J. E. WEAVER AND W. E. BRUNER, 1927

As to demands on [photosynthetic products], roots apparently have
the last call, yielding priority first to fruit and second to shoot growth. . . .
Composed almost entirely of primary tissues, and harboring parasites
and saprophytes capable of invading in turn, and exposed to
microbial toxins, products of incomplete decomposition, corrosive enzymes,
and oxygen and nutriment deficiencies, feeder rootlets . . . are transient,
but remarkably adaptable to their physical surroundings.
. . . Nowhere, as in the soil, are plant parts and organisms bathed
in each other's metabolites, those secreted by the plant providing
conditions selective for certain soil organisms, and those
of the soil organisms thus selected having injurious
as well as salutory effects upon the plant.

—S. WILHELM, 1959

The host plays a double role in the biological control of plant pathogens. It provides a meeting ground for the pathogen and its antagonists, and it reduces the incidence or disease-producing activities of the pathogen directly, through physiological processes that constitute host resistance. Host resistance is an ideal biological defense against pathogens. However, to the extent that this control is ineffective or unstable, as it obviously has been for many of the less specialized parasites, we logically turn to less direct methods and employ third-parties, that is, antagonistic microorganisms. Nevertheless, the host is still involved through modification of the environment that affects the outcome between pathogen and antagonist. Effects such as exudation, gaseous exchange, ion and water uptake, stimulation of saprophytic organisms, and light and temperature modification are all important, Much of the so-called field resistance or field tolerance used commercially today for control of certain root diseases may involve effects of the host on the microenvironment, which then affects the pathogen, the antagonists, or both. Some forms of resistance are a combination of direct and indirect effects of the host on the pathogen.

Of practical significance to biological control is the relative ease with which the host lends itself to genetic or tillage manipulations. Man has used this factor to advantage by breeding for improved crop varieties and by experimenting with date and method of seeding, watering, fertilizing, weeding, or harvesting to improve crop yields. He has generally ignored the fact that each change in crop variety or husbandry can effect significant or drastic changes in the microbiological balance, and hence also affect biological control. Of the three living agencies involved in most biological control, the host offers the greatest latitude for management. By breeding and selecting for crop yield and quality, man has sometimes inadvertently used practical biological control. We may never unravel the many intricacies of even present-day systems of biological control brought on by features of the host, but this should not prevent us from trying.

ROOT DYNAMICS

The ability of roots to adapt to the environment plays a role in susceptibility of plants to disease. Features of the root system relevant to disease and biological control are outlined here.

Structure

The tip of the root is protected by a cap and lubricated by copious mucigel as it pushes through the soil. This gel is utilized by soil microorganisms as the root pushes forward, and contributes to the rhizosphere effect. Behind the cap is the delicate meristematic region, 0.2–0.4 mm in length; there is little absorption of water through this zone. The region of elongation extends a few tenths of a millimeter behind the meristematic region, and above that is the region of maturation in which the vascular tissues are differentiated. "Supported behind by older, more rigid tissue, and on the sides by soil particles, root tips are pushed forward through the soil by their elongating cells, sometimes at rates of 5 cm or more per day" (Kramer, 1969). The extent of roots produced is frequently enormous.

Thus, a winter rye plant 4 months old had 623 km of roots, with a surface of 236m^2, and 14 × 10^9 root hairs, with a combined length of 4681 km and a surface of 401 m^2. The plant must have added 5 km/day of roots (Weaver, 1926).

Plants produce perhaps 50% more roots than are needed for survival, and therefore are able to withstand a good deal of root injury without obvious yield reduction (tolerance). The distribution of roots in soil varies with the plant, the environment, and the soil. Although largely in the surface half meter of soil, roots may reach depths of 10 m (apple, alfalfa). The soil mass is generally permeated by roots, and under pastures and cereal crops it may be almost wholly rhizosphere (Bowen and Rovira, in Whittington, 1969). *Phytophthora cinnamomi* spreads from tree to tree through the sandy soils of the eucalyptus forests of Western Australia in the dense root mats formed by susceptible understory shrubs.

Breeding crop varieties for specific patterns of root distribution as a means of reducing disease, although little studied, has merit. As an example, Acala SJ-1 cotton, susceptible to wilt caused by *Verticillium albo-atrum,* forms many lateral roots in the surface 46 cm of soil, whereas the field-resistant Waukena White forms its major laterals below the 46 cm zone, in which most of the microsclerotia are found (Phillips and Wilhelm, 1971).

Roots are in a state of subordinated competition with tops, even though, like an iceberg, much of the plant may be beneath the surface. Heavy crops of coffee berries, cotton bolls, corn ears, tomato fruit, cranberries, or strawberries deplete the carbohydrate reserves of roots, decrease their growth rate, and may hasten their death. Also, vigorous shoot development decreases root growth of apple and pear (Kramer, 1969). The ability of cranberries to withstand winter flooding, and of cereals to withstand winter injury, is closely related to the carbohydrate reserves in the plant. The implications of this in relation to biological control were discussed in Chapter 2.

There is much natural root grafting among woody plants. In pure stands of cultivated pine, such as are grown commercially in the southeastern states, root grafting may be so extensive as to produce an interconnected root system for an entire field. Stumps and root systems interspersed in the stand may survive for considerable periods on carbohydrates translocated through grafts from intact trees. *Fomes annosus* may infect through a stump and thereafter infect the root system of surrounding trees with little or no exposure to soilborne antagonistic organisms (Chapter 6). An effective biological control in this case is to plant mixed tree species that will not root-graft, thereby interrupting the avenues of spread for the pathogen (Koenigs, 1960; Hodges et al., 1971).

The longevity of roots varies from a few weeks for the transient noncambial feeder roots of strawberry (Wilhelm, in Holton et al., 1959) or seminal roots to many years for the main roots of trees. Some perennial grasses apparently produce a new root system each year (Kramer, 1969). The senescent and dead roots may provide a means of entry of saprophytes into the vascular tissue of healthy plants, as shown by Wilhelm for *Cylindrocarpon radicicola* in strawberry. This fungus may be another example of a saprophyte or weak parasite that occupies its host, causing little apparent damage except under conditions of age or stress of the host (Chapter 2).

Root Hairs

Root hairs, extensions of epidermal cells that increase the area of contact with soil particles, are important in water and nutrient absorption. Water stress in the root induces absorption through root hairs. Cailloux (1972) found that "there is an intimate correlation between the intensity of metabolism and the velocity of absorption of water." He postulated that a root hair acted as an "apparent metabolic water pump." There is one definite zone of absorption in root hairs, usually at the tip in young ones.

The absorption capacity of a root hair increases under aerobic conditions, but shifting to anaerobic conditions will cause it to excrete water and nutrients (Cailloux, 1972). The rate of this exudation is apparently limited by the amount of water supplied by neighboring root cells. Following such excretion, the absorption capacity of a root hair is reduced because of the decreased metabolism from loss of nutrients. A root hair can be made to alternately absorb or excrete by varying the amount of water in contact with it or by changing the carbon dioxide level around it; sudden flooding may increase exudation. Excretion also increases with the age of the hair. Under normal soil conditions, root hairs will thus frequently leak nutrients favorable to the growth of microorganisms.

Effect of Soil Moisture and Oxygen

Roots are geotropic, and are hydrotropic up to water potentials approaching zero (saturation). Plants grown in a bed with constant-water-level subirrigation, with a decreasing water potential upward from the water level, tend to have roots concentrated in the zone of their optimal moisture and aeration. However, under field conditions of fluctuating water potential, roots are more uniformly distributed in a pattern determined by their genes and the soil type. Newman (in Kramer, 1969) found that root growth of flax in each soil layer was independent of the water potential in other layers or in the shoot. When, for example, root growth in the upper dry soil layers was reduced, it might not be at all decreased in lower moister layers. ". . . apparently many kinds of plants survive and even grow with only part of their root system in soil above the permanent wilting percentage." This situation is characteristic of Pacific Northwest dryland wheat (Papendick et al., 1971).

The reduction of fusarium root rot of bean in the arid irrigated Columbia Basin of Washington by fracturing the plow sole hardpan illustrates how increased root depth may be used in biological control (Chapter 9).

Root injury from oxygen deficiency is a form of host predisposition to pathogenic attack (Chapter 9). The degree of root injury caused by oxygen deficiency from saturation or flooding of soil is determined by:

a. Resistance of the plant. Cypress, willow, mangrove, rice, waterlily, and cattail are able to grow in water-saturated soil because oxygen moves through the stems to the roots (Greenwood, 1967), which perhaps also have some capacity for anaerobic respiration. Roots of most plants die, however, if submerged suddenly and for more than 12–24 hours.

b. Rapidity of submersion. Plants form roots in water culture, which may sustain normal plant growth, but if seedlings grown in soil are transferred to water culture the roots die and are replaced by adapted ones before top growth is resumed.

c. Duration of flooding. Brief submergence may cause no serious injury, but submergence for 1 day early in plant growth may markedly reduce yields of tomatoes and peas. The longer the flooding, generally the more severe the injury. Snapdragon roots may be injured by submergence for a half day.

d. Dormancy of the plant. Dormant trees may survive weeks of flooding in winter with little or no injury, but flooding for 1 day during the growing season, when transpiration and respiration are high, may seriously injure some species (Kramer, 1969). Cranberry bogs may safely be flooded for long periods in winter, but the plants may be injured by 20–30 hours of flooding in summer. Respiration is increased and solubility of oxygen decreased with rising water temperature. Lowered carbohydrate storage in the plants following a heavy fruit crop, for example, also increases flooding injury (Franklin, 1940).

e. Soil temperature. Flooding warm soil is usually more injurious to plants than flooding cold soil, because of greater transpiration and

> respiration. Furthermore, some plants (cucumber) may sustain severe wilting and injury from cold soils that greatly decrease water absorption, whereas warm air maintains high transpiration. Tobacco exposed to the sun when the soil is still saturated from a heavy rain wilts rapidly because the deficiency of oxygen in the soil has reduced the water intake by the roots.

It is now thought that much "of the absorption of water and solutes by many perennial plants occurs through roots which have undergone secondary growth and are covered with layers of suberized tissue" (Kramer, 1969; Last, 1971). The continuing growth of roots taps new soil areas for water and minerals.

Mycorrhizal Relationships

The mycorrhizal relationship, an effective mechanism for increasing water and nutrient absorption by roots, has been reviewed by Harley (1969; in Baker and Snyder, 1965), Bowen and Rovira (in Whittington, 1969), and Hacskaylo (1971). Marx (1972) showed that mycorrhizal fungi protect roots from infection by *Phytophthora cinnamomi* (Chapter 7).

Establishment of the mycorrhiza may begin with infection of the seedling root by certain soil fungi, particularly Basidiomycetes. Subsequently the lateral branches of the long mother roots may be infected by internal hyphae before they emerge through the cortex; this probably insures possession of the rootlet before it is exposed to competitors. The mycelia of both ecto- and endomycorrhizae radiate for some distance into the surrounding soil, essentially expanding the effective radius of the root. Although such infection generally reduces the extent of the root system, substantial increases in absorption and storage of phosphate, for example, by the plant result. ". . . the most spectacular increases in phosphate uptake . . . occur under conditions of low available phosphate." Since amino-acid leakage from roots of *Pinus radiata* was 2.5 times as great when phosphate was deficient as when adequately supplied, the increased phosphate absorption via mycorrhizae could indirectly affect the rhizosphere flora as a whole (Bowen and Rovira, in Whittington, 1969). Whereas the most active area of absorption in a growing root is only briefly in contact with a given bit of soil, the mycorrhizal fungus remains in contact for a much longer period.

The fungus sheath and its mycelia spreading through the soil have a bacterial flora that complicates the interpretation of experimental results (Bowen and Rovira, in Whittington, 1969). Mycorrhizal fungi are generally sensitive to antagonistic effects of other microorganisms, and Harley (1969), on this and other bases, considered them as having properties similar to those of obligate root-inhabiting pathogens under natural conditions.

Function and Adaptation of Roots

The functions of the roots, and particularly of the white root tips, are greater than the commonly mentioned anchorage, absorption of water and minerals, and storage organs of reserve carbohydrates. For example, nitrogen conversion to amino acids occurs there (Bollard, 1960), as "most of the nitrogen passing from the root to the shoot is in organic form" (Street, 1966). Furthermore, cytokinins and auxins are formed in the roots, and either gibberellins are synthesized there or the gibberellic $acid_{19}$ synthesized in the shoot is converted to gibberellic $acid_1$ in the root and retranslocated to the shoot (Vaadia and Itai, in Whittington, 1969; Crozier and Reid, 1971). Probably other growth-regulating substances are also formed in the roots. The nicotine of tobacco is synthesized there. Roots, in turn, depend on shoots for thiamin, niacin, pyridoxin, and perhaps growth regulators. The yellowing of basal leaves from waterlogging or poor aeration of soil may result from a reduction in their hormone supply from the roots (Kramer, 1969).

There is abundant evidence that roots may be modified anatomically and physiologically in adapting to environmental changes. Roots developed in poorly aerated media or in water culture are generally larger in diameter, have fewer root hairs and lateral roots, and contain more and larger cortical air spaces than those in well-aerated media. The former roots develop mature pericycle and secondary thickening of walls of xylem vessels closer to the root tip. However, the cell walls tend to be thinner than in well-aerated media. Well-aerated roots are lower in total and reducing sugars. Roots produced under conditions of low oxygen are able to withstand low oxygen levels without a decrease of their respiration or salt accumulation.

When soil dries or salinity becomes excessive, root elongation ceases, roots become suberized to the tips, photosynthates may accumulate for a time, and cell permeability decreases. ". . . plants subjected to severe droughts do not recover their full ability to absorb water for several days after the soil is rewetted" (Kramer, 1969).

Rice, grown in drained soil and suddenly flooded, may sustain root injury until adventitious adapted roots develop. "Raising the water table even temporarily by irrigation causes the death of the deeper roots in many plants and usually results in decreased yield" (Weaver, 1926). ". . . high water tables maintained at a constant level are less injurious to crops than occasional periods of complete flooding" (Kramer, 1969). Injury to the tops may then result from insufficient water because of decreased root absorption, either from toxins anaerobically formed in soil or tissues, or from cessation of essential syntheses. Root systems developed in dry soil are generally larger than those in wet soil. The effects of displacement of oxygen by water, carbon dioxide, or nitrogen gas are much the same. Microorganisms may play a decisive role in oxygen depletion in soil in time of stress. As pointed out by Greenwood (in Whittington, 1969), "Most oxygen will be required by segments of root that are furthest away from the stems and are thus least able to get oxygen from the aerial parts by transport through the plant" (Chapter 9).

At low soil temperature, root maturation is retarded and roots tend to be white, succulent, and of relatively large diameter, with few scattered laterals; the cortex remains alive for some time. Rapid maturation occurs at high soil temperature, with early loss of the cortex, nearly complete suberization, frequent branching even close to the growing point, and formation of brown nonsucculent roots of small diameter.

The continuing growth of roots during periods of rapid top growth clearly maintains active roots anatomically and physiologically adapted to constantly changing physical conditions (water potential, oxygen supply, temperature, salinity, and perhaps others) and thus able to carry on their essential functions. The evolutionary advantage to the plant of such adaptability is obvious. Like all adaptations, however, this has its limits, and roots may be killed or become nonfunctional under rapidly changing conditions.

Many plants have evolved a means of evading the terminal effect of vascular parasites that invade the xylem, restrict water flow, or produce toxins. Successive new sets of lateral roots are formed, whose xylem connects to each successive layer of secondary xylem as it is laid down, and which in turn connects to new leaves. Therefore, the new lateral roots connect not to old xylem vessels but to new ones as they form, providing successively new conducting tissue, bypassing old vessels plugged with tyloses and those invaded, for example, by *Verticillium albo-atrum*. The pathogen may thus be buried by new xylem tissue of a woody plant, and each new set of roots and layer of secondary xylem apparently must be infected, since the fungus does not appear to spread radially across annual rings (Wilhelm and Taylor, 1965). This effective disease escape is possible only because there is continuing root growth. Since "growing roots, travel to root-infecting fungi more commonly than fungi travel or are carried to substrates" (Garrett, 1970), this continued root growth is not an unmixed blessing.

PHYSICAL AND CHEMICAL FEATURES OF THE RHIZOSPHERE

To the soil microbiologist, the rhizosphere is the narrow soil zone surrounding living plant roots, which contains root exudates, sloughed root remains, and large populations of microorganisms of various nutritional groupings. To the plant physiologist it is the zone of ion uptake and exchange, of oxygen and carbon dioxide exchange, and of the mucigel matrix. To the soil physicist it is the zone of minimal porosity, of water diffusion and uptake, and of water-potential gradients. To the plant pathologist it is the zone where root pathogens are stimulated by root exudates, and where they swarm, grow ectotrophically, or form infection structures prior to pathogenesis. The zones near seeds, tubers, bulbs, corms, or rhizomes are comparable. All these facets operate in nature as part of a whole, not independently. The more we understand how each process and component of the rhizosphere, singly and collectively, affects root pathogens, the more successful and predictable will be our systems of biological control.

Roots growing through soil displace their own volume and thus increase the adjacent soil bulk density. They may deviate through crevices, but aside from this, roots press forward and the soil gives way. Soil nearest the root compresses first, and as all possible space is filled with particles, the compression is exerted at ever-increasing distances until the entire root volume has been accommodated. The epidermal and cortical cells may be crushed in this process. The compressed area immediately surrounding the root contains the displaced soil and is referred to as the minimal-voids zone. Bulk density here is maximal. Since this is also the zone of penetration by plant pathogens, the increased bulk density is especially significant to the plant pathologist. As bulk density increases, pore size decreases, and water per unit volume and its tendency to move (unsaturated conductivity) is increased. This may increase diffusion of exudates away from roots, as well as nutrients toward the roots. It may also directly affect the growth of microorganisms by decreasing the diameter of the pores and the tortuous channels that must accommodate or permit penetration of microbial structures (Griffin and Quail, 1968). This is not grossly limiting to root fungi, as evidenced by the extensive mycelial growth that occurs near and on the root surfaces. On the other hand, it could affect the final stages of swarming of aquatic Phycomycetes if water-filled pores of sufficient size to accommodate zoospores were in short supply. The minimal-voids zone could thus act as a barrier to aquatic Phycomycetes in some soils. Motile bacteria, on the other hand, being smaller, could pass through smaller openings and thus accumulate with greater ease on the root surface.

The rhizosphere also differs from the soil mass in quantity of oxygen and carbon dioxide. Respiratory carbon dioxide is evolved by root cells and also by the active microbiota, so that the concentration decreases away from the root surface. Oxygen, on the other hand, apparently exists in a reverse gradient, being progressively lower at points closer to the roots. Oxygen from aerial parts apparently can satisfy oxygen requirements of the roots, and possibly even those of microorganisms on the rhizoplane or in the rhizosphere. Although occasional and brief anaerobiosis in the rhizosphere undoubtedly occurs under certain soil conditions, the evidence that there is oxygen diffusion from the tops to the roots would raise serious doubts about the dependability of oxygen competition as a mechanism of antagonism and biological control in the rhizosphere.

The process of water uptake by plant roots has a profound effect on physical properties of the rhizosphere and on microbiological activities. This is best described by first discussing the dynamics of water in the soil-plant-microorganism system.

Water in soil, plant tissues, or within microorganisms can be expressed most simply in terms of its potential energy, that is, water potential (Slatyer, 1967). This potential energy may be quantified in joules/kg (free energy units), but more commonly as bars (1 bar = 100 joules/kg). Water under pressure or elevated above a reference point has positive potential energy (+ bars); that adsorbed to a soil or cell matrix, or with dissolved salts or sugars, has a negative potential energy (– bars) relative to that of the standard, pure free water. Soil water potential is determined mainly by adhesion and cohesion (negative), that is, forces associated with the matrix (matric water potential), but includes an osmotic component (negative) from the dissolved salts (osmotic potential). The two are additive. The total water potential of a plant consists largely of an osmotic component (negative), but includes some matric (negative) and turgor component (positive, but sometimes negative). Again, the values are additive. The water potential of a microbial cell likewise includes an osmotic, matric, and turgor component. Regardless of the individual components, however, and because they are additive, the water potential of one system can be readily equated with that of another (Cook and Papendick, 1972; Cook, 1973). Methods of measuring water potential of soils or plants have been reviewed (Brown, 1970; Wiebe, 1971).

Water, like heat, flows from areas of high to low potential energy, according to the second law of thermodynamics. The flow of water from one area of soil to another or to the roots (unsaturated conductivity), its movement into roots (absorption), and its movement up the xylem and passage into the atmosphere (transpiration) are examples of water flow from high to low energy status. The soil profile, when drainage has slowed to a negligible rate after rain or irrigation, is at *field capacity* (−0.3 to −0.5 bars). Air with a relative humidity of 50% and a temperature of 25° C has a water potential of −740 bars. Evaporative loss is the direct flow of water from high energy status in the soil to lower energy status in the atmosphere. The plant provides a less direct means by which equilibrium is approached between soil and air. The water potential of a microorganism situated along the path of water flow, whether in soil or in plant tissue, will be determined by the water potential of that specific site. Because of their small mass, microorganisms have a negligible effect on the water potential of their surrounding environment.

There is some evidence that the plant water potential may be fairly uniform from the uppermost leaf to the deepest root (Cook, 1973); the steepest water energy gradients thus occur at the root and leaf surfaces. Assuming a leaf water potential of −20 bars, the root may likewise be near −20 bars. If the soil away from a root has a water potential of −0.3 bars, then soil adjacent to the rhizoplane must be at some intermediate water potential. This water-potential gradient toward the root will steepen on hot dry days with high evaporative demand and transpirational loss, and it will lessen on cool moist days.

Thus, the water potential in the rhizosphere may be quite different from that of the bulk soil where most measurements are made. Moreover, it can fluctuate from day to day and even from hour to hour, depending on aboveground environment, soil type, or other factors. Water can diffuse out of roots into soil if a reverse gradient develops (Cook and Papendick, 1972).

Organisms differ in their optimal and minimal water potential requirements. Some may be slightly stimulated by a given water potential, whereas growth of others is inhibited at this same potential. As changes in temperature effect corresponding changes in microbial growth and balance, so variations in water potential effect fluctuations in microbial activity.

Water potential in the rhizosphere will also vary along the root length, being high at the tip where absorption has not yet begun, low in the root-hair zone, and probably slightly higher again along mature sections of the root where suberization has occurred and absorption is less. The apical meristem and region of elongation in newly invaded soil is thus still sufficiently moist to permit diffusion of exudates and the increase of microbiological activity, particularly of bacteria and aquatic Phycomycetes. In the slightly drier zone near the region of maturation, where root hairs exist, microbiological activity reaches a maximum.

The mature-root zone is relatively devoid of exudates unless wounds occur. With rains or irrigation, the soil becomes water-filled to the root surface, eliminating whatever water-potential gradient may have existed. When the soil drains, the cycle is repeated, except that soil may remain wet longer near mature root surfaces until lateral roots, and hence new absorptive surfaces, emerge.

One might expect greatest fungus and actinomycete activity on the mature area of the root because of their initial slow growth rates. Bacteria, on the other hand, tend to be abundant in the region of elongation because of their shorter generation time.

Ion exchange between the root surface and the soil solution produces still another change of importance to root-disease organisms and to biological control in the rhizosphere. This effect is perhaps best illustrated by root uptake of NH_4^+ and NO_3^- and corresponding decreases or increases in rhizosphere pH, which affect *Gaeumannomyces graminis* (Smiley and Cook, 1973).

The suppressive effects of nitrogen on the take-all disease are well known, but "Ophiobolus or Gaeumannomyces patch" of turf in western Washington (Gould et al., 1966) and take-all of wheat in southern Idaho (Huber et al., 1968) were more effectively controlled by ammonium than by nitrate nitrogen. Smiley and Cook (1973) showed that the effect is related to pH, partly of the bulk soil (pH_b) but mostly of the rhizosphere (pH_r). As roots remove ammonium ions, anions are left behind, but hydrogen ions are liberated to maintain the equilibrium. This lowers the pH_r by as much as 2 units from that of nearby bulk soil. Conversely, the uptake of nitrate leaves cations behind, is accompanied by root liberation of hydroxyl ions, and the pH_r rises. There was a direct correlation between pH_r and severity of root disease caused by *G. graminis;* disease severity was progressively less with each successive pH_r unit below 7, regardless of soil, form of nitrogen, or treatment. Moreover, the controlling effects of ammonium could be nullified with lime applications, and control was possible with nitrate if the soil was acidified by some means.

On the other hand, in soil treated with methyl bromide, then infested with *G. graminis* and planted to wheat, the pH_r again declined with ammonium and increased with nitrate, but disease was uniformly severe at all pH_r above 5. Only limited disease occurred at pH_r below 5. Hyphal growth of *G. graminis* on straw in autoclaved soils of different nitrogen treatments and pH values was prevented at 4.9 and below, and was optimal between 6.5 and 7.0. Garrett (1936) showed that growth of runner hyphae of this fungus is improved with a rise in pH of the root zone. Smiley and Cook concluded that pH_r below 5 controlled *G. graminis* directly, because the fungus cannot tolerate these pH values, but between 5.0 and 7.0 the control was indirect, probably through antagonistic rhizosphere organisms. Addition of ammonium to soil increased populations of *Pseudomonas putida* in the rhizosphere (R. W. Smiley, unpublished). Apparently, by changing its rhizosphere environment through ion uptake and exchange, the host selectively favors antagonists over the pathogen, or vice versa.

Root pathogens are subjected to a constantly changing chemical and physical microenvironment in the rhizosphere that may be completely different from that of bulk soil. Root exudates buffer the soil pH, but apparently not enough to prevent significant shifts caused by ion exchange, which in turn affects the equilibrium between carbon dioxide, carbonate, and bicarbonate. An increase in volumetric water content could increase the unit volume of dissolved carbon dioxide by comparison with areas apart from the root, and also affect ion uptake. Microorganisms, and especially the balance among them, respond to the ecological niche as it changes with time, as an ever-fluctuating microbiological equilibrium. Man will increase the chances of biological control of root pathogens in the rhizosphere when he learns to manage the physical and chemical environment of this relatively small zone of soil, while paying less attention to the environment of the soil mass, which may have little direct bearing on the active microbiological interactions.

ROOT EXUDATION AND THE RHIZOSPHERE EFFECT

Although ion uptake, water diffusion, and compression of the soil are important, there are no two factors of greater significance in root disease initiation than exudation and the resultant rhizosphere effect. Exudation stimulates pathogens as well as many other microorganisms, some of which facilitate nutrient uptake by the plant and some of which suppress pathogens. The universality of this phenomenon suggests that it is of net benefit to plants. Rovira (1969; in Baker and Snyder, 1965) has reviewed root exudation, and Bowen and Rovira (1973), Lochhead (in Holton et al., 1959), and Katznelson (in Baker and Snyder, 1965) have reviewed the rhizosphere effect.

The surface (rhizoplane) of a root growing in a soil is a complex consortium with diffuse margins (R. C. Foster, unpublished). The living root has some crushed cells of the epidermis, cortex, and perhaps the root cap, as well as live root hairs immersed in the mucigel secreted by the root. In this mucigel are embedded individual cells or discrete colonies of bacteria, surrounded by their capsular materials. A bit farther out there may be clusters of plates of clay minerals, bits of organic matter, and sand particles, all embedded in the mucigel (Figs. 8.1 and 8.2). Beyond the irregular

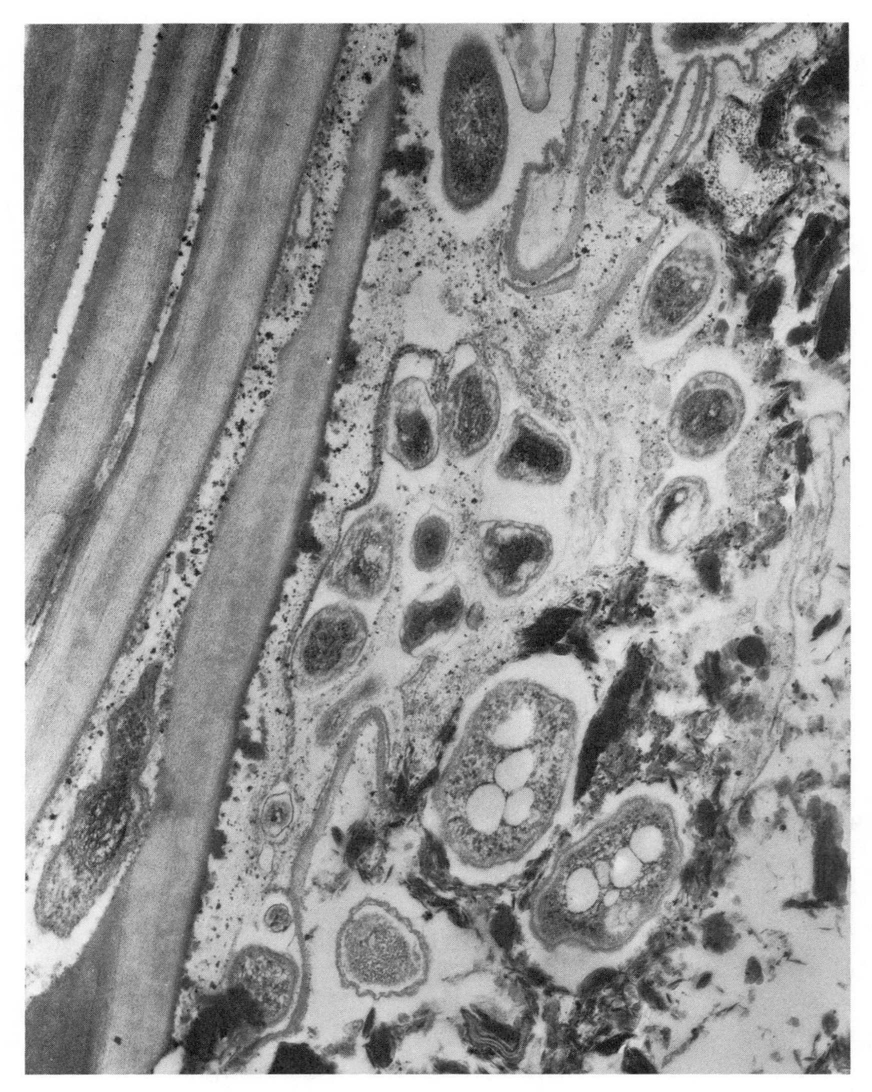

FIGURE 8.1
Ultrathin transverse section through rhizoplane of seminal root of 18-week-old wheat plant grown in field soil. Crushed cortical and epidermal cells at left. Several kinds of bacteria with their surrounding capsular gel are embedded in the mucigel of the root at right. × 19,-500. (Photo courtesy of R. C. Foster, CSIRO, Melbourne, Australia.)

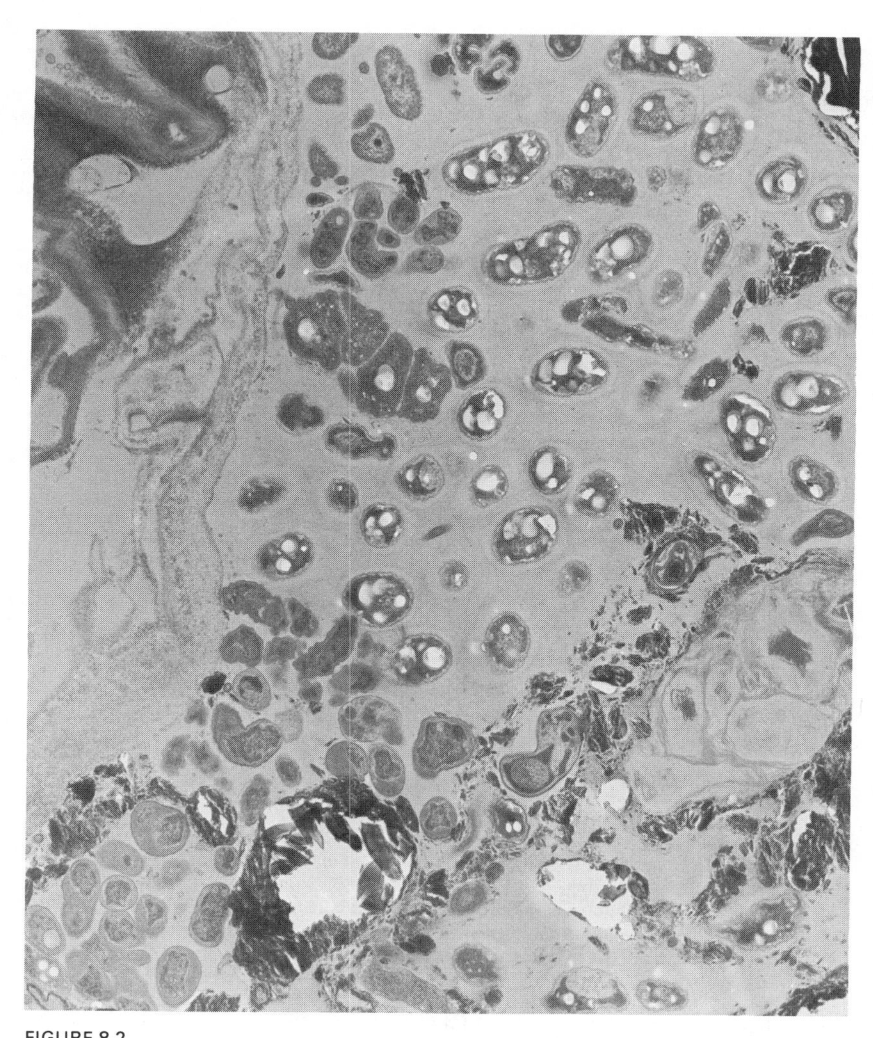

FIGURE 8.2.
Ultrathin transverse section through rhizoplane of 6-week-old *Trifolium subterraneum* root grown in field soil. Root surface is at left. Several kinds of bacteria are embedded in the mucigel of the root at right. Some cells are embedded singly surrounded by capsular gel, others are in colonies. Clay minerals at bottom (black group of plates), organic matter at upper right. × 7200. (Photo courtesy of R. C. Foster, CSIRO, Melbourne, Australia.)

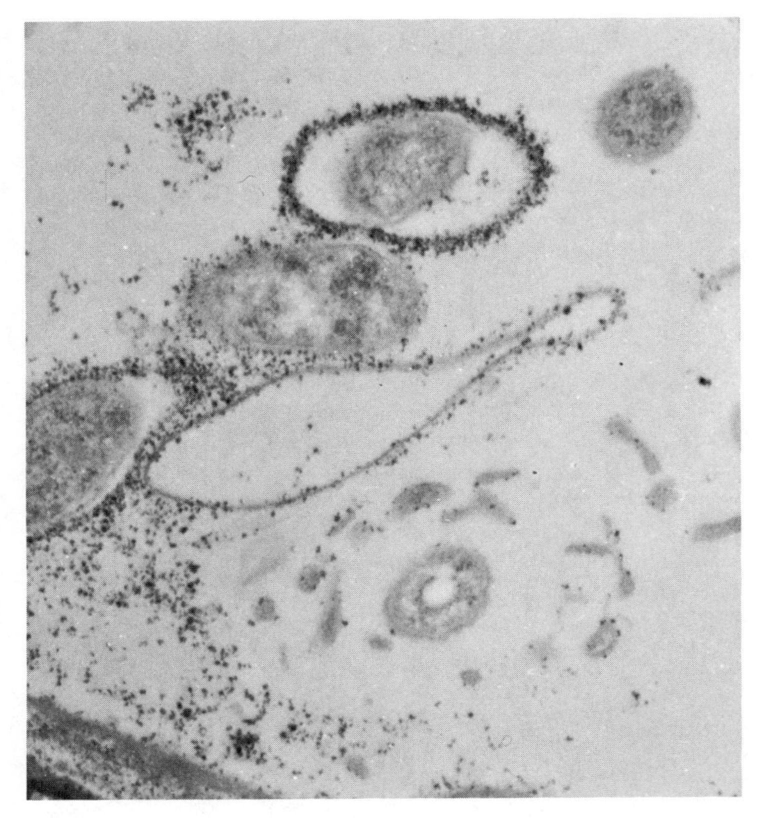

FIGURE 8.3.
Rhizoplane bacteria with sheath of small plates of clay minerals on surface of capsular gel surrounding cells. Empty sheath of a presumably dead bacterium in center; cells without sheaths at upper right and bottom center. Ultrathin section, with root of *Trifolium subterraneum* below. × 22,000. (Photo courtesy of R. C. Foster, CSIRO, Melbourne, Australia.)

boundaries of the mucigel, the bacterial cells may be attached to, or enclosed in, clay minerals or organic matter (Fig. 8.3). Actinomycetes, fungi, and other microorganisms also may be freely intermingled. Mycorrhizal fungi may form a dense mantle directly on the root surface; bacterial colonies then lie outside this layer. There is thus a gradual transition from what is clearly rhizoplane, to rhizosphere, to the soil mass, with indistinct boundaries. The number of bacteria in the mucigel is least at the root tip, increases in the regions of elongation and maturation, and continues along the mature root. Bacteria may cause loss of mucigel on unhealthy roots (Rovira and Campbell, 1974).

Microorganisms rapidly colonize growing roots within the first centimeter from the tip. Bowen and Rovira (1973) thought that bacterial colonies covered only 5–15% of the surface of 3-day-old *Pinus radiata* roots, but that this might reach 37% in 90-day-old roots. "Bacteria often occur as colonies which coalesce to form an almost continuous cover several bacteria deep on the root . . . . the mucilaginous material covering the root . . . may be heavily colonized by bacteria. . . ." (Bowen and Rovira, in Whittington, 1969). Such bacteria may stimulate plant growth.

Brown (1973) showed that wheat roots provide an environment where bacteria whose growth is normally restricted by bacteriostatic soil factors are able to multiply. This appears to be an effect from root exudates.

Increased leakage from roots caused by mechanical injuries, phosphate deficiency, infection by viruses (Chapter 10), or the feeding of insects, nematodes, or other animals may also provide nutrients favorable to microorganisms, including pathogens, in the rhizosphere. Leakage from mycelia may similarly be increased by virus infection (Hollings and Stone, 1971), by exposure to supraoptimal temperatures (Ward, 1971), and probably by other factors.

Related to this topic is the application to the foliage of a suitable systemic chemical, which is translocated to the roots, where it is excreted, possibly controlling root pathogens. Several workers have demonstrated the feasibility of modifying root exudates through foliar application of chemicals (Mitchell and Livingston, 1968; Rovira, 1969; Stǎnková-Opočenská and Dekker, 1970). So far the greatest success with this method has been achieved in nematode control. Root-knot nematode has been controlled on tomato, tobacco, pumpkin, and sweet potato, and *Pratylenchus scribneri* on bean by experimental nematocide D-1410. Root-knot nematode has also been experimentally controlled by ethyl 4-(methylthio)-*m*-tolyl isopropylphoramidate (Nemacur). This general method seems to represent a meeting ground between chemical and biological control, but is closer to the former.

Pertinent researches will be discussed here to illustrate the damping of the rhizosphere effect on root-disease organisms and how this may account for field resistance and tolerance to pathogens.

Some varieties susceptible to pathogens have been reported to exert a greater effect on the rhizosphere population than resistant varieties of the same crop, as reviewed by Lochhead (in Holton et al., 1959). The rhizosphere effect for fungi, bacteria, and actinomycetes was greater for Bison than for Novelty flax, the former being susceptible and the latter resistant to fusarium wilt. The same correlation has been reported for tobacco resistant, susceptible, or intermediate in reaction to *Thielaviopsis basicola;* the ratio of microorganisms in rhizosphere to those in soil was greatest with the susceptible tobacco. Since one would expect such a difference in rhizosphere population between two genetically different lines, the greater rhizosphere effect for the susceptible than for the resistant variety in a two- or three-variety study may be coincidental.

Resistance in plants to the wilt fusaria is fairly specific for a given race, and is an inherited character. Histological studies indicate that the wilt fusaria can invade the cortex and vascular tissues of resistant as well as susceptible varieties, and that resistance is expressed within the vascular system of the host (Beckman and associates, 1962–72). As discussed later in this chapter, there seems little doubt that host resistance to wilt fusaria involves internal direct, rather than external indirect, biological effects of the plant on the fungus.

Phytophthora citrophthora invades roots of both susceptible and resistant citrus in the regions of elongation and maturation, but mature cells of resistant species seem to inhibit development of the pathogen (Broadbent, 1969).

On the other hand, the rhizosphere effect of a susceptible banana variety may be related to whether the soil is conducive to the establishment of *F. oxysporum* f. sp. *cubense.* The ratios, rhizosphere to soil, for spore-forming bacilli in a wilt-conducive banana soil from the West Indies (Rombouts, 1953) were 10 for Congo (virtually immune to *Fusarium oxysporum* f. sp. *cubense*) and 25.3 for Gros Michel (highly susceptible). They were 5.3 and 42.4, respectively, in the wilt-suppressive soil, about a fourfold difference in varietal distinction between the two soils. The reverse was true for fungi, which were more suppressed in the rhizosphere of Gros Michel than of Congo, and which were suppressed more in the wilt-suppressive than in the wilt-conducive soil. Harper (1950) had reported earlier a markedly greater bacterial rhizosphere stimulation for Gros Michel and Silk Fig (both susceptible) than for Congo and Guindy (both resistant), and found a bacterium highly antagonistic to *F. oxysporum* f. sp. *cubense* and present in greater frequency in the rhizosphere of Congo than of Gros Michel. Rombouts was unable to repeat this work.

Smith and Snyder (1972) observed that in California soils suppressive and conducive to various fusarium wilts, the rate of bacterial multiplication (total count) in a given period was greatest in the suppressive soil. This is further circumstantial evidence for a role of bacteria in the suppressive nature of some soils to wilt fusaria (Chapter 4).

Lysis of some *Fusarium* germlings (*F. solani* and *F. roseum*), or their conversion back to chlamydospores, occurs commonly near exuding plant parts, particularly near cotyledons, root tips, and wounds, where the exudates are rich in amino acids and sugars (Chapters 6 and 7). In spite of this tendency, formae speciales of *Fusarium oxysporum* apparently infect through the region of root elongation and through wounds, where exudation is high. In so doing, they invade the cortex and enter the partially differentiated vascular system directly through broken or undifferentiated tissue, where they do not have to cross the endodermis. Presumably, germlings of *F. oxysporum* are subject to rapid lysis in competition and antibiosis, like those of *F. solani* and *F. roseum,* but information on their survival rate in the root tip region is not available. Assuming that lysis has an impact on the penetration activity of this species, and that the pathogen must penetrate through the region of elongation or wounds to infect, this could account for the apparent susceptibility of *F. oxysporum* to soil factors favoring soil bacteria. The host would thus play a major role in suppressive soils by supplying the nutritive substances that shift the balance against *Fusarium* and in favor of antagonists.

Resistance of monocotyledonous plants to phymatotrichum root rot represents a complex situation probably not attributable to any single factor. Eaton and Rigler (1946) demonstrated that corn was resistant to *Phymatotrichum omnivorum* when grown in nontreated field soil, but not in sterile soil. The pathogen grew in the soil and over the corn roots in both sterile and nonsterile cultures, but only the plants under sterile conditions developed lesions and died. The non-inoculated plants, and inoculated plants in nonsterile culture, remained healthy for the duration of the test. It was concluded that "The antibiotic protection afforded the maize plant by its root-surface organisms is apparently so great that *Phymatotrichum omnivorum* is unable to exert any effect upon the epidermal tissues." These studies with corn were repeated by Black (1968), and the results confirmed in sterile and nonsterile sand inoculated with *P. omnivorum* (Fig. 8.4). Histological studies showed almost total absence of the pathogen from roots in nonsterile soil, although the fungus mycelia grew parallel to them. In the sterile soil the fungus formed mycelial wefts on the root surface, and hyphae penetrated the cortex and even invaded the vascular system. He thought that the surface microorganisms were involved directly through antibiosis, or indirectly by affecting host physiology. Bloss and Gries (1967) have studied the chemical bases of the hypersensitive reactions of resistant and susceptible cotton plants.

Cotton in the seedling stage is resistant to phymatotrichum root rot, but when boll production begins, the plants become highly susceptible. Unlike corn, however, the cotton seedlings are resistant to the disease whether grown in sterile or nonsterile culture. Eaton and Rigler (1946) concluded that "any protection afforded by the root-surface saprophytes of young cotton plants was secondary to chemical or other resistance factors." On the other hand, older cotton plants may vary greatly in root-rot suscepti-

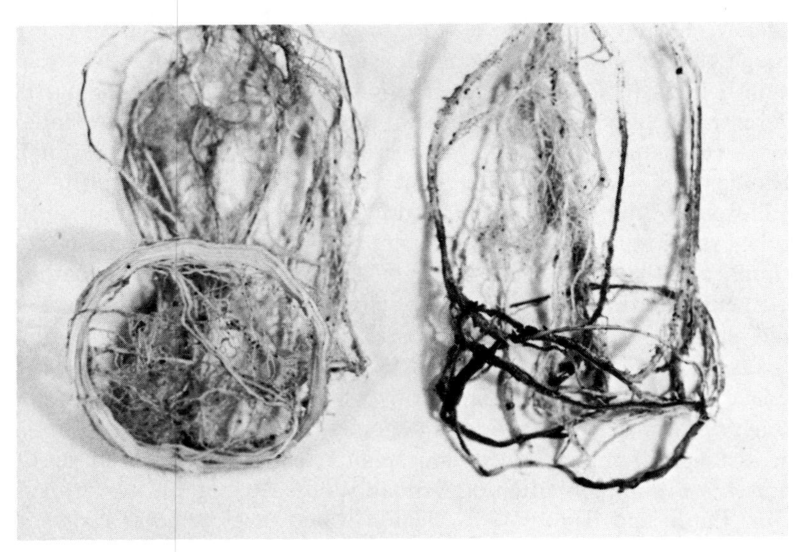

FIGURE 8.4.
Root systems of corn grown axenically for 45 days in sand. Nutrient solution had been added to the sand, which had then been inoculated with *Phymatotrichum omnivorum* and planted with surface-sterilized seeds. This approximated the methods of Eaton and Rigler (1946). Left: Root system grown in nonsterilized sand. Right: Root system grown in sterilized sand. (From Black, 1968.)

bility, and this variation could reflect antagonism to *P. omnivorum* by root-surface saprophytes, possibly promoted by a high root-bark carbohydrate content. By treating cotton plants in different ways, they were able to produce plants with a high, low, or intermediate root-bark carbohydrate content. The higher the content of root-bark carbohydrate, the slower the onset of severe root rot. Although plants with high root-bark carbohydrate became infected, subsequent progress of the disease was greatly restricted. In some instances, the root bark grew over the initial infection sites.

These workers also pointed out that carbohydrates do not inhibit *P. omnivorum;* the fungus thrives on such substances in pure culture. Other organisms will also thrive on the high carbohydrates in nature, and may in turn reduce ectotrophic parasitic activities of *P. omnivorum*. They also pointed out that *P. omnivorum* is commonly observed to die out in decaying organic substrates, even though food exhaustion is not complete. In sterile organic substrates, however, the fungus does not die out. Dilution-plate counts of the root-surface bacterial flora revealed higher total numbers of bacteria, including *Bacillus* spp., on roots of low bark-carbohydrate content. Although environments favorable to bacteria generally support low populations of fungi and actinomycetes, *P. omnivorum* may be an exception. Alternatively, it may be that saprophytic fungi or actinomycetes are more antagonistic to *P. omnivorum* than to bacteria (Chapter 9), and only when they were suppressed could *P. omnivorum* cause severe disease. As stated by Eaton and Rigler, "The fact that the total bacterial number was decreased [with a high root-bark carbohydrate content] suggests that the organism or organisms responsible for checking or destroying *Phymatotrichum omnivorum* may also have been unfavorable to other groups."

In contrast to *P. omnivorum, Gaeumannomyces graminis* attacks Gramineae, but not dicotyledonous plants in field soil. Müller-Kögler (1938) found that 73 dicotyledonous plants were susceptible in sterile soil, producing slight cortical infections. Perithecia were even formed on *Chrysanthemum segetum, Chenopodium album,* and *Convolvulus arvensis.* Rape and flax were also attacked in nonsterile soil, but more slowly than in sterile soil. Peas were found by Zogg (1969) to develop extensive black root necrosis from *G. graminis* in sterile soil, but remained healthy in the presence of the rhizosphere flora.

The genetic control by the host of its rhizosphere microflora has been studied by Atkinson et al. (1974), Larson and Atkinson (1970), Neal (1971), and Neal et al. (1970, 1973), using wheat varieties resistant or susceptible to *Cochliobolus (Helminthosporium) sativus* root rot. With one set of varieties, kinds and numbers of microorganisms and percentages of microorganisms antagonistic to the pathogen on agar plates were studied in the rhizospheres of three spring wheats: Apex (relatively resistant to *C. sativus*), S-615 (a susceptible variety), and S-A5B (identical to S-615 except resistant to *C. sativus* because of substitution of the chromosome pair 5B from Apex). As would be expected, each variety supported a unique rhizosphere flora made up of differing percentages of ammonifiers, nitrate reducers, starch hydrolyzers, sporeformers, and organisms requiring various growth factors and amino acids. The total bacterial count was about twofold greater for S-615 than for the two resistant varieties (Table 8.1). It was noteworthy, however, that about 20% of the bacteria in rhizospheres of two resistant varieties were antibiotic to *C. sativus,* whereas none were in those of susceptible plants. It would have been interesting to have grown the resistant lines under sterile conditions with *C. sativus* to see whether resistance would, in fact, disappear.

With a different set of varieties involving another disomic substitution with chromosome 5B, the role of antibiotic rhizosphere microorganisms in root-rot incidence was not evident (Atkinson et al., 1974, in Bruehl, 1974; Neal et al., 1973). In this study, the resistant variety Cadet again had a

TABLE 8.1
Microorganisms in the rhizosphere of spring wheats resistant and susceptible to root rot caused by *Cochliobolus sativus*. (Data from Neal et al., 1970.)

Rhizosphere type	Total organisms/g soil		Percentage of bacteria antagonistic to *Cochliobolus sativus*
	Bacteria $\times 10^6$	Fungi $\times 10^3$	
Resistant Apex	251	343	20
Susceptible S-615	577	124	0
S-615 on which resistance has been conferred by substitution of Apex chromosome 5B	266	81	19
Nonrhizosphere	45	119	7

lower total rhizosphere bacterial population than did susceptible Rescue or C-R5B (Cadet rendered susceptible by substitution of its chromosome 5B with 5B from Rescue). Instead of more antagonists in the rhizosphere of resistant Cadet, there were greater numbers of bacteria capable of producing cellulases, pectinases, and amylases in the rhizosphere of the susceptible Rescue. Moreover, transfer of chromosome 5B to an otherwise Cadet genotype changed its rhizosphere microflora. Although the role, in these studies, of the various bacterial groups is uncertain in the occurrence of wheat root rot, it is clear that the quality and quantity of the rhizosphere microflora is under host genetic control. This work thus opens a new aspect of plant breeding—specific alteration of the host genotype to selectively manipulate the rhizosphere flora and fauna, and thus perhaps influence host susceptibility to disease.

The question arises whether loss or acquisition of resistance of a variety to root pathogens occurs frequently with changes in the rhizosphere flora. If so, a variety planted in an area where the flora necessary for the protective rhizosphere is lacking could be erroneously discarded as being susceptible. The same variety planted in an area where components of the necessary rhizosphere microflora were present would be considered resistant.

The extent to which antagonists occur naturally on seed used for planting could also be important, as shown by Tveit and Moore for resistance of oats to *Helminthosporium victoriae* (Chapter 3). Thus, Ledingham et al. (1949) showed that severity of *H. sativum* on wheat seedlings is greater for surface-sterilized than nonsurface-sterilized seeds. Bacteria were the principal organisms involved, and when reapplied to the surface-sterilized seeds, they reduced the severity of disease. Varieties differed in abundance and effectiveness of their seed-surface flora, and correspondingly in their susceptibility to *H. sativum*. This effect could explain why Harding (1971), in a study of resistance to common root rot in wheat, obtained a poor correlation between varietal reaction in growth chambers and that in the field. It could also explain why Cohen et al. (1969), in a genetic study of resistance in barley to common root rot, found that resistance was lost when glasshouse-produced seed was used. In both examples, the proper antagonists may have been lacking. Ledingham et al. (1949) suggested that seed of test varieties should be surface-sterilized for a more accurate test of their "basic resistance."

These observations also emphasize the necessity of testing varieties at several field locations rather than at only one. Multilocation testing has the added value of preventing release of a variety resistant in the Experiment Station soil, but not in some grower's field. Perhaps the usefulness of such varieties could be extended with an application of the necessary organisms on the seed or soil (Chapter 4). Breeders might also be able to select for specific rhizosphere floras.

CROPPING HISTORY AND THE MICROBIOLOGICAL BALANCE OF SOIL

Different degrees of biological buffering against plant pathogens can be achieved through the host in different soils. The enforced monoculture of forests or orchards may have as important an effect in favoring a pathogen as does an annual crop. Each crop species selects for a specific saprophytic and parasitic microbiota, especially by the kind of nutrients supplied by the crop. Monoculture of a crop in a given field will favor the perpetuation of a specific and eventually stable flora and fauna. Cultivation of a different crop each season, on the other hand, will maintain the microbiota in a greater state of flux, and result in different dominant organisms each year. As stated by Zogg (1969), "During the vegetation period [of a single crop species] a quite typical association of microorganisms is built up in the soil, but is again conversed in another direction by cultivation of another crop plant species the following year." Crop rotations may in this way result in a type of microbial rotation.

Crop Sequences and Combinations Unfavorable to the Pathogen

The microbial makeup of any given field at any given time essentially is the product of all previous crops and their sequence in that field, much as the genetic makeup of an individual is the product of his ancestors. The immediate parent (or crop species) may make the greatest contribution in terms of genetic traits (or active microorganisms), but the input of all previous parents (or crops) is also manifest, the magnitude being less with each more distant generation. Some types of crops will undoubtedly have more effect than others.

No two fields will ever possess identical microbiota, although they may be fairly uniform within a given field. Can the microbiological makeup of a field be controlled by cropping so that a pathogen will not establish, or only with difficulty? This kind of situation may produce pathogen-inimical soils (Chapter 4).

A single crop not susceptible to *Gaeumannomyces graminis* inserted into an otherwise wheat monoculture will reverse the decline of take-all, and the disease may again be severe in the next wheat crop (Shipton, 1969; Vojinović, 1972). Possibly the flora and fauna favored by cereals, and antagonistic to *G. graminis,* are replaced by microorganisms less antagonistic to it. Vojinović found that antagonists of *G. graminis* functioned as secondary invaders of take-all lesions, but became less effective if the wheat or barley monoculture was interrupted by a *Gaeumannomyces*-resistant crop for 1 year. Apparently, disease in subsequent wheat was more severe if the monoculture was broken by sugar beets rather than by corn. Moreover, total number of microorganisms antagonistic to *G. graminis,* based on tests of random isolates on agar media, was lowest after sugar beet, intermediate after corn, vetch, or fallow, and highest after wheat.

Some host varieties probably increase suppressiveness more than others. This is suggested by the work of Rombouts (1953) on fusarium wilt of banana, discussed earlier in this chapter. Perhaps varieties could be bred for greater efficiency in bringing about disease suppression. Alternatively, antagonism may be enhanced by specific combinations of varieties grown in a particular sequence. As suggested by Kommedahl (in Baker and Snyder, 1965) for control of root rot of wheat caused by *Fusarium* and *Helminthosporium,* "varietial sequence in a given crop may be as effective as crop sequence."

Although disease control through crop rotation is commonly viewed as a passive process (disinfestation by attrition during absence of the host), some of the benefits may involve stimulation of organisms that partially or totally suppress the pathogen during its parasitic or saprophytic phases. Peas may in this way increase antagonism in soil to *G. graminis* (Zogg, 1969). Zogg suggested that wheat after peas will, therefore, show little or no injury from *G. graminis.* Schroth and Hendrix (1962) observed that population increase of *Fusarium solani* f. sp. *phaseoli* is favored by tomato, corn, and lettuce, but not onion. Goss and Afanasiev (1938) found fusarium wilt of potato more severe after corn than sugar beets.

Crop sequence to manipulate the microbiological balance probably will prove more difficult in areas of established monoculture where microbiota are stabilized than in virgin desert areas being brought into cultivation. The Palouse region of eastern Washington and northern Idaho, for example, has been planted to wheat and peas with occasional barley, oats, or fallow for several decades, and attained microbiological stability and complexity years ago. Fumigation or some comparable shock (Chapter 3), because it drastically upsets the microbiological balance, would be a helpful preliminary to accelerate the desired microbiological shift resulting from a change of crop. An abrupt change to crops other than cereals or peas will also accomplish a shift, but only after many years. The long-term influence of wheat and peas on the microbiota, therefore, will not be easily overcome.

The vast areas of desert soils used for irrigated agriculture are ideally suited for soil microbiological manipulations (Chapter 7), as are lands fumigated with methyl bromide each year. Practically no research has been conducted on crop sequence of new lands as a means of biologically buffering against the establishment of plant pathogens. Instead, man has brought desert areas directly into irrigated monoculture (potato, sugar beet, or wheat) only to find it necessary after 2 or 3 years to move on to still newer land because that just reclaimed contains some pathogen. Perhaps cultivation of certain plants for 2 or 3 years, solely for their influence on the soil microflora, would make possible long-term disease-free production of commercial crops. Through plant introduction from specific habitats elsewhere, it may be possible to find antagonist-promoting plants for use as the first crop in new or fumigated lands. Crops may be bred specifically for this effect. In any event, research should be undertaken to determine how new lands, like a new automobile, may be broken in gradually.

Crop Sequences That Favor the Pathogen

The host can also increase favorableness of the soil biotic environment to root pathogens by suppressing antagonists, or by selectively favoring growth of the pathogen. Remove the host and the pathogen may die. Stover (in Holton et al., 1959) has described this effect in relation to bananas and *Fusarium oxysporum* f. sp. *cubense* in long-life, short-life, and flood-fallowed soils. The long-life soil was most suppressive to *Fusarium*, but cultivation of the banana plant made possible multiplication of the fungus and the disease. When the banana plant died or was removed, the *Fusarium* population declined. In short-life soils, cultivation of banana also increased the *Fusarium* population, but since the microflora was less suppressive, the *Fusarium* multiplied more rapidly, and the planting was quickly destroyed. With plant death or removal, the *Fusarium* population declined more gradually. The suppressive flora was largely eliminated in flood-fallowed soil; surviving contaminants or inoculum established and multiplied, particularly when the host was reestablished, even in soil that was highly suppressive in the virgin state.

Nematode-transmitted plant viruses provide further examples of crop sequences favorable to the pathogen (Cadman, in Baker and Snyder, 1965). Species of ectoparasitic stylet-feeding *Xiphinema, Longidorus,* and *Trichodorus* can acquire and transmit a number of viruses by feeding on roots of infected plants. Since these viruses are not transmitted to nematode progeny through the eggs and the nematodes are short-lived, elimination of source hosts will break the virus cycles. These viruses unfortunately are carried through a high percentage of seeds of many hosts, and germinated weed seeds thus provide the principal source of infection of nematodes and crop plants. In addition, grape roots may remain alive in soil for extended periods after the planting is removed, so that neither fanleaf virus nor its nematode vector are quickly eliminated. Use of herbicides and clean cultivation to eliminate weeds may provide an effective means of biological control of virus. *Olpidium*-transmitted viruses may be diminished in a soil by rotation with crops not susceptible to the virus (for example, sugar beets for lettuce big-vein virus). Breeding for host resistance to or tolerance of big-vein of lettuce appears promising. The prospect for success with other "soilborne" viruses is favorable, since resistance to either the virus or the nematode or fungus vector would be effective.

The use of herbicides for weed control, however, because of their selective effect on the weeds present, may increase the occurrence of other viruses. Compositae, especially sow thistle, have recently increased in the Salinas Valley, California, and this has been reflected in a greater incidence of the beet yellow-stunt and sow thistle yellow-vein viruses on lettuce. The control of rabbits in Australia by myxomatosis has also selectively affected the weed populations. Sow thistle has become more prevalent, and this virus reservoir near lettuce fields in Victoria is thought to have led to an increase in lettuce necrotic yellows there (Matthews, 1970).

PLANT RESIDUES

The role of crop residues in the biological control of plant pathogens has been recognized, studied, and made use of for decades. The effects achieved with proper management of the residues can be spectacular. Unfortunately, they can also be unpredictable and undependable, as has been shown repeatedly by research such as that of the Western Regional Project W-38 (Cook and Watson, 1969). This variability is caused by second- and third-order interactions among soil microorganisms and between them and the pathogen, each factor of which is independently affected by the environment. **The less direct the control, the more it varies in relation to fluctuations in environment** (Chapters 4 and 9). Nevertheless, this postmortem effect of the host is an extremely important one, and should attract more attention as man increases his effort to dispose of agricultural wastes in the most beneficial manner.

The role of plant residues is basically one of stimulating microorganisms antagonistic to plant pathogens. This antagonism may involve rather specific forms of antibiosis, such as that of soybean green manure in the suppression of potato scab (Weinhold and Bowman, 1968). More commonly, it involves nonspecific antagonism, some of which undoubtedly is antibiosis and hyperparasitism, but most of which is probably competition. This agrees with the common observation that control with amendments does not depend on stimulation of specific species of microorganisms, but rather on high numbers of organisms, especially bacteria that use carbon, nitrogen, and oxygen as rapidly as available. This general effect explains why, in control of *Phymatotrichum omnivorum* with organic amendments, the important feature is not what is used, but that some amendment is used. Rich substrates seem to work best, probably because they support the highest number of organisms. The conclusion of Clark (1942) that the environment best for maximum microbial activity after adding amendments was also best for control of *P. omnivorum* supports this competition theory.

Johnson (1953) found that where sclerotia of *Sclerotium rolfsii* were destroyed with amendments, bacterial numbers were proportional to the amount of alfalfa meal added, and to the rate of destruction of sclerotia. He showed that all sclerotia were destroyed within 6 weeks if 4% of meal was added, 2 months with 2% of meal, and 3 months with 1 or 0.4% of meal. The work on volatiles from alfalfa (Gilbert and Linderman, 1971) may explain this result (Chapter 2).

The degree of control of *Rhizoctonia solani* on beans from fresh-green or dry-plant materials added to soil was greatest with materials that stimulated the highest total number of organisms and that decomposed the fastest (Papavizas and Davey, 1960). Control was poorest with materials that stimulated the lowest total microbial activity and were slowest to decompose. **The greater the variety and number of microorganisms supported by the organic amendments, the greater the probability of suppressing the pathogen.** As previously outlined (Chapters 6 and 7), intensified microbial activities will, in effect, steepen the nutrient gradient outward from an exuding plant part and thus lessen the chances for growth of a pathogen. Bacteria and actinomycetes antagonistic to the fungus pathogen may have the added effect of increasing exosmosis from mycelia, reducing oxygen, and favoring other processes that result in endolysis (Chapters 2 and 6). The pathogen may in such situations be swamped by the competing microflora.

Plant residues also control root rots and lower the populations of pathogens by propagule germination-lysis. Two different mechanisms seem to be involved. In some cases, germination apparently is triggered by something other than a nutritional stimulus. This could be the case for ascorbic acid, shown to stimulate germination-lysis of conidia of *Helminthosporium sativum* (Chinn and Ledingham, 1957), and for volatile compounds released from decomposing organic matter. The germling consumes the endogenous reserves of the spore and dies. Most residues supply nutrients that stimulate germination and sustain growth, but because the nutrients are quickly utilized by other microorganisms, the germlings starve. Death results from endolysis primarily through starvation, unless new propagules can be formed (Lockwood, in Gray and Parkinson, 1968), or unless they have the ability to germinate repeatedly, which *H. sativum* apparently does (Hsu and Lockwood, 1971).

Germination-lysis was first reported by Mitchell et al. (1941), following their observations on the elimination of *Phymatotrichum omnivorum* sclerotia after organic amendments. It was also described by Chinn et al. (1953) to explain the control of *H. sativum* by soybean residues. The soybean meal stimulated conidial germination, but germ tubes subsequently lysed, and the propagules were killed.

The germination and subsequent lysis of chlamydospores and germlings of *Fusarium solani* f. sp. *phaseoli* in nutrient-amended soil can also be explained in terms of nutrition (Cook and Snyder, 1965). Glucose alone stimulated chlamydospore germination to the extent that nitrogen was available in the soil solution, but lysis occurred very slowly with this substance, replacement chlamydospores were formed, and the population remained steady or increased slightly. The addition of inorganic nitrogen with the glucose stimulated greater germination, but also more lysis than with glucose alone. If the glucose was supplemented with organic nitrogen (asparagine), lysis was still more rapid, and the population declined slightly. Finally, glucose + asparagine + yeast extract stimulated the *Fusarium,* but greatly intensified the microbial activity, and lysis occurred the most rapidly of the four treatments. High carbon substrates may stimulate chlamydospore germination, but in the absence of available nitrogen, neither the pathogen nor the competitors can sustain normal growth, and the fungus slowly returns to the dormant stage. If nitrogen does not become limiting, available carbon probably does, and dormancy or death soon follows. *When carbon is ample but nitrogen depleted, growth is slow and autolysis does not occur; when nitrogen is ample but carbon depleted, growth stops and autolysis occurs.*

Papavizas et al. (1968) obtained similar results, and showed that with many plant materials, lysis might not be sufficiently rapid to prevent new chlamydospore formation, and hence propagule numbers increased. It is commonly observed that this fungus multiplies when certain plant residues are added, because chlamydospore formation is not prevented by lysis.

Most experimental control of pathogens with plant materials depends critically on the time of planting relative to the time the materials are added. Control is generally best if planting is done within a few days after amendment, when germination-lysis is complete and microbial activity and antagonism is still intense. Planting at the time of amendment can result in increased disease, because of the increased propagule germination. Planting too long after amendment can also result in increased disease, because microbial activity has subsided and in the meantime the pathogen has multiplied. Disease control by amendments is therefore temporary unless it results in a drastic reduction in inoculum density, as apparently happens for conidia of *Helminthosporium sativum* in soil amended with soybean meal (Chinn et al., 1953).

A suppressive effect against *Fusarium solani* f. sp. *phaseoli* that lasts for about 28 days results from adding spent coffee grounds to soil (Adams et al., 1968a). The germination-lysis process was completed within the first 48 hours after amendment, and control was optimal at 7–14 days after amendment, but ultimately populations of the pathogen increased. Control of this same pathogen with materials of high C:N ratios (barley straw, sawdust, cellulose, or glucose) is also optimal a few days after the amendment (Chapter 4), when nitrogen available for pathogen growth is maximally immobilized, but subsides after a few weeks, and disease again may be severe (Snyder et al., 1959; Maurer and Baker, 1965). The period of control varies with the readiness with which the material is decomposed. Control from a single application of glucose may last only a few days, whereas control with mature barley straw may last a month.

The propagule germination and subsequent lysis triggered by volatiles from decomposing plant residues, first reported by Menzies and Gilbert (1967), have considerable potential in biological control, because it may lower the population of the pathogen, and also because it apparently changes the biological balance qualitatively. The volatile substances included acetaldehyde, isobutyraldehyde, isovaleraldehyde, methanol, and ethanol, which stimulated microbial growth at low concentrations but inhibited growth at higher concentrations. Some fungi, actinomycetes, and bacteria may increase, whereas others may decrease. Apparently the vapors trigger propagule germination and abortive growth of some organisms, including some pathogenic fungi. The killed organisms are replaced by others that grow on nutrients released by their dead predecessors, resulting in qualitative and quantitative microbiological changes. Legumes, and especially alfalfa hay, are excellent sources of the volatile substances. Does this perhaps account for the effectiveness and popularity of legumes as organic amendments for disease control? Fries (1973) has recently reviewed the effect of volatiles on fungi. The balance between ethylene and soil nutrients has an important role in fungistasis (Chapter 6).

Root-disease control by volatiles may also depend on time of planting relative to time of amendment. Thus, control of aphanomyces root rot of peas by crucifer amendments (cabbage, kale, mustard, turnip), which may result from isothiocyanates and sulfides normally released from crucifer tissues during decomposition, persisted for no longer than about 15 weeks (Lewis and Papavizas, 1971). During this period, formation, motility, and germination of zoospores of *Aphanomyces euteiches* were inhibited. Once tissue decomposition reached a certain stage, however, growth and activity by this fungus was again normal.

There are numerous examples in which crop residues may also yield, on microbial decomposition, materials toxic to crop plants (Patrick and Toussoun, in Baker and Snyder, 1965). For example, corky root rot of lettuce is caused by *m*-ethyl-aminobenzoic acid and other compounds formed during decomposition of fresh lettuce residue (Amin and Sequeira, 1966).

THE HOST AS A RESERVOIR OF INOCULUM

When the host functions as a reservoir of pathogen inoculum, it may be both an impediment to and an opportunity for application of biological control. Pathogens that depend on host tissue for survival may be controlled by accelerated decomposition of the plant remains (Chapter 6).

Pathogens internally seedborne are assured of perpetuation within their hosts and are unlikely to be controlled biologically other than through host resistance. Because they can persist indefinitely within their host, they offer little or no opportunity for suppression by external antagonists. One of the few opportunities for biological control would be at the time of seed infection by airborne inoculum—for example, spores of *Ustilago nuda* as they land in barley florets. Many pathogens could be more easily controlled except for their ability to be transmitted through seed (Baker,

1972). This is particularly true for seedborne viruses, which persist systemically in the plant and have a narrow host range. Heat treatment of such fungus-, bacteria-, and nematode-infected seed to free it of the pathogens (Baker, 1962, 1969, 1972) might increase the effectiveness of biological control in some cases.

Weeds, especially perennial species, are important reservoirs of virus inoculum for diseases in cultivated plants. Epiphytotics of curly top in tomatoes and sugar beets in the western United States originate from leafhopper-borne virus acquired from wild hosts on nearby noncultivated land (Bennett, 1971; Duffus, 1971). Outbreaks of wheat streak mosaic originate from virus harbored in wild grasses and volunteer wheat along roads, waterways, and fence lines, and transmitted to cultivated wheat by eriophyid mites. Cadman (in Baker and Snyder, 1965) suggested that soilborne plant viruses are pathogens primarily of wild plants. When man eliminates the wild hosts of viruses, he is accomplishing biological control of virus diseases of his crops.

Buddenhagen (in Baker and Snyder, 1965) outlined how wild hosts harbor inoculum of *Pseudomonas solanacearum,* which subsequently attacks banana. The virgin jungle in Central America contains species of the related *Heliconia* infected with *P. solanacearum.* Apparently the predominant strain on *Heliconia* is only weakly pathogenic to banana, but a more aggressive strain exists within the population and readily increases with banana cultivation. The strain virulent to banana is not so virulent to *Heliconia,* nor can it compete effectively with the common wild strain of *P. solanacearum,* but is maintained nevertheless in low populations throughout the jungle and only needs the banana to flourish.

Wild hosts are also important reservoirs for plant-pathogenic fungi, as *Fomes annosus* in native stands of oak and pine subsequently cleared for pulpwood plantations in southeastern United States, or *Gaeumannomyces graminis* on wild grasses in virgin areas subsequently brought into wheat production. The unsatisfactory control of *Verticillium albo-atrum* with rotations lasting 8–12 years has been attributed to maintenance of this fungus on roots of nonhost plants, including weeds (Evans et al., 1967). The wilt fusaria also infect nonhost plants, and in this way persist for long periods, once established in a field. Watson (1970) and Van der Plank (1968) have suggested that the lasting control of the wilt fusaria by monogenic resistance may relate to the fact that such pathogens must compete with soil saprophytes, which slow or prevent population shifts such as occur in rusts. If so, wild hosts of such a *Fusarium* may enable it to survive as a forma specialis that would not carry over on the more usual host crop cultivar. This would make weed control a type of biological control of a plant pathogen.

Leach (1939) showed that sudden outbreaks of root rot of tea on newly cleared land in Africa resulted from inoculum harbored on native tree species. Many of these species were resistant to the fungus, *Armillaria mellea,* and had only restricted lesions on their roots. When these trees were felled, however, resistance ceased, the fungus colonized the remainder of the root system, and massive inoculum resulted. Leach suggested a system of biological control whereby the native trees were ringbarked in advance of felling. This apparently caused reduction in root carbohydrate reserves, permitted root colonization by saprophytic soil fungi, and restricted *A. mellea* to the localized root areas parasitically colonized earlier. *Armillaria* would not colonize tissues occupied by other fungi. This is one of the first observations on possession as an important mechanism for survival of fungi in tissues. The phenomenon aids survival of pathogens within the host (Chapters 1 and 6), but as reported by Leach, it can also work to advantage in biological control. Ringbarking has also been used in Ceylon to control the spread of *Ustulina deusta* from shade trees to adjacent tea. This prevents extension of dormant lesions on the shade-tree roots and thus limits the opportunity for saprophytic increase of the fungus on the felled tree. According to Shanmuganathan (in Toussoun et al., 1970), "There is no satisfactory alternative at present to ringbarking shade trees in order to eliminate infection by *Ustulina.*"

Armillaria mellea can also invade uncolonized dead wood buried in soil. Wood prunings in African tea plantations were sometimes buried during culture, and armillaria root rot of tea was consequently severe. Leach (1939) showed that leaving prunings on the soil surface for a while permitted their colonization by airborne saprophytes, and that when subsequently buried, the wood was not colonized by *A. mellea.* This practice has since been suggested by Cook (1970) to control saprophytic increase of *Fusarium roseum* f. sp. *cerealis* 'Culmorum' in wheat straw; straw allowed to mold slightly on the soil surface are not colonized by Culmorum when subsequently buried, whereas clean bright straw is readily colonized by this fungus saprophytically.

The general resistance of living plant tissue when it is plowed under prevents invasion by saprophytes. It is, therefore, selectively invaded by pathogens such as *Pythium, Phytophthora,* and *Rhizoctonia.* Fresh plant refuse of peanuts is also selectively invaded by *Sclerotium rolfsii* when turned under. *Rosellinia pepo* similarly invades fallen leaves in areas of high rainfall in the tropics, and then attacks the trunks of cacao from this food base (Waterston, 1941).

Prevention of colonization of the host remains by pathogens is thus one approach to biological control. Another is to eliminate the pathogen from the dead host residue after its establishment. This generally involves replacement of the pathogen with saprophytes, but is not readily accomplished because of the advantages of prior possession by the pathogen. Many pathogen-infested host tissues are known to decompose more slowly than noninfested tissues, suggesting a mechanism for exclusion of other organisms. *Armillaria*-infested wood, for example, is more resistant to termites than is noninfested wood, which, under the same conditions, is readily destroyed by these insects. Wheat straw infested with *Cercosporella herpotrichoides* also decomposes more slowly than noninfected straw (Macer, 1961). To supplant a pathogen with saprophytes will generally require a shift in environment that will suppress metabolism of the pathogen relatively more than that of saprophytic colonists (Chapter 9). This has been accomplished experimentally for *Cephalosporium gramineum* in wheat straw by increasing soil pH to the alkaline range, or by allowing the soil to dry to a water potential below −150 bars, which apparently limits metabolism by *C. gramineum* and allows colonization by saprophytes (Bruehl and Lai, 1968b; Chapter 9). It has been accomplished for *Armillaria mellea* and certain other pathogens in roots in soil by treatment with carbon disulfide or some other fumigant which apparently weakens the pathogen and encourages its replacement by soil saprophytes (Chapter 4). It has also been accomplished by Garrett (1940) for *Gaeumannomyces graminis* in wheat straw by allowing the soil to become nitrogen-deficient, thereby reducing nitrogen available for metabolism by *G. graminis* in the straw, which then permitted entry by saprophytes. Because wheat varieties differ in the amount of nitrogen retained in straw, perhaps biological control of *G. graminis* during the saprophytic phase could be improved by breeding for low nitrogen content of the straw. This would be a less important feature of cereal roots, which are already relatively quick to decompose, leaving the occupant to perish.

The exhaustion of food reserves remains the best method of biological elimination of many pathogens in host tissues (Chapter 6). Shanmuganathan (in Toussoun et al., 1970) has reviewed this method as used in Ceylon for control of *Poria hypolateritia* on tea. Although fumigation with D-D kills the fungus to a 46-cm depth, it persists at depths up to 76 cm. If Guatemala grass is planted after soil treatment with D-D, the fungus exhausts its root food supply within 2 years, and tea can again be planted. This is similar to the control outlined by Fox (in Baker and Snyder, 1965) for root diseases of rubber in Malaya with creeping legumes grown among the rubber trees (Chapter 4) to encourage microbial antagonism and the decomposition of infected roots, exhausting the pathogen's food supply. It has also been one of the main purposes of crop rotation (Chapter 4) to provide time for decomposition of the residues of previous crops that contain plant pathogens.

HOST RESISTANCE

Man has long been interested in the phenomenon of host resistance. He has learned to transfer resistance from one variety to another to achieve disease-free production of high-quality food and fiber, and has used it to advantage in trap and inhibitory plants (Chapter 4). In spite of these applications, however, the phenomenon itself is still largely unexplained.

Some forms of host resistance result from mechanical barriers to microorganisms, such as thick cell walls or cuticle. Others are associated with: suberin and lignin; fungitoxic phenolics such as chlorogenic acid, caffeic acid, catechol, or tannins; glucosides that give rise to phenolics; quinones and other toxic oxidative products of phenolics. Plants may also synthesize fungitoxic compounds (phytoalexins) in response to attempted invasion by microorganisms. Indeed, it seems clear that most resistance is a response process involving the entire host-cell protoplast, and includes increased respiration, protein and RNA synthesis, manufacture of proteins different from those of healthy cells, changes in enzyme systems, and shifts in metabolic pathways (Stahmann, 1967). These processes collectively are associated with exclusion or inhibition of the pathogen, but we are still uncertain which are cause and which are effect. Nevertheless, resistance usually is a dynamic rather than a passive process, which explains why dead plants have no resistance to microbial invasion, but living plants do.

A high carbohydrate content improves resistance in cotton to phymatotrichum root rot (this chapter), and in winter wheat to snow molds (Chapter 9). Other examples are given in Chapter 2.

We should not expect any one phenomenon to explain all host resistance or all successful parasitism. Each host-parasite interaction is undoubtedly unique in some way. What constitutes a physiological barrier to a parasite in one potential host-parasite interaction may be irrelevant in another.

C. H. Beckman and associates studied the nature of resistance in different plants to saprophytic and would-be vascular parasitic organisms, with emphasis on resistance of certain banana varieties to *Fusarium oxysporum* f. sp. *cubense,* and tomato varieties to *F. oxysporum* f. sp. *lycopersici.* Resistance relates to the development of vascular occlusions by the plant, which physically seal off the intruding organism (Beckman and Zaroogian, 1967). The occlusions include gelation of perforated plates and end and side walls of vessels, and formation of tyloses within them (Beckman and Halmos, 1962). In a susceptible interaction of banana and *F. oxysporum* f. sp. *cubense,* gel formation is less extensive, and that which does form is temporary and breaks down quickly. Some partially dissolved gels shear under the transpirational water pull. The gels also break down in the resistant plants, but timing is such that the more permanent occlusion by tyloses occurs first. A single dominant gene for resistance controlled the rapid formation of tyloses in tomato plants inoculated with *F. oxysporum* f. sp. *lycopersici,* whereas in a near-isogenic line lacking this resistance gene, tylose formation was retarded beginning about 2 days after inoculation (Beckman et al., 1972). It was suggested that the susceptible line was sensitive to growth-inhibiting substances produced by the pathogen, which then limited formation of tyloses and permitted the pathogen to spread through the plant.

The gel consists of pectins, calcium pectates, hemicelluloses, and proteins (Beckman and Zaroogian, 1967). Gels and tyloses apparently are of host origin, but form in response to microorganisms. They did not form, or did so rarely, in banana roots inoculated with sterile suspensions of tracer-particles, but formed extensively when microconidia of *F. oxysporum* f. sp. *cubense* or spores of random rhizosphere microorganisms were mixed with the particles (Beckman, 1966). Particles passed through vessels approximately 20–25 mm in 5 minutes, 48 hours after aseptic banana roots were cut and incubated at 21°C, but moved almost no distance in 24 hours if the roots were cut in a cell suspension of root microorganisms. Microorganisms normally present on the root, which presumably challenge the host plant by attempted invasion, are in this way localized in the plant within 1–2 days and prevented from systemic distribution. Vascular occlusions may, therefore, serve as a general host-protective response (Beckman, 1968). They are also of possible fundamental significance as a basis for resistance to wilt pathogens stimulated by certain nonpathogens (Chapter 7 and below) or susceptibility induced by root-knot nematodes (Chapter 6).

An earlier hypothesis was that pectic enzymes of a microorganism in the vascular system caused cleavage of host pectins, plugging of the vascular system, and hence wilting. The studies of Beckman et al. (1962) leave little doubt that pectin accumulation in vascular elements is the product of resistant rather than susceptible response. Fusarium wilts of plants are generally much more severe at 27–28°C than at 21° or 34°C. Occlusions by gel and tyloses in vessels of the wilt-susceptible Gros Michel banana were adequate to cause some restriction of pathogen sporulation and upward spread at 21°C (where some disease occurred) and this was especially rapid and continuous at 34°C (where little or no disease occurred). At 27°C, however, the gel disappeared most rapidly, tylose formation was delayed, and the pathogen spread upward unchecked. When the pathogen was systemic, having moved in advance of the protective response, the delayed vascular occlusions eventually became systemic as well, and wilt ensued.

Genetics and Evolution of Host Resistance

Types of resistance have been classified into those controlled by a single gene (monogenic) and those controlled by several genes (polygenic) (Van der Plank, 1958). Monogenic resistance is often sufficiently effective to qualify as immunity. It is stable under a wide range of environmental conditions, but is usually specific for a certain race or virulence gene of the pathogen. In contrast, polygenic resistance, which is more sensitive to environmental fluctuations, does not result in immunity, but is more uniformly effective against variants of the pathogen. Moreover, it is primarily responsible for whatever resistance exists in host plants to the less specialized parasites such as *F. roseum, F. nivale, F. solani,* and certain *Phytophthora* spp. Host species show considerable variation in expression of polygenic resistance, providing attractive opportunities for genetic studies and varietal improvement.

In addition to the clear-cut examples of monogenic and polygenic resistance in plant species, there is good evidence that all plants have a general resistance to pathogens that cannot yet be defined genetically, but which is controlled by the collective action of multiple physiological factors associated with active host metabolism and growth. As suggested by Van der Plank (1968), this is a form of polygenic resistance that "is governed by genes that are not special resistance genes, but are simply genes that occur ordinarily in healthy plants and regulate ordinary processes." Garrett (1970) touched on the same theory in stating "a decline in root resistance to infection is likely to follow any occurrence that reduces vigour of the plant as a whole." The existence of a general resistance to parasites in all plants helps explain: why pathogenicity is the exception rather than the rule; why, as pointed out by Van der Plank, many plant species have resistance to parasites to which they have never been exposed; why the great majority of microorganisms is relegated to a strictly saprophytic or weakly parasitic life. The theory is also consistent with the observation that weaker and nonspecialized parasites are troublesome primarily to plants exposed to marginal growing conditions (Chapter 1). Of all the forms of resistance, this general form is subject to greatest modification by environment. As implied by Garrett, a suppression of the ordinary physiological processes of the host plant as a whole increases its vulnerability to disease. Any practice that improves growth and biological efficiency of the plant also improves the general or broadly horizontal resistance of the plant, and thus qualifies as biological control.

A general (indirect) resistance to pathogens probably developed during early evolution of plants, and persists in their genetic constitution to this day. These plants would certainly have needed resistance for protection

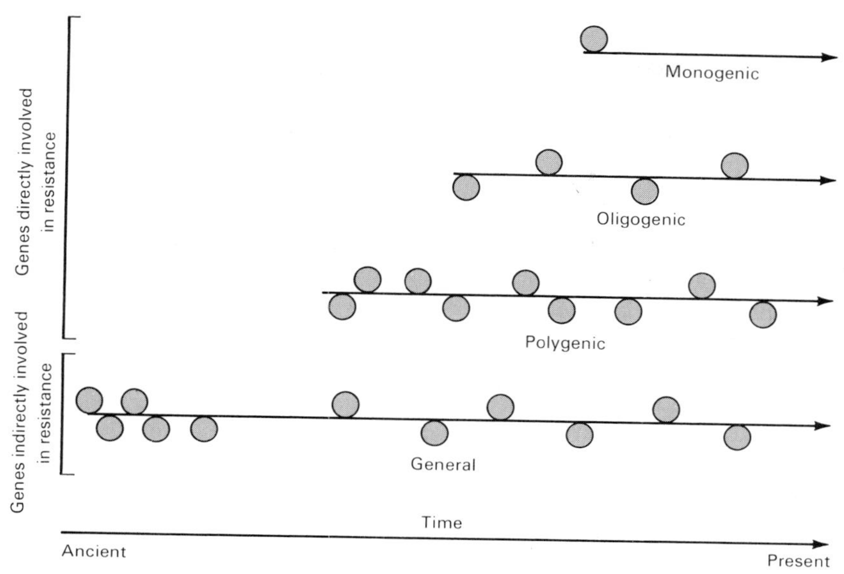

FIGURE 8.5
Diagram illustrating a possible development of direct (monogenic, oligogenic, and polygenic) resistance to plant pathogens. General resistance, of possibly even greater antiquity, provides a background against which this direct resistance may have evolved.

against parasites with which they grew. As higher plants became more complex, modifications in the resistance probably occurred. This general resistance was modified as plant families, genera, and species became more specialized. Monogenic resistance, as seen today, would then be a more recent evolutionary development than polygenic resistance within any given plant taxon. The complexity of polygenic resistance, in turn, would indicate antiquity within the taxon; it may initially have been monogenic but has accumulated additional genes directly affecting resistance (Fig. 8.5). Indeed, it seems probable that **the greater the complexity of genetic factors that condition resistance, the longer the evolutionary development of that resistance and the greater its spectrum of effectiveness against pathogens, including unspecialized types.** Conversely, the simpler the genetic control of resistance, the more recent its evolutionary development, and the more specialized the parasite that it controls.

Although we commonly consider evolution within plant populations as having been, through selection pressure, toward greater resistance, the reverse may also have occurred. Such events could account for the number of parasites restricted to one species, genus, or family of plants. The genetic loss of a portion of the general (indirect) resistance could have resulted in susceptibility to some pathogen. The probable rarity of such an event could account for the scarcity of compatible host-parasite interactions. Presumably, the origin of a susceptible line would then be followed by a natural selection for more resistant individuals.

Resistance Stimulated by Nonpathogens

There are several lines of evidence to suggest that some microorganisms function as antagonists of pathogens by stimulating a resistance in the host (Chapter 7). The production of phytoalexins and related compounds has recently been reviewed (Ingham, 1972).

Hadwiger and Schwochau (1969) have shown that many microbial metabolites, including actinomycin D, mitomycin C, phytoactin B, gliotoxin, cycloheximide, puromycin, chloramphenicol, and culture filtrates of several different fungi, are potent inducers of physiological processes associated with the resistance response within pea cells, including production of the phytoalexin pisatin. They suggested that in interactions of the gene-for-gene type, the ability of a host cell to alter or disorganize its own metabolism, to the detriment of the parasite (hypersensitive reaction), is present but repressed in uninfected tissues of resistant varieties. They further suggested that susceptible varieties lack the genetic potential for this reaction, whereas resistant varieties possess a dominant gene that directs the physiological alteration. A parasite is avirulent when it possesses a dominant gene for production of a metabolite that can activate (derepress) the host gene for cellular alteration. A virulent parasite would thus lack the potential to derepress the host gene for hypersensitivity.

Littlefield (1969) succeeded in inducing resistance in flax to an otherwise virulent race of *Melampsora lini* by prior inoculation of the leaves with an avirulent race. The possibility exists, therefore, that certain saprophytic organisms likewise could induce resistance in a host, perhaps by producing the proper metabolite to activate the host gene for hypersensitivity.

Resistance has also been induced in bean hypocotyls to *Colletotrichum lindemuthianum* by prior inoculation with an avirulent race of the fungus (Rahe et al., 1969). Hypocotyls of the variety Perry Marrow, normally susceptible to the gamma but resistant to the beta race, when inoculated with the beta race followed 18 or 36 hours later by the gamma race, were resistant to both. Substitution of water alone for the beta race as a check permitted the usual fully susceptible reaction when plants were subsequently inoculated with the gamma race. Moreover, simultaneous inoculations with the two races on a stem usually permitted expression of the susceptible reaction to the gamma race.

One particularly striking example of resistance to a pathogen induced by prior inoculation with a nonpathogen is with sweet potato varieties susceptible to *Ceratocystis fimbriata* (Weber and Stahmann, 1966). Susceptible tissues of the sweet potato variety Jersey Orange, inoculated with a nonpathogenic isolate of *C. fimbriata,* were immune or resistant when inoculated 2 days later with a pathogenic isolate of *C. fimbriata.* Tissues not inoculated first with the nonpathogen were fully susceptible. Tissues of a resistant variety were resistant with or without prior inoculation with a nonpathogen. All of the eight isolates of *C. fimbriata* pathogenic to various hosts (cacao, oak, almond), but nonpathogenic to sweet potatoes, induced resistance to the sweet potato isolate. In contrast, none of several other species of *Ceratocystis, Verticillium albo-atrum,* or *Phytophthora infestans* induced resistance. Induction of resistance by a nonpathogenic isolate of *C. fimbriata* was possible in all the susceptible sweet potato varieties tested. The resistant zone was only a few millimeters deep.

Host resistance under field conditions may be far more complicated than normally indicated by the usual tests in highly artificial laboratory situations. Christiansen (1971) demonstrated, for example, that peas produce considerable amounts of pisatin in the lower epicotyl-upper radicle region when in nonsterile but pathogen-free soil. Few plant organs, tissues, or lesions are completely free of saprophytic organisms under natural conditions. Presumably these saprophytes invade tissue to the greatest extent possible, until prevented by the resistance response. The reduced severity of wilt symptoms in plants inoculated first with a nonpathogenic, then with a pathogenic, strain of an organism may relate to increased vascular occlusion, stimulated by the nonpathogen but subsequently effective against the pathogen as well (Chapter 7). Possibly then, the saprophytes assist in keeping the superficial or internal tissues of plants in a near-continual state of enhanced resistance to pathogens. Although the common view of plant-saprophytic organism interactions is in terms of the influence of the plant on the saprophyte, perhaps the saprophyte has an equally great influence on the plant. The known examples where a plant is resistant under nonsterile conditions, but susceptible under sterile conditions, should be restudied to determine the influence, if any, of the saprophytes on host physiology and resistance.

The influence of saprophytes on disease expression does not end, furthermore, with the pathogen's penetration of the host, as pointed out by Wilhelm (in Holton et al., 1959) and Flentje (in Baker and Snyder, 1965). The clubs caused by *Plasmodiophora brassicae* on cruciferous roots produced only slight stunting and plant failure in field soil treated by materials that controlled secondary invaders but not *Plasmodiophora* (Chapter 2). In nontreated soil the clubs decayed and produced toxins injurious to the tops. The same type of effect probably occurs with galls produced by crown-gall bacteria and root-knot nematodes. Although there are many secondary invaders that do not appreciably increase the severity of disease, functioning rather in a mopping-up capacity, it is probable that study would reveal many instances in which the disease is intensified by saprophytic secondaries.

DECOY, TRAP, AND INHIBITORY PLANTS

Creeping legumes are planted between rows of rubber trees as decoy crops for *Ganoderma pseudoferreum, Fomes noxious,* and especially *F. lignosis* (Chapter 4). *Crotalaria spectabilis* may be planted as a trap crop in peach orchards to reduce the population of root-knot nematodes in the soil and in peach roots (Chapter 4). MacFarlane (1952) found that planting perennial rye grass between cruciferous crops reduced the population of *Plasmodiophora brassicae* through abortive zoospore infection of root hairs. *Tagetes* spp. produce terthienyls that inhibit nematodes, reducing their numbers; asparagus roots have a similar effect (Chapter 4). Native savannah grass in southern Africa produces an exudation inhibitory to nitrifying bacteria and to rhizomorph formation by *Armillaria mellea* (Chapter 4). White (1954) found that planting *Datura stramonium* densely and turning it under at time of flowering, caused the resting spores of *Spongospora subterranea* to germinate and die. The subsequent potato crop had only 7% of the tubers with powdery scab and an average disease rating of 1, whereas 37% were scabbed, with an average rating of 4 in the fallowed control plots.

Host resistance may thus serve in either of two ways in biological control: Incorporated into an economic crop species, it provides reduction of disease. Incorporated into a decoy, trap, or inhibitory plant species that is grown in mixture or in sequence with the economic crop, it may indirectly reduce disease.

9

ROLE OF THE PHYSICAL ENVIRONMENT IN BIOLOGICAL CONTROL

Because environmental factors are interrelated and dynamic, and because they often exhibit delayed effects, an alteration of one factor frequently initiates a series of adjustments of far-reaching and often unpredictable consequences.

—R. DAUBENMIRE, 1959

Environment controls the outcome of all host-pathogen-antagonist interactions. Indeed, plants and microorganisms are captives of their environment, in that completion of their life cycles depends on environmental stimuli. Because of this cause-effect relationship, biological activity and biological control could not be discussed in previous chapters without including the role of environment. This chapter treats biological control specifically in relation to the abiotic environment.

The role of environment in biological control is, by definition, indirect (Chapter 2). When it is direct, as in killing the pathogen with cold, heat, or desiccation, it is ecological, rather than biological, control. *Environment in biological control operates by increasing the ability of the host to resist, tolerate, or escape the pathogen, through increasing the capacity of the antagonist as a competitor, parasite, or antibiotic producer, or through weakening the ability of the pathogen to affect the host or resist the antagonist.*

ENVIRONMENT OPERATIVE THROUGH THE HOST

The general and even moderately specific (horizontal) forms of resistance in plants to microorganisms are greatly affected by environment. Plants subjected to inadequate water, excessively low or high temperature, prolonged darkness, or a mineral deficiency have lowered resistance to microorganisms, and commonly succumb to facultative types of pathogens (Chapter 1). Within this group, some (*Botrytis cinerea*) require senescent or predisposed host tissue for infection; others are able to infect healthy tissue, but the disease they cause is accelerated by environmental stress. As metabolism of the host slows because of age, disease, injury, or stress from an unfavorable environment (Levitt, 1972), previously latent or subdued infections may quickly kill the plant. This effect is in contrast to that on obligate parasites, which generally decline in activity as the host declines (Fig. 1.3:B). It is also in contrast to the effect on obligate saprophytes, which may inhabit healthy tissues of plants in substomatal chambers, in intercellular spaces, or even in vascular tissues (Chapter 10), but cause no decay until death of the tissues.

Snow mold of wheat, caused by *Fusarium nivale* and *Typhula idahoensis,* are examples of diseases that occur when normal host resistance ceases to function (Bruehl et al., 1966). Beneath deep snow at about 0.5° C, photosynthesis and many other physiological processes within the plant presumably are restricted or stopped. Respiration continues at a slow rate, and the plant utilizes the carbohydrate reserves of leaves, crowns, and roots. As the reserves diminish, resistance apparently declines, and in 6–8 weeks pathogens capable of growth at −1° to 1° C kill the plant. Death would occur eventually with snow cover alone, but is earlier if the pathogens are present. Early melting of snow in northeastern Washington will permit almost complete plant recovery, but a prolonged snow cover can be disastrous. When the snow melts, the organisms are again limited to very slow parasitism, mostly on senescent basal leaves. Applications of certain fungicides before snowfall will largely protect the plants under prolonged snow cover, indicating that death is not caused by snow alone. Wheat varieties have been developed that apparently conserve more carbohydrate reserves in their crowns and thus survive, even though the pathogen may destroy the foliage completely (Bruehl and Cunfer, 1971).

The fruiting of morels in early spring may be a result of the ability of the antagonist-sensitive *Morchella* to grow at 2° C, when other micro-organisms cannot (L. C. Schisler and K. F. Baker, unpublished).

Certain ill-defined crown rots of clover, alfalfa, and other legumes associated with *Fusarium* spp. probably are also products of a weakened host combined with prior establishment of generally weak parasites. The repeated removal of tops from the plants for hay provides one obvious source of stress on the roots and crowns (Chapter 2). Improved fertility, proper watering, or altered time of cutting, which improve general plant vigor and hence resistance of the plant, alleviate this disease problem (Leath et al., 1971).

Towers and Stambough (1968) observed greater progression of *Fomes annosus* implanted as wood blocks of inoculum into roots of loblolly pine if the trees were subjected to water stress. Six months after inoculation, the fungus had progressed an average distance of 8.1 cm through roots of stressed trees and 2.7 cm in roots of nonstressed trees. In the glasshouse, root infection of 2-year-old loblolly pine seedlings was significantly greater after 10 months when inoculated plants were subjected to frequent but temporary wilting, compared with other inoculated plants grown in soil maintained near field capacity. Apparently, the host had less resistance to *F. annosus* when subjected to water stress (Chapter 10).

The results of Ghaffar and Erwin (1969) show clearly that water stress increased the susceptibility of cotton to root rot caused by *Macrophomina phaseoli*. Root rot was severe and sclerotial formation abundant when plants were allowed to wilt briefly before inoculation. Plants treated in the same manner but without the fungus, or those inoculated without prior stress, remained healthy. Adding Czapek's solution plus sucrose with the inoculum did not increase pathogenicity of the fungus in the nonstress treatments, indicating that increased exudation did not cause the increased fungus activity on desiccated plants. Charcoal rot of sorghum caused by this fungus apparently also increases in severity when the sorghum plants are subjected to water stress (Edmunds, 1964). In the absence of such stress, plant resistance to the fungus functions, and disease does not develop.

Crown and foot rot of wheat caused by *Fusarium roseum* f. sp. *cerealis* 'Culmorum' in Washington is another example where water stress lowers host resistance (Papendick and Cook, 1974). Disease is most obvious in old fertile fence rows, near edges of the field in a pattern that reflects overlap of the fertilizer application during implement turning, and in early-seeded fields. High fertility and early seeding both increase demand of the plant for water, by increasing the rate of photosynthesis, transpirational leaf area, and plant size. The dryland wheat areas (25 cm annual rainfall) have only limited water in the soil profile, conserved during a year of fallow. Rainfall is nonexistent or ineffective from mid-May onward, and as the plants deplete the finite water supply, they ripen early. Time of maturity ordinarily permits yields to reach or exceed 40 bushels/acre (35 kl/ha). Where the soil contains *Fusarium*, however, and plants are stressed from lack of water, severe disease develops and yields may be 25 bushels/-acre (17 kl/ha) or less.

Papendick and Cook (1974) monitored the rates of water use from the 1.8 m profile in wheat plots, and also the leaf-water potentials of wheat in plots of different row spacings (plant densities) and rates of nitrogen application. Percentages of plant infection were not affected by the treatments, in that plants of all plots were uniformly infected, but these remained latent or progressed slowly through May and early June. During early May, however, the rate of use of soil water was subsequently greater in high-fertility or narrow-row plots; this was reflected in consistent slightly lower average leaf-water potentials and more rapid onset of fusarium foot rot. Infection progressed most rapidly in plots having the greatest rate of water use and the lowest leaf osmotic water potential. Thus, whereas leaf-water potentials of plants of all plots were −18 to −20 bars in April, regardless of treatment, by mid-May plants from plots of high fertility and with narrow rows were consistently 3 to 4 bars lower than those of low fertility and wide rows. It was significant that some varieties apparently used soil water more slowly than others, which is reflected in generally higher leaf-water potentials (less stress) and less severe crown and foot rot. Some club wheats of the Omar type (Fig. 1.4) were especially conservative in water use, stressed very slowly, and showed essentially no severe disease, whereas adjacent plots of Gaines had 18–20% severe foot rot. This more conservative use of water by some varieties can be likened to the miler who paces himself for the first three laps in order to finish the fourth. Varieties planted in *Fusarium*-infested soil and which use soil water at a high rate early in the season do not cross the finish line. The present approach in Washington to develop wheat resistant to *F. roseum* f. sp. *cerealis* 'Culmorum,' is to select lines that use water efficiently. Such plants should permit the soil to remain moist for a longer period. This is an example of breeding for one character (efficient use of water) to achieve another (resistance). Development of winter wheats with a high carbohydrate level to achieve resistance to *F. nivale* and *Typhula idahoensis*, explained earlier in this chapter, is another example.

The greater disease severity with plant water stress in wheat is probably similar to the influence of temperature on seedling blight of wheat and corn caused by *F. roseum* f. sp. *cerealis* 'Graminearum' observed by Dickson (1923). The fungus grew on agar at 3–32° C, with optimum at 24–28° C. Wheat emerged and grew fastest at 8–12° C, whereas corn seedlings developed best at 16° C and above. Wheat seedlings showed little or no blight at a soil temperature of 12° C or less, because the host was then most resistant, and because food reserves for the fungus were minimal. Corn seedlings showed no blight above 24° C for the same reasons. Blight was severe in wheat at about 28° and in corn at 8–20° C, because these plants were most susceptible under these contrasting conditions.

Perhaps resistance to *Fusarium* in corn could be achieved by breeding varieties adapted to colder soils. Dickson also observed that low soil moisture favored fusarium seedling blight of wheat, caused by Graminearum, even at low temperatures, which supports the suggestion that resistance in wheat to *Fusarium* can be lowered because of moisture or temperature unfavorable to the plant.

Hoppe (1957) used cold testing of corn seed to screen hybrids for resistance to *Pythium*. Seeds and *Pythium*-infested soil were wrapped in moist paper towels and incubated 4–5 days at 7–10° C, and then transferred to 20–24° C. Hoppe (1949) had earlier shown that 7–10° C was ideal for seed decay caused by *Pythium*, and although infection occurred at 20–24° C and seedlings were stunted, they showed no conspicuous lesions at the warmer temperature. This supports the concept (Chapter 2 and this chapter) that the influence of environmental stress is on damage caused by facultative microorganisms rather than on infection itself. Apparently corn has little or no resistance to *Pythium* at a temperature unfavorable to corn growth. Could resistant hybrids have been selected in the cold test, even without *Pythium*, simply by selecting those lines capable of fastest seed germination and seedling growth at 7–10° C? This would be another example of breeding for one character (cold-tolerance) to achieve another (resistance to *Pythium*).

Leach (1947) showed that incidence and severity of damping-off of seedlings caused by *Pythium ultimum* and *Rhizoctonia solani* at different temperatures related directly to the relative influence of temperature on the growth rate of the plant and of the pathogen (Fig. 9.1). Thus, a low-temperature crop (spinach) showed greatest injury from *P. ultimum* at 12–20° C and escaped preemergence damping-off at 4° C. *Pythium* grew in pure culture at 4° C, but the spinach plant grew relatively much better than the fungus, as compared with 12–20° C where growth of *Pythium* was relatively better than for spinach. Seedling injury to this same host by *R. solani* was greatest at 20° C and above, and was mild below 12° C. *Rhizoctonia solani* in pure culture grew over the range, 8–35° C, but grew optimally at 25–30° C. A high-temperature crop (watermelon) emerged fastest, grew best, and showed little injury from either *P. ultimum* or *R. solani* at soil temperatures above 20° C, and was affected most severely by the two pathogens at soil temperatures below 20° C. In all host-pathogen combinations, infections were most severe at temperatures not necessarily optimal for the pathogen, but relatively less unfavorable for the pathogen than for the host. Conversely, temperatures that favored the host

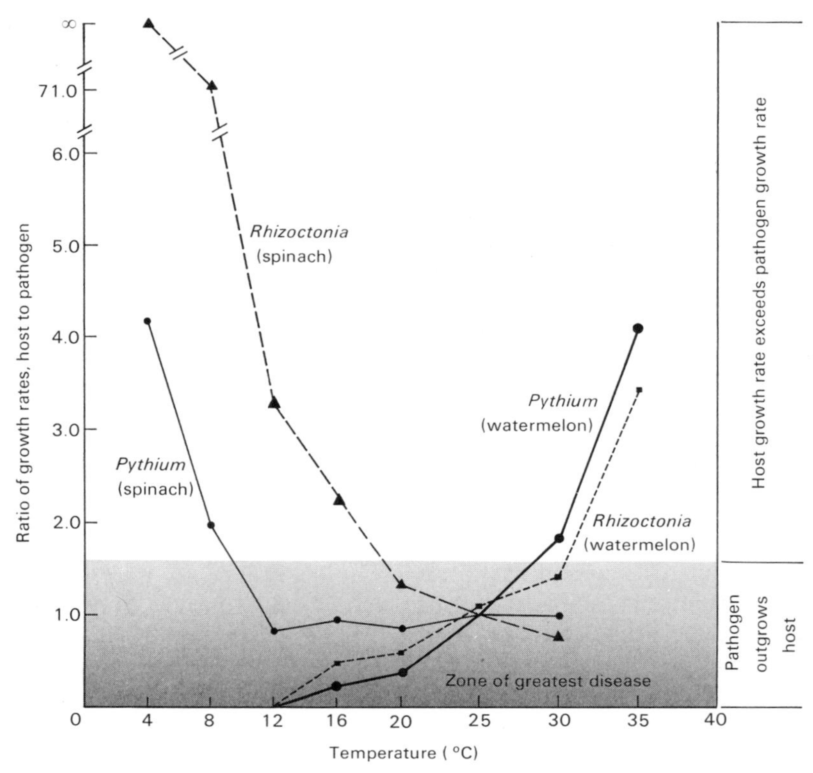

FIGURE 9.1.
Ratios of emergence rates of spinach and watermelon seedlings to growth rates of *Pythium ultimum* and *Rhizoctonia solani* in relation to temperature. (Data from Leach, 1947.)

more than the pathogen permitted expression of the resistance that plants generally have to facultative pathogens (Fig. 1.3:B; Chapter 8), and hence resulted in little or no disease.

This influence of growth rate of the host versus that of the pathogen, as influenced by environment, has been shown by Malalasekera and Colhoun (1968) to be useful in the control of seedling blight of cereals caused by *F. roseum* f. sp. *cerealis* 'Culmorum.' Incidence of preemergence death of wheat seedlings was greatest in dry soil, where the wheat seed was slow to sprout but the pathogen grew well. A 3-hour water soak of the seeds before they were sowed in dry soil hastened sprouting and emergence and reduced seedling mortality.

This does not imply that organisms which cause diseases of weakened plants are any less important than those which parasitize vigorous healthy plants. Water stress, an impermeable soil layer, prolonged snow cover, or lack of a mineral nutrient can kill plants without assistance from organisms (Fig. 1.3:B). The concern here is for the common intermediate levels of stress that can be tolerated by the plant except when combined with pathogens. If the pathogen is controlled, the plant is better able to tolerate the adverse environment. Conversely, modifying the environment or breeding for better-adapted varieties allows the plant to resist the pathogen, and biological control comes into play.

Environment also affects the expression of multigenic resistance in plants to the more specialized parasites. Some forms of resistance in wheat to stripe rust (*Puccinia striiformis*) and stem rust (*Puccinia graminis* f. sp. *tritici*), for example, function at some temperatures but not at others. The extensive studies of Walker (1969, 1971) on the influence of temperature on polygenic resistance in cabbage and tomatoes to *Fusarium oxysporum* f. sp. *conglutinans* and f. sp. *lycopersici,* respectively, provide other examples of this effect. Polygenic resistance in cabbage (Type B, or horizontal resistance) was effective against the *Fusarium* at 22° C, less so at 24° C, and ineffective at 28° C. Susceptible varieties were killed by the pathogen at all three temperatures, but those with monogenic (Type A, or vertical) resistance were unaffected at all three temperatures. An intermediate-resistant tomato variety (comparable to Type B) became susceptible after several weeks exposure to higher temperatures before inoculation with *F. oxysporum* f. sp. *lycopersici,* even if the plants were returned to a cooler environment immediately before inoculation. Predisposition of plants by exposure for several weeks to dry soil, low light intensity, short day lengths, low nitrogen and phosphorus, high potassium, or low pH (4.5) all lessened or nullified the effectiveness of the intermediate form of resistance to the wilt *Fusarium* when the plants were subsequently inoculated and incubated under uniformly ideal conditions. Tomato plants with monogenic resistance, on the other hand, were resistant in all treatments, regardless of conditions of predisposition (Chapter 8).

Walker (1971) pointed out that **monogenic resistance is more stable with variation in environment than is polygenic resistance.** Perhaps this generalization may also apply to the more general resistance of plants to facultative microorganisms, which apparently is not regulated by specific resistance genes (Chapter 8), and to host resistance conferred by the presence of rhizosphere or epiphytic microorganisms, which are even more environment-sensitive. The converse is true if stability of host resistance is assessed in terms of genetic changes toward virulence in populations of the pathogen. **The more complex and indirect the form of host resistance, the more stable it is against genetic diversity in the pathogen.**

The effect of poor aeration on a root disease has been shown by Stolzy et al. (1965) for citrus root rot caused by *Phytophthora citrophthora.* Plants exposed to water-saturated soil for 8 hours three times per month showed more root decay than plants exposed to water-saturated soil conditions for a few minutes every 4 hours during the 6-month test. Plants exposed from the beginning of the experiment to a constant 15% soil water content, comparable to that to which they had become accustomed prior to application of the treatments, showed little root rot, unless the oxygen supply was curtailed. Lack of oxygen, either by 8 hours of water saturation, or by experimental curtailment of the supply, thus predisposed the roots to greater decay caused by *P. citrophthora.* Had the plants been reared from the outset under submerged conditions, thereby allowing for root adaptation and possibly the means for oxygen diffusion downward from the tops, predisposition and the concomitant acute root decay of the citrus seedlings might have been less. Lowered oxygen levels in the soil also reduced growth and regeneration of roots, and decreased recovery and tolerance of the plant (Chapter 8).

Concentrations of carbon dioxide such as occur in soil apparently enhance initiation of lateral roots, as may ethylene in very low concentrations (Abeles, 1973). The extent to which soil saprophytes or plant pathogens may produce soil volatiles capable of stimulating lateral-root development is unknown. The effect of these conditions on root exudation is not defined either, but is undoubtedly marked.

Any situation that reduces respiration and growth of roots generally may also reduce the capacity of plants to resist pathogens. An inadequate supply of oxygen is only one way by which this may be accomplished. Root growth may be slowed by inadequate water at the growing points caused by (a) insufficient soil moisture, (b) low relative humidity aboveground, or (c) sunlight, which, through increasing photosynthesis, lowers the osmotic potential (solute potential) of leaves and thereby increases the demand of the tops for water. This steepens the water-potential gradient between tops and roots, and as equilibrium is again approached, the root osmotic potential drops. Transpiration and photosynthesis may create root stresses in this manner, and probably facilitate infection by facultative types of pathogens.

Still another factor is the unsaturated conductivity of the soil (water-supplying power), determined by pore-size distribution. On days of high evapotranspirational demand and hence of high rates of water absorption by roots, capillary flow in a clay soil may keep pace with absorption, but that in a sandy soil may not. The water-potential gradient will then steepen greatly in the sandy soil, and water potential of the rhizosphere will approach that of the internal root (Chapter 8). This would greatly favor pathogens, such as *Fusarium,* that prefer slightly reduced soil-water potentials, but would be unfavorable to pathogens that grow optimally at water potentials near zero. Moreover, as water uptake fails to keep pace with the evaporative demand, leaf-cell turgor drops, stomata close, transpiration and photosynthesis are sharply reduced, and leaf temperatures rise slightly. Osmotic potential within the plant will then also rise, but meanwhile the plant has been stressed. How this relates to the generally greater incidence and spread of *Fusarium* pathogens in sandy than in clay soils, including those formae that cause wilt, is unknown but deserves serious consideration.

ENVIRONMENT OPERATIVE DURING DORMANCY OF THE PATHOGEN

Environment can directly kill the pathogen during dormancy, as with death of sclerotia of *Phymatotrichum omnivorum* at temperatures below −25° C, of cells of *Pseudomonas solanacearum* by desiccation, or of chlamydospores of *Fusarium* by high soil temperature. However, for each example of this type, there are many where the pathogen propagule is weakened by environment, but could still function were it not for the action of soil microorganisms (Coley-Smith and Cooke, 1971). This role of environment in biological control is very important because it enables many microorganisms to play a part when they would otherwise be ineffective. It also increases the effectiveness of sublethal environmental conditions—*the environment starts the process and microorganisms finish it.*

Stress Effects on the Propagule

Marshall (1969, 1971) suggested that "most microorganisms in soil exist in a sorbed state. . . . selective sorption of certain microorganisms on soil particles is one of the most important factors whereby the solid phase of the soil exerts some control over the balance between populations of microorganisms in soils." Sorption of bacteria is thought to occur in three ways:

a. Sorption onto the surface of large particles of clay or organic matter, but not sand. Bacteria attach to the surface on their ends or sides, either reversibly or permanently.
b. Aggregate formation with soil particles of bacterial size.

c. Sorption of small clay particles by bacterial cells. An envelope of clay may form around the cell (Fig. 8.3) and may restrict or prevent diffusion of nutrients, water, or oxygen to the cell, or the escape of carbon dioxide from it. Substances antibiotic or inhibitory to the bacteria may also be sorbed by the clay and reduce growth. On the other hand, cells may be able to multiply quite readily and even be stimulated by sorption; nutrients may be selectively sorbed and thus concentrated for the bacteria from dilute solutions. Furthermore, the clay envelope may protect bacteria from a dry external environment by modifying the rate of water loss from the cells. Slow-growing *Rhizobium* spp. are remarkably resistant to dry heat even in the absence of clays, and legumes with these species may occur in arid or semiarid areas in sandy soils. Legumes with fast-growing *Rhizobium* spp., however, seem to occur in moist areas or in soils containing montmorillonite or illite clays. Nodulation of clover by *R. trifolii* in sandy soils of Western Australia was successful when inoculated seed was planted, but the bacteria did not survive in the soil into the second year because of lack of protection from desiccation (Marshall et al., 1961). It would be expected that similar relationships on survival of antagonistic or pathogenic bacteria in soils would exist, exerting a profound influence on the level of suppressiveness shown.

Observations of Smith (1972) on effects of desiccation of sclerotia of *Sclerotium rolfsii, S. cepivorum, Sclerotinia sclerotiorum,* and *S. minor* provide an excellent example of the way environment, together with soil microorganisms, affects pathogen survival. Sclerotia air-dried for 4 hours lost about 90% of their water. When returned to moist soil they leaked amino acids and sugars, germinated, and decomposed within 2 weeks. The organisms that colonized and decomposed the sclerotia (apparently fungi, including *Trichoderma* spp.) appeared to be stimulated by the nutrients that leaked from the sclerotium. Smith showed that the breakdown was not caused by drying alone, but by the combined action of drying and colonization by the soil microorganisms; without prior drying, decomposition did not occur. Moreover, if after drying they were washed to remove nutrients and then placed in moist soil, they no longer leaked nutrients and were not decomposed, but did germinate. Sclerotia not predried did not germinate or infect a host.

Smith suggested that in the field, drying is an important trigger for sclerotial germination and infection. This perhaps explains why the disease caused by *Sclerotium* is often inconsistent in fields uniformly infested with sclerotia, and why disease is commonly most severe when wet weather follows a dry spell. He suggested a possible method of biological control of

S. rolfsii in which the field is rapidly dried, at least at the surface where most sclerotia occur, then irrigated and kept moist for 2 or more weeks, and finally planted. This assumes, of course, that controlled drying and watering is possible, as in irrigation districts, nurseries, or gardens. The method probably will not work where rainfall is more evenly distributed or less seasonal (Smith, 1972).

Other factors such as ethylene (Chapter 6) act in concert with the above drying phenomenon in determining the germination of sclerotia.

Staten and Cole (1948) reported that postemergence damping-off of cotton caused by *Rhizoctonia solani* in New Mexico could be lessened significantly if the soil was irrigated 3 or more weeks before planting. By comparison, if the soil was irrigated at time of planting, postemergence seedling blight was severe. The difference occurred in either naturally or experimentally infested field soil. They suggested that "the pathogen is partially inhibited by antagonistic or competitive organisms when the soil is given a pre-planting irrigation." Perhaps sclerotia of this pathogen respond like those of *S. rolfsii* and decompose during the 3-week period between irrigation and planting. The desiccation requirements would readily be met by field conditions of the arid southwest before irrigation. Planting at time of irrigation, on the other hand, would permit infection before decomposition.

Sclerotia of *Typhula idahoensis* apparently survive very well in the dry tillage layer in Washington through at least 1 year of fallow. G. W. Bruehl (unpublished) observed with the scanning electron microscope that these mature sclerotia have a waxy surface not found on immature or culture-grown ones. Perhaps this surface in some way protects the sclerotium from adverse environment and microorganism attack.

Predisposition of sclerotia to infection and decay is similar to that of some seeds that leak nutrients and stimulate pathogens. Seeds with small cracks or fissures in their coats leak greater quantities of nutrients, and decay in higher percentages, according to Flentje and Saksena (1964) for peas, and Schroth and Cook (1964) for beans. Flentje and Saksena showed that presoaking pea seeds for 24 hours to remove the exudates largely averted preemergence damping-off of seeds when subsequently planted in *Pythium*-infested soil. In a sense, the improved seed survival paralleled that of sclerotia with washing. Sclerotia may in this way be analogous to seeds, and much information about the biology and pathology of seeds (Baker, 1972) might apply to them.

The microsclerotia of *Verticillium albo-atrum* will germinate several times in soil that is repeatedly air-dried and remoistened, and apparently do not die until the cycle has been repeated several times. Some germinated up to nine times, although percentage of germination and number of germ tubes per microsclerotium declined with each dry-wet cycle. Since *Verticillium* microsclerotia consist of aggregates of thick-walled surface cells or hyphae surrounding hyaline cells, which germinate, each germination may arise from a different cell or hypha rather than from the same cell. This could account for an observed gradual decline in percentage of germination and number of germ tubes per microsclerotium with each dry-wet cycle (Farley et al., 1971); eventually all the cells would have germinated and lysed.

Verticillium dahliae microsclerotia also germinate to produce conidiophores and microconidia in moistened soil, but in so doing tend to lose resistance to desiccation (Menzies and Griebel, 1967). After several weeks of sustained sporulation in moist nonsterile soil, the microsclerotia were apparently weakened and became progressively more sensitive to drying; eventually they could be killed by drying the soil to a water potential at which fresh sclerotia would have survived indefinitely. Perhaps the weakening of these structures by sporulation would also make them more vulnerable to decay and thus shorten their life expectancy even in soil not dried.

Sclerotia of *Botrytis convoluta* are able to produce six successive crops of conidia (Jackson, 1972).

Chlamydospores of *Thielaviopsis basicola* and some fusaria lose their rapid germinability with desiccation, in what apparently is a reversible process. Linderman and Toussoun (1968b) showed that, if *T. basicola* was maintained in air-dry soil for several months, chlamydospores would not germinate, and cotton seedlings would be infected by the fungus only after the infested soil had been wetted for 5–9 days. In contrast, chlamydospore germination and seedling infection promptly occurred in soil kept continuously moist. However, Papavizas and Lewis (1971) found that germinability of chlamydospores of *T. basicola* remained high after several months storage in air-dry soil, and also in soil at 4 and 15% moisture-holding capacity, but declined at increasing rates over a 20- to 60-day period as the water content was increased to 30, 45, and 60% of moisture-holding capacity. Rewetting the air-dry soil improved germination somewhat, but the effect was not nearly as great as that observed by Linderman and Toussoun. They suggested that the different results reflected isolate differences. A more likely explanation is that "air-dry" (equilibrium with the atmosphere) was considerably different in the two tests. The water potentials to which *T. basicola* chlamydospores were exposed could have been as different as −750 bars (50% relative humidity or RH in Berkeley) versus −150 bars (90% RH in Beltsville), certainly not comparable values. This illustrates how meaningless the term air-dry is unless the relative humidity of the air is specified.

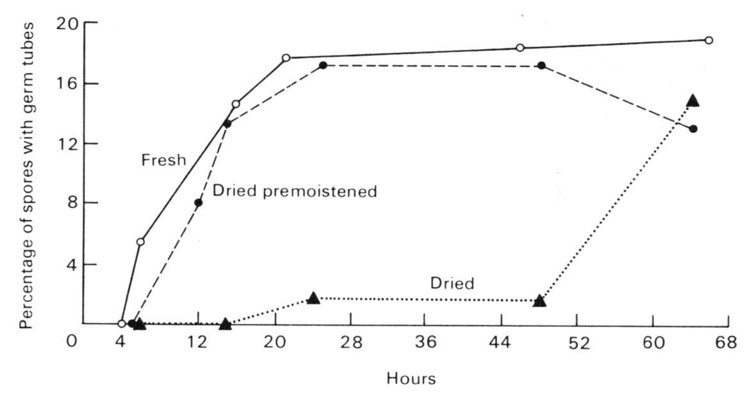

FIGURE 9.2.
Percentage of chlamydospore germination of *Fusarium solani* f. sp. *phaseoli* in three different soil samples. A. Recently infested (fresh). B. Stored air dry for 90 days after infestation, then premoistened for 1 week (dried-premoistened). C. As in B, except not premoistened (dried). Glucose (0.1M) used to test for germinability. Note the long lag phase in soil dried but not premoistened before applying glucose. (From Cook, 1964.)

Chlamydospores of *F. solani* f. sp. *phaseoli* respond to desiccation in much the same way as those observed by Linderman and Toussoun for *T. basicola.* Germinability near bean seeds was 60% when chlamydospores were new, but declined steadily after 60 or more days of storage in air-dry soil (average RH near 50%). By 5–6 months, only 10–20% or less of the chlamydospores would still germinate. However, when the soil was wetted for 2–3 days, and then amended with a solution of glucose, glucose + asparagine, or a bean seed, chlamydospore germination again readily occurred. Germination also occurred in air-dry soil amended with a glucose solution, but 48 hours elapsed before it began (Fig. 9.2). In contrast, germination began after only 4 hours in continuously moist or air-dried and remoistened soil. Apparently, the effect of desiccation is on the lag phase in germination, extending it from 4 to about 48 hours (Cook, 1964).

The longer lag period for chlamydospore germination in *F. solani* f. sp. *phaseoli* was detectable with a sugar solution (Fig. 9.2), but not when a mixture of amino acids or bean-seed exudates was used as the stimulus for germination. Sugars without nitrogen do not support as rapid an increase in microbial numbers as does a more complete nutrient source (Cook and Schroth, 1965). When a combination of streptomycin and penicillin was added to air-dried soil along with a mixture of amino acids, chlamydospores germinated near the end of the second day (Cook, 1964).

The extended lag period was detectable with richer sources of nutrition only if bacterial competitors were subdued. Since the bacteria would not be inhibited in nature, germination and disease would probably be less in soil that was dry for long periods, even though the propagules might be alive.

The extent to which "desiccation dormancy" develops in desiccated chlamydospores is unknown, but is probably common, at least in *Fusarium*. Chlamydospores of both *F. solani* f. sp. *pisi* and *F. roseum* f. sp. *cerealis* 'Culmorum' are less germinable after the soil has been air-dried for long periods (2–3 months or more), but germinate readily if the same soil is first wetted, and then amended with nutrients. French and Neilsen (1966) reported that 91% of the chlamydospores of *F. oxysporum* f. sp. *batatas* germinated overnight in a nutrient solution after storage for 11 months in water, but did not germinate in 48 hours if they had been stored air-dry. They took the absence of germination after 48 hours as evidence of death, but perhaps in another few hours germination would have occurred. Nevertheless, propagules in this condition are less functional, even if not dead, because the fungus is placed at such a competitive disadvantage.

The significance to biological control of delayed germination is that by the time the propagules revive from desiccation, the level of fungistasis is high, and germination may be prevented.

Environmental Effects on Antagonists of the Propagule

Thin-walled conidia commonly lyse quickly in soil. Such observations are generally made with test soils in the laboratory, with water and temperature ideal for microbial growth. However, the soil surface is commonly dry in nature to the extent that lysis is retarded or prevented. J. W. Sitton (unpublished) showed that in soil at −30 to −60 bars water potential, conidia of *F. roseum* f. sp. *cerealis* 'Graminearum' may persist for more than a year at temperatures from 9° to 28° C. The conidia died quickly (could not be revived with soaking) in air-dry soil (30–40% RH), and they endolysed in soil at −1 bar water potential, although slower at 9° than at 25° C. Perhaps thin-walled conidia of *Fusarium* play a role in survival and infection in soil too dry for the activity of other microorganisms. The observation suggests further, however, that a timely irrigation or rain in advance of planting could promote lysis and reduction in populations of conidia, unless the conidia were converted into chlamydospores.

The results of Papavizas and Lewis (1971) on survival of endoconidia and chlamydospores of *Thielaviopsis basicola* in soils of different moisture levels also suggest that propagule death caused by soil microorganisms depends on soil moisture and temperature. In very dry (but not air-dry) nonsterile soil, chlamydospore germinability remained near 100% for 9 months at 10° C, and declined only to 60% during the same time period at 26° C. Chlamydospore germinability also remained high in moist soil maintained at 10° C, but declined quickly in moist soil maintained at 26° C, unless the soil had been autoclaved, in which case propagule survival was high. Endoconidia survived for up to 8 months in dry soil at 5° C, but declined to nearly zero within 3 months in wet soil at 18° C or higher. The conditions that favored rapid loss of spore germinability are those that favor the greatest microbial activity, and consequently are those that result in biological control of the dormant propagules.

Actinomycetes in soil digest the chitin between the endwalls in chlamydospore chains of *T. basicola* and cause the chains to break apart (Christias and Baker, 1967). Individual chlamydospores only then can germinate, but are also more vulnerable to antagonism from soil microorganisms. The poorer survival of *T. basicola* in warm moist soil could relate to increased fragmentation of the chains, followed by destruction of the propagules.

Forms of nitrogen, pH, and possibly interactions between the two also affect propagule survival by affecting multiplication of soil microorganisms. One example is sclerotia of *Sclerotium rolfsii*, which germinate poorly or not at all if placed in soil amended with various nitrogen compounds. Apparently some forms of nitrogen stimulate antagonists of the sclerotia of this fungus (Henis and Chet, 1968). Only ammonia caused complete loss of germinability in sterile soil, whereas in nonsterile soil urea, chitin, peptone, calcium nitrate, and ammonium acetate, but not ammonium bicarbonate and ammonium sulfate, caused loss of germinability of the sclerotia when subsequently removed and tested on agar. That ammonia is directly toxic to cells and tissues has been shown many times (Cochrane, 1958). Henis and Chet suggested that the other forms of nitrogen favored certain soil microorganisms that could attack the carbonaceous sclerotial wall. They demonstrated high numbers of bacteria, actinomycetes, and fungi in crushed sclerotia and in washings of sclerotia for each of the nitrogen treatments that reduced germination, except for calcium nitrate, where numbers of organisms were less than in the control. Ethanol extracts of sclerotia so handled reduced germination of other sclerotia from 96 to 36–50%, suggesting that antibiosis was involved. The observations of Henis and Chet probably were with sclerotia stored dry before use, in which case their subsequent burial in moist soil might result in decomposition regardless of nitrogen applications (Smith, 1972), as discussed earlier in this chapter. Perhaps the ammonium bicarbonate and ammonium sulfate helped protect sclerotia against an otherwise inevitable decay. It is known from field observations that nitrogen reduces the sugar beet disease caused by *S. rolfsii* (Leach and Davey, 1942), but the mechanism whereby the disease is reduced is still uncertain. Avizohar-Hershenzon and Shacked (1969) have suggested that the effect may be, at least partially, directly on the fungus, suppressing its ability to compete with other microorganisms. Further study is needed to determine whether the control of *S. rolfsii* in sugar beets with nitrogen is a type of biological control.

Soil fungistasis is generally associated with sites of high microbiological activity—the top 15 cm or areas of high moisture, organic matter content, or temperature. It has also been shown to increase with increasing soil pH (Schüepp and Green, 1968; Schüepp and Frei, 1969). Soil organisms, especially bacteria, are probably more active in alkaline than in acid soil. This greater microbiological activity would, in turn, limit the germination of fungus spores, whether by depriving them of nutrients, by liberation of fungitoxins or ethylene (Chapter 6), or both.

Not all environmental effects on propagule antagonism need be directed at decomposition of the propagule in soil, or at prevention of its germination. The stage in the life cycle of the pathogen when propagules are formed presents a point of attack; an example is chlamydospore formation in *Fusarium*.

Macroconidia provide *Fusarium* with a means of aboveground dispersal by water splashing, but like thin-walled spores of other fungi, these structures are short-lived when washed into soil. In at least three species, however, *F. solani, F. oxysporum,* and *F. roseum,* conidia are converted into thick-walled, long-lived chlamydospores within 1–2 days if moisture and temperature are favorable. Conidia not converted into chlamydospores lyse. Chlamydospore formation in pure cultures of these species can be related to depletion of the carbon substrate (Griffin and Pass, 1969).

Chlamydospore formation may be particularly rapid and profuse if conidia in liquid culture with a carbon substrate are allowed 10–12 hours to germinate, and then are washed and transferred to a basal salts solution without a carbon substrate (Griffin and Pass, 1969; Meyers and Cook, 1972; Stevenson and Becker, 1972). This treatment more or less parallels the natural situation: the *Fusarium* conidium commonly germinates in soil within 10–12 hours, presumably with nutrients supplied by the slime coating acquired originally from the sporodochium. This spore is, however, quickly robbed of all exogenous nutrients by soil bacteria and other organisms. In addition, there is evidence that one or more substances of microbial origin act as morphogens and help to induce chlamydospore formation in *F. solani* f. sp. *phaseoli,* even in the presence of high carbon substrate (Ford et al., 1970). Possibly carbon starvation, combined with high levels of a morphogen in the *Fusarium* cell, is the key to chlamydospore formation.

Regardless of the mechanism, soil bacteria are involved in this process, and any condition that influences bacterial activity should, therefore, affect chlamydospore formation. A mechanism for biological control is thus provided. Any environment favorable to increased microbial activity will also increase the rate of carbon substrate use and consequently hasten onset of chlamydospore formation by these fungi. Delay in their formation is commonly observed in soil habitats where nutrients are abundant, and rapid formation is observed in habitats of limited, quickly expendable carbon substrates (simple sugars) and of intense microbial competition. A warm moist environment could actually increase the rate, and thereby aid survival of the fungus unless microbial activity was extremely intense, when germ tubes and conidia might lyse before completing chlamydospore formation. In certain *F. solani* f. sp. *phaseoli* "resistant soils" (Chapter 4) in Washington, the pathogen forms chlamydospores slowly or not at all, whereas in other soils it forms them quickly and survives (Burke, 1965b).

Undoubtedly both carbon starvation and the presence of certain substrates are involved in chlamydospore formation in *Fusarium.* **Most biological processes may be initiated by more than one factor or agent;** this has survival value for microorganisms. The success of the manned space program has demanded the provision of such back-up support systems.

Environment obviously has a very great influence on effects of plant residues on pathogens during dormancy, as for example the germination-lysis process (Chapter 8). Moisture, temperature, pH, and aeration that favor multiplication of the soil microflora supplied with plant residues will hasten the depletion of nutrients, encourage other forms of antagonism, and thus accelerate lysis. As reported by Clark (1942) for elimination of sclerotia of *Phymatotrichum omnivorum* with amendments, *the more ideal the environment for microbial growth, the faster the rate of amendment breakdown and hence the more rapid the death and decomposition of sclerotia* (Chapter 8). The procedure outlined by Streets for control of *P. omnivorum* (Chapter 6) involved liberal application of inorganic nitrogen with the residue to hasten microbial breakdown of the organic matter, including sclerotia and infected cotton roots. Papavizas (1968) showed that nitrogen prolonged the survival of endoconidia of *Thielaviopsis basicola* in soil and permitted the population to stabilize at a higher density. Survival was better in alkaline than in acid soils, which correlates with soil reactions that favor pathogenesis by this fungus. Apparently, biological elimination of endoconidia of this fungus is thus best accomplished with moist, warm, slightly acid soils.

Environmental Conditions that Help the Pathogen Withstand Antagonists

A propagule in the center of a particle of organic matter, a soil crumb, or clod is better protected from a harsh environment and probably from antagonists than one in a pore between soil particles. Thus, J. H. Warcup and K. F. Baker (unpublished) found that hyphae of *Rhizoctonia solani* picked out of soil had at least 1° C higher thermal tolerance to aerated steam when in particles of organic matter than when free of it. Morris and Winspear (1957) found that a buried clod 3¾ inches in diameter reached 100° C 36 minutes after steaming began. A clod 5¾ inches in diameter required 50 minutes to reach 100° C at the center, and a 7-inch one took 120 minutes to reach 85° C at the center. In all instances, the surface layers of the clods reached 100° C in 20 minutes. A similar heterogeneity of environment exists throughout soil and clods of soil in relation to drying and wetting, distribution of oxygen and carbon dioxide, and radiant heat. Propagules are also protected from antagonists if they are situated where antagonists cannot grow.

The soil environment is extremely important to survival of *Cephalosporium gramineum* in wheat straw, because it affects production by the organism of the antibiotic needed to resist soil antagonists (Bruehl and Lai, 1968a). At soil water potentials of –150 to –210 bars, *Penicillium* spp. overran *C. gramineum* in the straw. At -75 bars or higher, or -270 bars or lower, *C. gramineum* retained posession of the straw. Antibiotic production by *C. gramineum* does not occur below -75 to -100 bars (Bruehl et al., 1972). Below this water potential, *Penicillium* can enter the straw without resistance. Under the driest conditions both *C. gramineum* and *Penicillium* are inactive, and hence *C. gramineum* is preserved, free of antagonists. A similar effect has been demonstrated for soil pH: antibiotic production by *C. gramineum* occurs and the pathogen survives best under acid conditions, but in alkaline soils antibiotic production is minimal and the pathogen is more quickly overrun in the straw by saprophytic soil fungi (Hopp, 1972).

ENVIRONMENT OPERATIVE DURING GROWTH OF THE PATHOGEN

Optimum conditions for pathogen growth in nonsterile soil are often different from those in pure culture. Where a host is involved and pathogenesis is the criterion for measurement of pathogen growth, the effect may be through the host, as already discussed. However, some of these shifts in environmental optima are obviously related to differential effects on the relative competitive advantage of the pathogen versus the antagonists. An explanation is that an environment suppressive to growth of a pathogen probably is never equally suppressive to growth of all surrounding potentially competitive organisms. It would probably make little difference to a pathogen, for example, whether an environmental change that reduced its growth likewise reduced growth and metabolism of all other organisms in the community by the same proportion. The more likely occurrence is that some will be affected more than the pathogen, some less, and some may even be favored in growth by the environmental change. The net effect is a change in growth pattern of the pathogen that is greater than can be explained by the environmental change alone.

Indeed, the more we learn of the life cycles of microorganisms in relation to their environment, the more apparent it becomes that **the reaction of an organism to its abiotic environment, unless extremes are involved, is usually not the direct consequence of that environment, but rather results from environmental effects on surrounding microorganisms, which in turn affect the pathogen.** As stated by Snyder (1963), "although biological control . . . could be and at times is accomplished by direct attack on the pathogen by some element of the soil fauna or flora, perhaps the major mechanism is of an indirect nature, or through the creation of greater stresses for the pathogen in its competition for moisture, air, food, and space." These stresses may be on the dormant propagule as discussed above, but more commonly they are on the pathogen during its growth, thus weakening it to the antagonism of other microorganisms.

The relative effect of the environment on metabolism and growth of a pathogen and of its antagonists will largely determine their relative competitive abilities. Therefore, **antagonists should be able to grow over as broad a range of environmental conditions as the pathogen or be exceptionally well adapted in a critical range of the pathogen.** It is possible, however, that an antagonist could produce enough antibiotic to inhibit the pathogen, even under unfavorable conditions. This illustrates the desirability of using several antagonists rather than one, as one may become effective when another is ineffective.

The influence of water stress on antibiotic production by soil microorganisms has been studied for *Cephalosporium gramineum* (Bruehl et al., 1972) and *Streptomyces* sp. (Wong, 1972). Antibiotic production was inversely related to the growth rate of the microorganism, and was greatest in each instance under moderate water stress. *Cephalosporium* produced maximum antibiotic between −27 and −55 bars osmotic water potential. The vigor of this fungus in straw on water-saturated soil shows

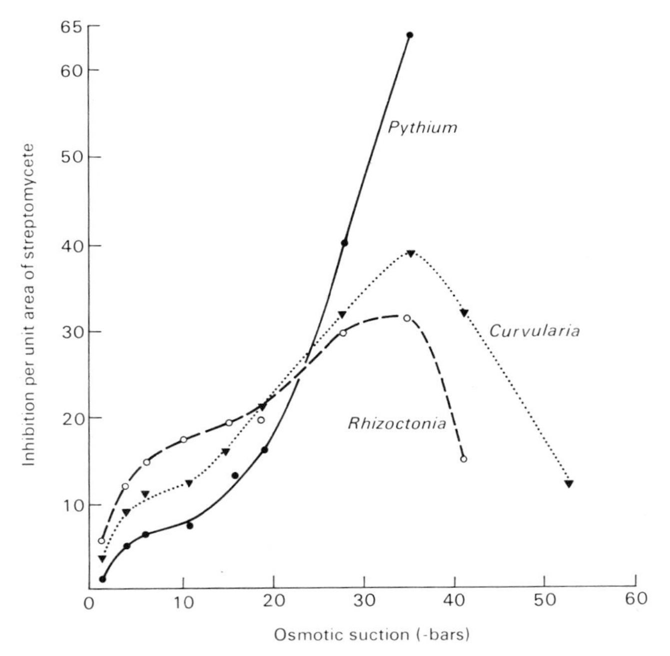

FIGURE 9.3.
Relationship between growth inhibition of *Pythium ultimum*, *Rhizoctonia solani*, and *Curvularia* sp. by *Streptomyces* sp. at different osmotic suctions, controlled by potassium chloride. *Pythium* was inoculated 8 days after the *Streptomyces*, the others 10 days after. (Photo courtesy of T.W. Wong, New South Wales Department of Agriculture, Rydalmere.)

its tolerance of bacteria, and its performance between −10 and −137 bars indicates that relatively xerophytic fungi are the most severe antagonists to this fungus in nature. The *Streptomyces* produced maximum antibiotic between −30 and −40 bars, based on its inhibition of *Rhizoctonia solani*, *Curvularia* sp., and *Pythium ultimum* (Fig. 9.3). There was also an interaction between antibiotic production by the antagonist and an increased sensitivity of the pathogens to the antibiotic, caused by the growth reduction that resulted from water stress. *Pythium* was particularly sensitive to the antibiotic at −20 to −30 bars, and the most sensitive of the three pathogens to water stress in the absence of antibiotic. Inhibition at a given water potential is probably the net effect of many factors, including antibiotic production by the antagonist and pathogen sensitivity.

One of the earliest demonstrations of environment affecting a pathogen through effects on soil microflora was that of Henry (1932) on the influence of temperature on root rot of wheat caused by *Gaeumannomyces graminis* in sterile and nonsterile soil. Growth of *G. graminis* in pure culture occurs over the range, 4–33°C and is optimal between 19–24°C, depending on the isolate (Davis, 1925). Disease caused by *G. graminis* is generally most severe at 12–16°C, which is also the optimum temperature for the host (McKinney and Davis, 1925). In contrast to results of Dickson for fusarium seedling blight of wheat and corn, and of Leach (this chapter) for damping-off of various crops, disease was more severe when the temperature was more favorable for the host than for the parasite. In the words of McKinney and Davis, "the explanation of this will require the study of factors . . . far more basic than the phenomena of growth rate and vigor" of host and parasite. Henry (1932) showed that the occurrence of maximum disease at 14–18°C was characteristic of nonsterile soil, and that in sterile soil disease was uniformly severe at all temperatures over the range, 14–26°C (Fig. 9.4). If the optimum temperature of 14–16°C for disease was simply a result of relative effects on the host and pathogen, the effects should have been expressed in sterilized as well as in nonsterile soil. He concluded that disease suppression in warmer soils, even though the fungus grows optimally there, is caused by increased antagonism from soil microflora.

Since Leach (1947) and Dickson (1923) both used pasteurized soil in their studies, any differences in temperature optima for disease versus pathogen growth was probably, as they concluded, related only to host resistance and vigor affected by the temperatures. Moreover, both worked

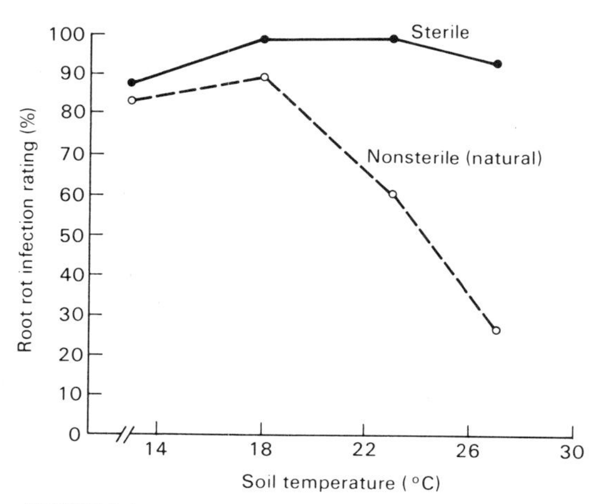

FIGURE 9.4.
Influence of soil temperature on root rot of Marquis wheat seedlings caused by *Gaeumannomyces graminis* in sterile and nonsterile soil. (Data from Henry, 1932.)

with relatively unspecialized parasites whose pathogenesis is generally favored by conditions unfavorable to the host. *Gaeumannomyces graminis* is more specialized and would be more likely, therefore, to be virulent at temperatures ideal for growth of the host and also at those optimal for its own growth. Because of its vulnerability as a rhizoplane occupant (Chapter 6), however, Henry's conclusion seems reasonable that the greater microbial activity above 18–20° C, compared with that of cooler soils, suppressed parasitism by *G. graminis.*

The optimum water potential for disease severity caused by *G. graminis* may also reflect relative effects of environment on the pathogen versus the antagonists. *Gaeumannomyces graminis* in pure culture or in sterilized soil grows progressively more slowly as the water potential is lowered below −1 bar, and ceases at −45 to −50 bars. Growth rate at −20 bars is about half that at −1 bar water potential. Disease, on the other hand, is severe almost exclusively in wet soils and rare in drier soils, including those that if sterile, permit excellent growth of the pathogen. As suggested by Henry (1932) for temperature, the suppression of *G. graminis* on wheat in drier soils may reflect an advantage conferred on antagonistic microorganisms. A water potential of −15 bars or drier, although unfavorable to *G. graminis,* is not sufficient to prevent its growth, but does place it at a competitive disadvantage relative to other organisms not suppressed at −15 bars (Cook et al., 1972).

The suppressive effects of rhizosphere pH values between 5.0 and 7.0 on severity of root infections caused by *G. graminis* in nonsterile soil, and the occurrence of severe disease in sterile soil at these same pH_r values, is probably another example of partial suppression of this pathogen, but not of its antagonists (Chapter 8). The early decrease in take-all severity in acid, but not in limed soil at Puyallup, Washington, also exemplifies greater suppression of the pathogen than the antagonists by environment (Chapter 7).

Phymatotrichum root rot of cotton and other hosts, like take-all of wheat, is more severe in wet than in dry soil. Again, the question arises whether this reflects an effect of soil water only on the fungus, on the host, or on antagonists of the fungus. Eaton et al. (1947), in a study concerning the rarity of cotton root rot in dry summers, showed that fungus growth from a substrate into dry soil was substantial at above −30 bars. They concluded that dry summers did not directly affect activity of the fungus. Root-bark carbohydrate levels (Chapter 8), which had been previously suggested as affecting *P. omnivorum* through effects on saprophytic organisms on the root surface, were on the other hand much higher in dryland than in irrigated cotton. Except for this work, the significance to disease of the effect of soil water on root-surface organisms or the host is still to be demonstrated.

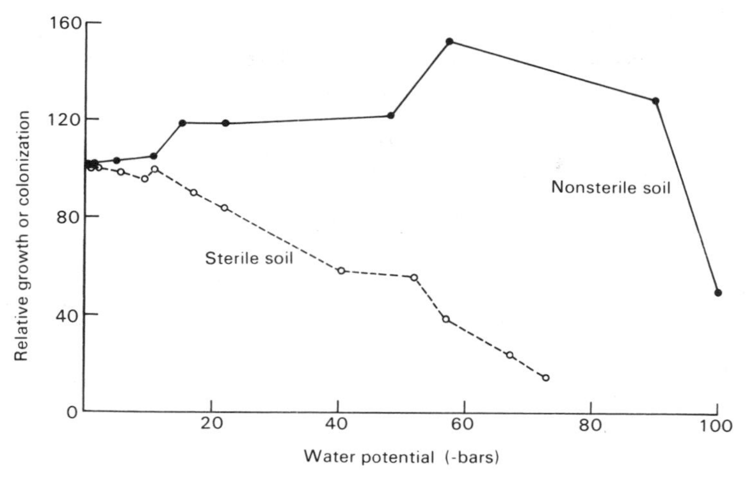

FIGURE 9.5.
Influence of soil-water potential on relative activity of *Fusarium roseum* f. sp. *cerealis* 'Culmorum' in sterile soil (activity measured as linear mycelial growth in concavity of straw; Cook et al., 1972) versus nonsterile soil (activity measured as amount of straw colonized; data from Cook, 1970).

Environment greatly affects the relative competitive advantage of *Fusarium roseum* f. sp. *cerealis* 'Culmorum'. Burgess and Griffin (1967) found more straw colonization by Culmorum at 10° than at 20° C, yet in pure culture the fungus grew about three times faster at 20° than at 10° C. Cook (1970) similarly found little change in colonization over the temperature range, 5–15° C, even though the fungus grows much better at 15° than at 5° C. Results on the influence of water potential on growth and colonization of straw are even more revealing. Culmorum colonized a progressively higher percentage of straws as the soil water potential was lowered to −90 bars even though, based on pure culture performance (Cook et al., 1972), activity of the fungus should have been less, not more, down to −90 bars (Fig. 9.5). The influence of temperature and water potential on straw colonization in nonsterile soil cannot be explained in terms of direct effects on the fungus. Neither can it be readily explained in terms of effects on the straw itself, since a dead substrate probably changes relatively little with altered soil environment. Reduced antagonism in the cooler or drier soils more than offset the handicap of reduced growth of Culmorum under these conditions. Plowing wheat stubble under when the soil is warm and moist should result in less colonization of straw by Culmorum than plowing during cool dry periods. Plowing after the straw has been colonized by airborne saprophytes is preferable (Chapter 8).

Biological factors modulated by the abiotic environment probably exert a major influence on foot rot and related fusarium diseases of cereals in various parts of the world, and may even determine the kinds of symptoms expressed by this host-pathogen interaction. For example, in humid cereal-producing areas (England, Europe, and parts of the midwestern United States), *Fusarium roseum* f. sp. *cerealis* is pathogenic primarily on aboveground parts, causing head blight, stem rot, and scab. Culmorum is the main pathogenic cultivar in England and Europe, whereas in the Midwest Graminearum dominates. Much of the early literature (Atanasoff, 1920) discusses "foot rot," but this apparently was caused by aboveground infections near the soil line, often initiated by conidia or ascospores caught in the angle between leaf blade and stem. In the drier cereal-producing areas, aboveground infections are essentially nonexistent. *Fusarium* there infects from soilborne inoculum (chlamydospores or mycelia in host refuse) and causes a crown rot that eventually spreads up the stem as a brown decay, referred to as dryland foot rot (this chapter). Culmorum is the predominant type in the northwestern United States and Canada, whereas in California and Australia, Graminearum predominates. The existence of these two types of foot rot, distinguished by mode of infection, has been a major source of confusion in the literature; workers in humid regions report severe foot rot in wet years, whereas those in semiarid regions report the opposite relationship. Obviously, both may be correct.

Foot rot initiated by aboveground infections can be recognized by discrete aboveground stem lesions and by the fact that outer leaf sheaths turn brown sooner than does the internal culm tissue in response to inward advance of the pathogen. With belowground infections, culm tissues display a uniform brown discoloration extending upward from the base where crown-root infections occur; outer leaf sheaths commonly remain free of the pathogen. The aboveground type is as rare in the Northwest as the belowground type is in England (J. Colhoun, personal communication). Isolates of Culmorum from England and Washington both produced the belowground type foot rot (but not seedling blight) on wheat grown to maturity in pots of Ritzville silt loam infested with 20,000 chlamydospores/g. Both isolates produced the aboveground-type foot rot when they were atomized onto plants as conidia, and the plants were subsequently incubated in a humidity chamber (R. J. Cook, unpublished). Seedling blight is virtually nonexistent in the drier areas (Cook, 1968), apparently because this phase of the disease is produced by seed- but not soilborne inoculum. Chlamydospores germinate primarily near crown roots, which are absent from seedlings. Both the Washington and English isolates produced seedling blight, however, when introduced as seedborne conidial inoculum.

Different types of foot rot and the presence or absence of seedling blight in various regions are thus probably not related to isolate differences. It seems likely that biological antagonism, such as is associated with moist soils in the Northwest, is at least as effective in moist soils of humid regions in other parts of the world, thereby helping to prevent the belowground type of foot rot. When such areas have high humidity, however, the pathogen operates aboveground, thereby escaping antagonism. That seedborne inoculum is more apt than soilborne to escape antagonism can account for the prevalence of belowground foot rot in humid areas, even though dry soil is more favorable than wet for this phase (Colhoun and Park, 1964; Dickson, 1923). There can thus be no single biological control of *Fusarium* on cereals, since one effective against soil- or seedborne infections may not work against an aboveground epiphytotic, nor will a control effective against soilborne chlamydospore inoculum necessarily work against water-splash or aerial conidia or ascospores.

USING ENVIRONMENT FOR PREDICTION

Predicting the Kinds of Antagonists

We suggested in Chapter 7 that *Pythium* spp. during their long evolutionary association with soil bacteria may have developed a tolerance of them, and thus have been able to occupy wet alkaline soils. *Fusarium* spp. were offered as examples of organisms that prefer dry acid conditions and have relatively little tolerance for soil bacteria. The question might now be raised whether preference for one set of environmental conditions over another has any prediction value for the kinds of antagonists to which the pathogen is most or least sensitive.

The preference of *Gaeumannomyces graminis* and *Phymatotrichum omnivorum* for wet alkaline conditions might indicate a tolerance for bacteria, like that of *Pythium.* It might further indicate that saprophytic fungi are the important antagonists of these two pathogens. Most saprophytic fungi seem to prefer lower water potentials, and certain species of *Penicillium* and *Aspergillus,* as reported by Griffin (in Toussoun et al., 1970), may be "addicted to" or able to grow only at low water potentials. Eaton and Rigler (1946) indicated that antagonism to *P. omnivorum* in their corn trials (Chapter 8) could not be explained in terms of the soil bacteria they studied. Gerlagh (1968) and Shipton et al. (1973) eliminated antagonism of *G. graminis* from suppressive soils with heat treatments as low as 60° C, treatments known to eliminate most fungi, nonspore-forming bacteria, and many actinomycetes (Chapter 4). R. J. Cook (unpublished) tested 200 random soil bacteria from suppressive soil, none of which were antibiotic to *G. graminis* on agar media. This is similar to the situation found with *Phytophthora cinnamomi* (Chapter 4). Considering the marked preference of *Phymatotrichum omnivorum* and *G. graminis* for a wet alkaline environment, they must have some tolerance to soil bacteria, which are most active under these conditions. Considering the extent of suppression of these two pathogens in dry acid soils, and considering that these conditions are favorable to saprophytic fungi, most of which are eliminated by steaming at 60° C, by inference the biological antagonism of the two would most likely involve fungi, to some extent actinomycetes, or both. These microorganisms have been suggested (Chapter 4) as the active agents in the general antagonism to *G. graminis.* Diseases caused by *Verticillium albo-atrum* and *Thielaviopsis basicola* also are associated more with wet alkaline than dry acid soil. Perhaps these two pathogens also have a greater tolerance for bacteria than for fungi or actinomycetes.

Soils suppressive to *Phytophthora cinnamomi* have high microorganism populations because of the favorable environmental conditions (see below). That portion of the total microflora that withstands treatment with moist heat (60° C/30 minutes) confers suppressiveness to these soils, but individual isolates from it do not. Sporeforming bacteria and some actinomycetes are the active agents involved (Chapter 4).

Actinomycetes are most active in dry, neutral to alkaline soils. Accordingly, potato scab caused by *Streptomyces scabies,* and soil rot of sweet potato caused by *S. ipomoea,* are most severe in dry soils and are suppressed by adequate watering or soil pH values below 5.2–5.5. The effect of pH is probably direct because, like bacteria, actinomycetes grow poorly in pure culture on acidified media. On the other hand, the lack of disease in wet soil may also reflect a form of biological control performed by soil bacteria. Lapwood and Hering (1970) controlled potato scab in field plots by maintaining the soil water near field capacity during tuber formation. Timing of the irrigation was extremely important, because if it was stopped before all the internodes on the tubers were formed, those internodes formed subsequently in dry soil became infected. *Streptomyces scabies* infects tubers through stomata and newly formed unsuberized lenticels. Once the corky layer develops in lenticels, infection no longer occurs. By varying the time of irrigation in relation to the stage of tuber development, they produced tubers with scab at the stolon end, the apex, both stolon end and apex, or as a band around the middle (Fig. 9.6). They pointed out that those who failed to control scab by irrigation probably were applying the water too late or too early. A dry period after the tubers are completely formed has considerably less influence on scab than one when tubers are forming and are still susceptible.

Does control of *Streptomyces scabies* by irrigation exemplify biological control by soil microorganisms? As reported in Chapter 4, scab was low

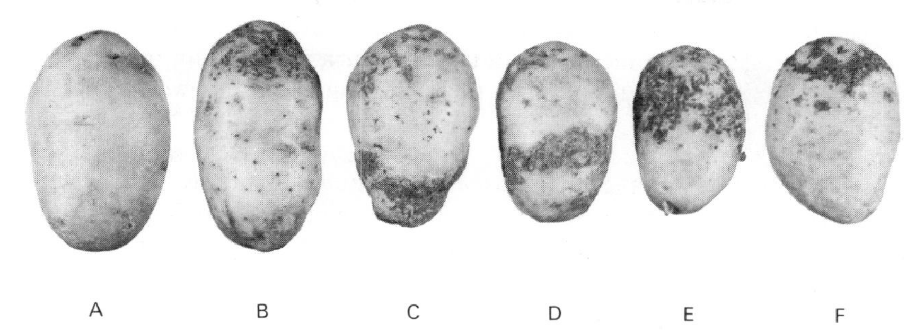

FIGURE 9.6.
Relationship between time of irrigation and position of scab lesions on tubers. A. Control irrigated to maintain soil water potential of –0.2 bars. B, C, D. From plots where irrigation was stopped twice during tuber formation, but progressively later each time. Note that scab is progressively farther away from stolen end on B–D, which corresponds to progressively later periods during which watering was discontinued. Note scab at apex end of B–D, which corresponds in each instance to the second time watering was discontinued. E, F. Watering discontinued once; E, beginning about halfway through the period of tuber formation, and F, later than E. Scab developed when watering stopped. (From Lapwood and Hering, 1970.)

after many years of irrigated potato monoculture in Washington, Nebraska, and California, but remained severe in dryland-potato monoculture in Nebraska (Werner et al., 1944). Scab was also suppressed with soybean green manure on potatoes grown in irrigated plots (Weinhold et al., 1964). Control of potato scab with soybean residue may be less dependable in dryland areas. In axenic culture of potato tubers, scab infections are favored, not suppressed, by moisture (Barker and Page, 1954). Lewis (1970) found that bacteria predominated in the lenticels of immature tubers in wet soil, and actinomycetes in those of immature tubers in dry soil. Where bacterial numbers were high, actinomycetes generally occurred in low frequency. He suggested that scab control with high water potentials might result from an increase in bacteria antagonistic to *S. scabies*. Labruyère (1971) also obtained scab control by irrigation, and implicated soil bacteria in the suppression of *S. scabies*. Alternatively, the effect may be more on the increased competitive advantage of bacteria over actinomycetes in wet soil, rather than on increased population of bacteria (D. H. Lapwood, personal communication). Nevertheless, the implication is clear that soil organisms, not just water alone, are involved in control of potato scab by irrigation.

Predicting the Suppressiveness of Soil

There is insufficient information on naturally suppressive soils (Chapter 5) to permit more than suggestions as to how they may be found, that is, predicted. Like weather prediction, this must be thought of as probability or betting odds. With increasing data, the accuracy should improve.

The odds are good that soil in the following situations will prove suppressive:

1. Areas free of a certain disease despite a favorable abiotic environment and repeated introduction of the pathogen (Chapter 4). One may grow susceptible pea varieties in California, for example, without fusarium wilt, or wheat in the Palouse area of the Pacific Northwest without concern about loss from fusarium foot rot. It is probable that such soils may be characterized sufficiently to permit improved prediction.
2. Areas free of a certain disease despite a favorable abiotic environment and the presence of the pathogen (Chapter 4). The soils suppressive to *Phytophthora cinnamomi* in Queensland have now been characterized sufficiently that one having abundant organic matter, calcium, ammonium and nitrate nitrogen, and phosphate, and a pH of 6.0–7.0 is likely to have minimal losses; it is still undetermined whether this will apply in other areas. Banana soils in Central America that are high in clay and alkaline in reaction are also likely to prove "intermediate" or "long-life" in relation to Panama disease.
3. Soils cropped in monoculture for a number of years (Chapter 4). A soil continuously cropped to wheat is much more likely to give minimal losses from take-all than is a virgin soil or one in which wheat has been intermittently planted. Soils cropped continuously to potatoes will give less common scab than those in which they are grown intermittently, especially if irrigated. It has also been found that the decomposed litter in established pine forests inhibits *Fusarium* spp. These types of suppressiveness are microbiological in character, and with further study, hopefully the level of inhibition can be predicted.

USING ENVIRONMENT TO NUDGE THE BIOLOGICAL BALANCE

Although the change in environment must sometimes be drastic to permit the establishment or function of an antagonist, biological control probably is accomplished extensively in agriculture through slight shifts in microbiological balance caused by subtle but significant and reasonably permanent changes in environment that affect the activity of a pathogen. As pungently phrased by A. M. Smith, "You don't need much freeboard on a boat if the waves are small." Biological control attempts to keep the waves small. The process may be likened to that which occurs in a river, lake, or sea where the water temperature is elevated slightly but permanently by an industrial activity of man. A slight temperature change may have little direct effect on any given aquatic species, but may exert a profound impact on that species by affecting the balance of organisms active in the water. Although the effect on reproduction and growth of each species in the food chain may be slight, the net ecological effect can be significant.

A change in almost any factor in the physical or chemical environment can alter the balance of active microorganisms. Indeed, *it is probably impossible to change one component of the soil environment without causing microbiological change.* The direct effects of water potential, temperature, pH, and gases on growth, reproduction, and relative competitive advantage of individual species are best known, but their effects go far beyond this. Water increases the heat capacity of soil and hence probably influences the heat tolerance and temperature optima of organisms. It also serves as a solvent for carbon dioxide and other volatiles, and as a medium for solute and ion diffusion, but is a poor medium for oxygen diffusion and commonly contributes to oxygen deficiencies within soil microsites. Higher temperatures increase vaporization and hence evaporative water loss from soil, which lowers the water potential. Carbon dioxide lowers the pH of the soil solution (Whitney and Gardner, 1943). Changes in soil pH, on the other hand, can shift the equilibrium relationship of carbon dioxide to bicarbonate to carbonate; increased soil pH shifts the equilibrium toward higher carbonate concentrations. These and other salts can accumulate in soil with evaporative water loss, and as their concentration increases, may provide an osmotic stimulus to growth of some organisms, and depress growth of others (Cook and Papendick, 1972).

This covers only part of the interactions among major components of the soil environment. Equally important are nutrients needed for microbial growth, especially forms of nitrogen and organic carbon, which vary in availability and concentration in the soil solution according to the environment. The number of distinct abiotic situations is infinitely large, and to understand each exceeds current technology and comprehension. Consideration of the equally large number of biotic patterns that can occur with each slight shift in environment also illustrates how really complex the soil is. It reemphasizes the futility of awaiting complete knowledge of each ecological situation before attempting to make biological control work (Chapter 3). It is better strategy, through informed judgment, first to make biological control work, and then to unravel as many of the mechanisms and factors as possible to make it work more predictably.

This approach was suggested in Chapter 5 as potentially useful in utilizing specific antagonists or groups of antagonists from pathogen-suppressive soils, in Chapter 8 for finding a crop sequence that will encourage development of a pathogen-suppressive soil, and is suggested here as the most practical approach to biological control through environmental nudging of the microbiological balance.

Irrigation Practices

Irrigation practices are an exceptionally versatile means of environmental manipulation because the daily mean and minimum water potentials of a field soil can be raised or lowered with relative ease. The proper timing and frequency of irrigation of potatoes for control of potato scab, discussed earlier in this chapter, is an example of an irrigation practice that controls a disease through a shift in the microbiological balance. Raised soil beds insure that basal stem tissues, as well as many roots and other infection courts are located in dry soil. This is another means of controlling soil water potential (Cook and Papendick, 1972). This is used to reduce verticillium wilt of cotton in furrow irrigation districts in the southwestern United States, although the reason for its effectiveness is not known. Soil in raised beds would probably not be too dry for growth of *Verticillium* because the fungus can grow at water potentials down to -100 to -120 bars (Manandhar and Bruehl, 1973), but it could reduce the competitive advantage of this fungus relative to actinomycetes and other fungi (Chapter 7). Another possibility is that the drier soil of raised beds prevents infection by other root pathogens such as *Pythium* sp., or nematodes (Chapter 2) that contribute to losses from verticillium wilt. Root-knot nematode combined with *Verticillium* is more damaging than *Verticillium* alone on many crops, including cotton (Chapter 6). Control of the nematode directly, or of *Verticillium* through reduced competitive ability would constitute a form of biological control accomplished by the drier soil in raised beds.

Fertilizers and pH

Certain fertilizers have an acidifying effect on soil that may not be striking the first year, but can be considerable if certain forms are used consistently year after year. This results because they saturate the cation exchange, and cause the bases to move downward in the soil profile. As introduced cations are removed from the exchange by root uptake, or in the case of ammonium, converted to leachable nitrates, the exchange becomes increasingly dominated by hydrogen ions, and the pH drops.

TABLE 9.1
Relative influence of various nitrogen forms of fertilizer on soil pH. (Data from Shaw, 1965.)

Fertilizer	Equivalent[1] acidity or basicity in pounds of calcium carbonate	
	Acid	Base
Anhydrous ammonia	147	
Ammonium chloride	140	
Ammonium sulfate	110	
Diammonium phosphate	75	
Urea	71	
Ammonium nitrate	62	
Monoammonium phosphate	58	
Aqua ammonia	36	
Nitric phosphates	15–29	
Calcium cyanamide		63
Sodium nitrate		29
Calcium nitrate		20

[1] Equivalent per 100 pounds of each material.

Some forms of fertilizer cause much greater pH changes than others. Table 9.1 gives the equivalent acidity or basicity of some common nitrogen fertilizers, using calcium carbonate as the standard for comparison. In general, ammonium forms of nitrogen have the greatest acidifying effect on soil, in contrast to nitrate forms, which have a tendency to raise the soil pH, at least initially. Anhydrous ammonia and the ammonia released by urea initially cause the soil pH to rise markedly, but this is temporary, and eventually the pH of soils so fertilized drops well below neutral (Smiley et al., 1970, 1972). Nitrification is an acidifying process, and together with the effects of ammonium on the eventual increase in hydrogen ions in the cation exchange, accounts for most of the pH drop characteristic of ammonium applications. It is not uncommon for the pH of a calcareous, sandy, poorly buffered soil to drop two full units with repeated applications of an ammonium form of nitrogen. Application of sodium nitrate to a clay loam will displace the calcium ions and deflocculate the soil, making it so impermeable to water that it may become impossible to use ditch irrigation. Gas exchange is then also restricted.

Lime application is a well-known means of counteracting pH drop, or of increasing the pH of acid soils, and has been used to control club root of crucifers, fusarium wilt of tomato, and certain other diseases suppressed by alkalinity or enhanced by acidity. As discussed in Chapter 8 and in more detail below, perhaps some form of nitrogen likewise could reduce a disease by creating a partially unfavorable soil pH.

Huber et al. (1965) pointed out that some root diseases are less severe and others more severe when specific forms of nitrogen fertilizer are used. In their examples, diseases caused by *Fusarium, Rhizoctonia,* and *Aphanomyces* are increased in severity by ammonium nitrogen and decreased by the nitrate form. Verticillium wilt was reduced by ammonium nitrogen (Huber and Watson, 1970). The take-all disease, caused by *Gaeumannomyces graminis,* was reduced with ammonium nitrogen, and unchanged or made more severe with nitrate nitrogen (Huber et al., 1968). The effect with take-all apparently relates to pH, especially in the rhizosphere, which is related to a shift in the biological balance, since take-all is severe with ammonium nitrogen at all rhizosphere pH values above 5 if the soil is treated first with methyl bromide to free it of microorganisms.

Smiley (1972) correlated the effect of different forms of nitrogen on various pathogens and diseases with the favorable pH ranges. Among the diseases favored by acid soils and by applications of ammonium nitrogen (or suppressed by nitrate nitrogen) are Panama disease of banana (Rishbeth, 1957), fusarium wilt of cotton (Tharp and Wadleigh, 1939), fusarium seedling blight of cereals (Shen, 1940), and fusarium root rot of bean (Weinke, 1962). Among those favored by alkaline soils and by applications of nitrate nitrogen (or suppressed by ammonium nitrogen) are verticillium wilt (Huber and Watson, 1970; Wilhelm, 1951), phymatotrichum root rot of cotton (Neal, 1935), thielaviopsis root rot of tobacco (Swanback and Anderson, 1947), and common scab of potato (Huber and Watson, 1970; Potter et al., 1971). This correlation is undoubtedly significant and suggests that the depression of soil and rhizosphere pH by ammonium nitrogen, and its elevation by nitrate nitrogen, are potentially important for disease control through fertilizer and pH management.

Maintenance of ammonium nitrogen in the ammonium form with nitrification inhibitors such as N-Serve or Telone could also affect disease by lowering soil pH. This could explain the reduction of verticillium wilt and increase of rhizoctonia black scurf of potato with Telone (Easton, 1964), the reduction of verticillium wilt and scab of potato with N-Serve (Huber and Watson, 1970; Potter et al., 1971), and the increase of fusarium root rot of bean with N-Serve (Maurer and Baker, 1965). Soil acidity created by a predominantly ammonium form of nitrogen would probably favor *Fusarium solani* f. sp. *phaseoli* and *Rhizoctonia solani,* but suppress *Streptomyces scabies* and *Verticillium albo-atrum.* Nitrogen stabilization by inhibitors of nitrification thus increases the opportunity for disease control through fertilizer applications.

It is apparent that the pH effect on root diseases operates through shifts in the biological balance. The pH change associated with ammonium or nitrate fertilizer in most cases is not large enough to inhibit the pathogen directly or to account for the observed magnitude of change in disease severity. The probable effect is to inhibit the pathogen sufficiently to reduce its growth in competition with others, that is, its inoculum potential *sensu* Garrett, and thus its capacity to cause disease. An effect that operates through other soil organisms could explain why the effects are not always clear-cut or consistent from experiment to experiment, or why, as pointed out for verticillium wilt (Wilhelm, 1950), the disease may be associated primarily with alkaline soils but occurs in some acid soils. **An environmental effect that is mediated through soil microorganisms will be more variable than one that operates directly against the pathogen.** The effect of different forms of nitrogen noted by Huber et al. (1965), and proposed by them for possible biological control of root diseases, is obviously real and should be exploited with those diseases where the pathogen is known to be pH-sensitive.

A different effect of ammonium-N was reported by Hornby and Goring (1972), in which take-all increased in severity when this form of nitrogen was added to a magnesium-deficient acid soil that had been leached to remove all nitrate and then uniformly treated with N-Serve to prevent nitrification. Their results may relate to impaired root metabolism (and hence predisposition to root rot) sometimes inherent in plants grown with ammonium alone. Ammonium toxicity, according to Thiegs (1955), is most likely to occur in acid soils devoid of nitrate-N, and may be greatly reduced or even eliminated by the presence of a small amount of nitrate-N. One may safely assume that the examples reviewed by Huber et al. (1965) and Smiley (1972) were based on a predominance of one form versus the other, rather than a total presence or absence of a given form.

Tillage Methods

The method of tillage and of handling crop residue affects the temperature and water potential of soil, and thus constitutes another practical means of changing the environment and perhaps nudging the biological balance in a desired direction. The effects of tillage on soil temperature are well illustrated by unpublished data of R. I. Papendick for plots near Pullman, Washington. These were mold-board plowed, chisel plowed, or nontilled (stubble left largely undisturbed on the soil surface). The presence of maximum residue on the soil surface slowed the rate of cooling of soil in the fall but kept the soil cool longer in the spring. In addition, the soil froze deeper and more frequently, and remained frozen longer in mold-board plowed plots than in those having maximum surface stubble. Differences between the tillage treatments was particularly pronounced on the north-

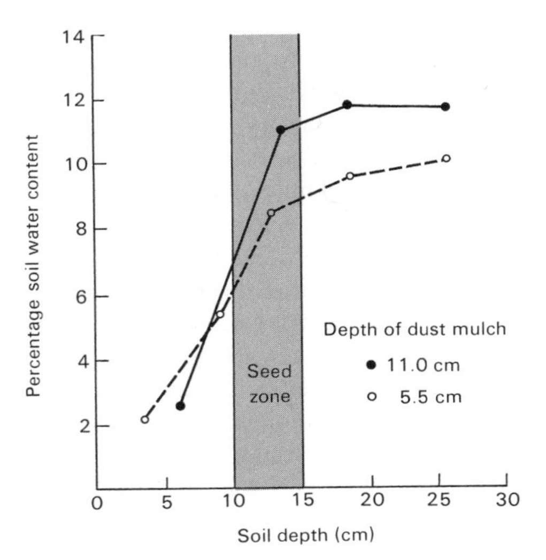

FIGURE 9.7.
Water content distribution at 10–15 cm depth at seeding time (September 20) under two depths of dust mulch in eastern Washington. (Based on data of Papendick et al., 1973.)

facing slopes. Differences in soil temperature in the fall and spring, and in rate of thaw, affect the growth rate of wheat and hence its tendency to contract or escape certain diseases. Freezing and thawing also affect the winter storage of water, and hence water available for plant growth (this chapter) and for pathogens and other microorganisms.

Tillage effects were also pronounced when abrupt weather changes occurred; mold-board plowed areas were quickest to warm and cool with daily changes in air temperature or cloud cover. In contrast, the presence of straw on the soil surface buffered against marked daily soil temperature oscillations.

The depth of dust mulch in the semiarid wheat-growing areas of eastern Washington affects both soil temperature and water potential in the wheat seed zone (10–15 cm), regardless of straw cover on the soil surface (Papendick et al., 1973). The soil at 13 cm during summer fallow was consistently 2–3° C cooler beneath a 11-cm dust mulch than where the dust mulch was only 5.5 cm thick. This caused less water loss from the seed zone and thus favored wheat seed germination at fall planting (Fig. 9.7). Mulching breaks the continuity of channels through which unsaturated water flow otherwise would occur. In addition, the deeper the mulch, the better the temperature insulation and the less the loss of water through vaporization. Seeds unable to germinate in dry soil are commonly destroyed by *Aspergillus* or *Penicillium* spp. and other weak pathogens capable of growth in it. Seeds in wetter soil may germinate and resist or escape decay. Conserving seed-zone moisture is extremely important to seedling establishment in semiarid eastern Washington, and can reduce the need for seed-protectant fungicides.

Perhaps plant water stress can also account for the severe fusarium root rot of bean under irrigation in the Columbia Basin of Washington. Burke et al. (1972) showed that subsoil tillage to break a hardpan at the plow sole reduced *Fusarium* damage by permitting better root penetration. Incidence of infection on roots and stems in surface soil containing the pathogen apparently was not affected by subsoil tillage, but proneness of plants to succumb to disease was reduced. The evaporative demand is exceedingly high in the arid Columbia Basin during the growing season, and crop demand for water is great. Plants with shallow and restricted root growth are undoubtedly less capable of meeting the transpiration demand than those with deep well-developed roots. Plant water potentials presumably will remain higher (less plant water stress) during the season when the absorptive capacity of roots can keep pace with transpirational loss. A lowered internal water potential of a bean plant may permit or even encourage a more aggressive attack by *Fusarium,* as shown for fusarium foot rot of wheat (Papendick and Cook, 1974). Loss of host resistance as a result of environment stress is discussed in Chapter 2.

Using tillage in a less subtle but very effective manner, the Dutch *rigolen* process involves turning the top 75–90 cm of soil upside down; this places many of the pathogens at a depth where they neither infect nor survive (Drayton, 1929). The buried material is decomposed before it is brought to the surface by the next plowing.

Planting Date

Planting dates, time of seedling establishment, and the geometry of plant distribution can greatly affect the microclimate near the soil surface at any given time or season, and thus have possibilities for changing the environment so as to nudge the biological balance. Winter wheat seeded near Pullman by September 15 on summer fallow usually develops severe cercosporella foot rot, whereas the same field seeded one month later has almost no foot rot (Bruehl et al., 1968). The effect is attributed to the difference in density of leaf canopy from late fall through spring, when infection by *Cercosporella* takes place. A dense canopy helps maintain a moist soil surface favorable to the pathogen. Curly top of tomatoes or beets, on the other hand, can be controlled by planting sufficiently early so that there are few open areas in the crop canopy by the time of leafhopper flights. The moister microclimate is unfavorable to leafhoppers, and thus provides biological control of curly top, acting through the leafhopper (Bennett, 1971; Chapter 2). Growing tomatoes in the hot interior valleys of California avoids infection by spotted wilt virus because the climate is unfavorable to the thrips vectors.

It may not always be necessary to create uniform environmental changes in a field to obtain a desired effect. In the work of Papendick et al. (1973) on depth of dust mulch, for example, water conservation at the 13–15 cm depth where the seed is to be planted is of greater importance than through the entire tillage layer. Indeed, it was necessary to sacrifice water in the surface 10 cm to insure adequate moisture in the seed zone. The soil-water potential, temperature, or pH are rarely uniform throughout the soil mass, but rather exist as gradients. The surface is subject to the greatest environmental fluctuations, and these vary less below 15–18 cm. The pH of the soil solution exists as a gradient away from residue where ammonia may be produced (for example, an alfalfa-meal amendment), or away from the bands of fertilizer application. By proper methods of fertilizer application and proper depth or method of tillage, it should eventually be possible to create desired microenvironments in zones that contain specific infection courts of a host or specific pathogen-antagonist interactions. The activity of most soilborne pathogens fortunately is confined to the tillage layer, where the greatest changes in environment are possible. The pH, water potential, or thermal conditions of this layer can be maintained at values different from those in deeper layers in the soil profile. Plant roots penetrate deeper than the depths of infestation of most pathogens, and therefore can be exposed to pH, nutrients, and water potential different from those of the pathogen (Chapter 8).

INTEGRATION BETWEEN BIOLOGICAL AND CHEMICAL CONTROL

Perhaps no approach to manipulation of the microbiological balance offers greater potential than the use of specific growth-regulating or biocidal chemicals. A major objection of many of the chemical controls currently in use is their indiscriminate effect on organisms other than the pathogen. **Rather than kill the pathogen, it may only be necessary to weaken it and make it more vulnerable to antagonism of the associated microflora.** Alternatively, biocides and biostats directed at some purely saprophytic element of the flora to shift the microbiological balance may indirectly control the undesirable microorganism.

There are several examples where fungicides controlled a disease, even though the materials or rates used were not considered directly inhibitory to the pathogen. McKeen (1949) observed, for example, that control of *Aphanomyces cochlioides* and *Pythium aphanidermatum* of sugar beets with thiram was partially the result of direct fungicidal action on the pathogen, but he thought that an indirect effect caused by a shift in the microbiological balance was also involved. Domsch (1959) also obtained control of damping-off with captan at rates not considered directly fungicidal to the pathogen, and suggested that biological control initiated by the chemical was involved. Although *Helminthosporium sativum* conidia were not affected directly by 5 ppm of methylmercury dicyandiamide, this rate controlled root rot of wheat caused by the fungus (Chinn, 1971). Certain microorganisms, especially *Penicillium* spp., were observed to have increased profusely in the treated soil, and were suggested as an explanation for the disease control obtained. *Pythium ultimum* was difficult to establish in soil treated with thiram even after the fungicide was largely gone 9 weeks after treatment and, even though residual amounts of the compound were low and probably sublethal, they delayed or prevented natural buildup of the fungus in flats recropped to peas (Richardson, 1954).

Anderson (1962–64) found that fumigation of soil with chloropicrin controlled pineapple root rot caused by *Phytophthora cinnamomi* in Hawaii for a period of 3 years, although the pathogen could be baited from the soil as early as 2 weeks after treatment. The population of *Trichoderma viride* was increased by the treatment. In contrast, PCNB increased disease severity over that in nontreated plots, and reduced counts of *Trichoderma, Penicillium,* and actinomycetes. Ethyl mercury phosphate at 50 pounds/acre foot (56 kg/ha 30 cm deep) also increased disease, but 150 pounds (168 kg) controlled it. As the *Trichoderma* population decreased in these trials, the disease increased, and vice versa. *Trichoderma viride* was an active antagonist in moist soil, but was suppressed under very wet conditions; it was also inhibited by a soil pH above 5.4. Application of coral sand to temporary sugarcane railroad tracks greatly increased soil pH and heartrot caused by *P. parasitica* in subsequent plantings of pineapple, a result of suppression of *T. viride* and the wide pH range of this pathogen (E. J. Anderson, unpublished). Disease control was thus thought to be indirect, through the effect of the treatments on organisms that ordinarily inhibit *Trichoderma* or other antagonists of *Phytophthora*.

Not all the indirect effects of fungicides bring about disease control. Where the chemical selectively inhibits antagonists more than the pathogen, or where competing pathogens are differentially affected by the chemical, the pathogen least affected may cause considerably more damage than without the fungicide. One well-known example is the increase in damping-off of seedlings of many different crops by *Rhizoctonia solani* when Dexon is used to control *Pythium* spp., or the increased seedling disease from *Pythium* and *Fusarium* spp. when PCNB is used to control *R. solani* (Garren, 1963). In still other instances, control of both *R. solani* and *Pythium* has increased seedling damage caused by *Fusarium*.

Although in some instances there is obviously competition among pathogens that cause damping-off, and control of one favors the other, the question may be asked whether Dexon and PCNB destroy antagonists of *Rhizoctonia* and *Pythium,* respectively. They may do so in the increased damping-off of conifer seedlings in Africa caused by *Pythium* (Gibson et al., 1961). The pathogen made better growth through soil incubated in the presence of PCNB vapors than through untreated soil. *Penicillium paxieli* was common in the nontreated, but not in PCNB-treated soil. Moreover, the fungus inhibited growth of *Pythium* on agar in the absence of PCNB, but not when the plates were incubated over PCNB. If *P. paxieli* is, in fact, the cause of the partial suppression of *Pythium* in untreated soil, then fungicides would be profitably selected on the basis of their preferential effect on *Pythium* over *P. paxieli.*

Vaartaja et al. (1964) emphasized that for many soil fungicides, the significant control is probably in part through changes in antagonistic flora, without which the fungicide would be only partially effective. In their work, Dexon alone gave significant control of damping-off of conifers, but stands were poor if Dexon was mixed with anilazine. The stand was also poor when Dexon was mixed with PCNB. The anilazine or PCNB, they suggested, interfered with biological control made possible by Dexon alone. They further suggested that this effect on the soil microflora can explain the great variation in disease control that workers obtain from year to year in fungicide trials.

Although specific antagonists may be involved in the indirect control or the accentuation of disease by fungicides, as suggested for some of the examples above, a large part of the effect is probably nonspecific and related to the total microbiota. PCNB, for example, has a broad-spectrum effect against fungi and actinomycetes, but a narrow-spectrum effect against soil bacteria. The results of Farley and Lockwood (1969) indicate that this suppression of PCNB-sensitive fungi and actinomycetes reduces competition for nutrients for less sensitive pathogens such as *Fusarium* and *Pythium,* which are then more active. Glucose was used more slowly in soil with than without PCNB, but without glucose as an energy source for microbial growth, PCNB had no effect on soil respiration or the number of organisms. The PCNB apparently thus selectively affected growth response to nutrients.

Probably the greatest role of fungicides in biological control will involve their effect on the relative competitive advantage of pathogens versus antagonists. **When any two microorganisms that are in competition are exposed to a differentially harmful factor, other conditions being equally favorable, the microorganism least affected will be most successful.** Many instances of control with sublethal doses of a fungicide undoubtedly involve this mechanism. Sublethal is not synonymous with harmless; even mild dosages are restrictive in some way to the pathogen, even if not measurable by reduced conidial germination or colony growth rate. Where an increase in dosage obviously inhibits the pathogen, it may be assumed that the milder dosage is also harmful, but in a more subtle way. On the other hand, it is common for a toxic compound to be stimulatory at very low dosages, which could also influence the relative competitive advantage of a pathogen or its antagonist.

The potential for selective inhibition or stimulation of organisms or groups of organisms with chemicals clearly exists, but awaits man's ingenuity to ascertain which compounds or rates to use. Clues on the indirect benefits of biocides have been found more or less by accident, in tests where direct kill of the pathogen by the chemical was the objective. Far more progress would be made if research were directed specifically at indirect control, enlisting the aid of antagonistic microorganisms.

10

BIOLOGICAL CONTROL OF PATHOGENS OF AERIAL PARTS

The surfaces of plants carry a non-parasitic flora apart from pathogens that may be deposited upon them. . . . Interactions occur between the organisms on the surfaces. They compete for nutrients and may secrete toxins harmful to others or substances that are mutually beneficial. A weakening of the surface structure . . . may assist attack by another organism. . . . There is mounting evidence that the phyllosphere microflora may be active in suppressing attack by pathogens. . . . Pathogens themselves interact on plant surfaces, the attack by a pathogen on its normal host may be suppressed by the presence of a pathogen of another.

—J. T. MARTIN AND B. E. JUNIPER, 1970

Soilborne plant pathogens have been emphasized in this volume, in part, because most plant pathogens spend some of their cycle on or in the soil. Many pathogens that cause diseases of aerial parts are also found in soil. Perithecia of *Venturia inaequalis* mature in apple leaves on soil; oospores of downy mildews survive for long periods in the soil; teliospores of stem rust of wheat overwinter on leaves on the ground; sclerotia of *Sclerotinia sclerotiorum,* cause of white blight of many plants, carry over between crops in refuse on soil; dodder seedlings start in soil, but quickly become exclusively aerial; nematode-borne viruses are transmitted in the soil environment, although the conspicuous symptoms are in the tops; infection by the fusarium and verticillium vascular-wilt pathogens occurs in the soil, but they occupy the xylem of aerial parts; nearly all leaf-spotting microorganisms spend a portion of their cycle on the ground; stinking smut of wheat is initiated with infection of the seedling from seed- or soil-borne spores through the coleoptile, although the symptoms are aerial. On the other hand, the causal agents of the following diseases, among others, have no known connection with the soil: fire blight of pome fruits; peach leaf curl; white pine blister rust; powdery mildews; the mistletoes; Dutch elm disease; chestnut blight; aphid- and leafhopper-borne viruses.

In this chapter we are concerned with pathogens that infect through aerial plant parts and that spend part or all of their life cycle on aboveground organs. An excellent general summary has recently appeared (Preece and Dickinson, 1971).

The aboveground abiotic environment is very different from that in soil. Temperature and moisture fluctuate more widely and quickly, and the microorganism is exposed to sunlight and ultraviolet radiation; there is greater exposure to weathering and risk of being washed away by rain. The osmotic potential, surface tension, and sorption phenomena of water on leaf surfaces certainly differ from those on the root. Although the microflora of the rhizosphere and the phylloplane are both nourished by exudations from the plant surface, these differ in kind and quantity. There is in both environments intense competition for nutrients, and exposure to lysing enzymes, antibiotics, and hyperparasites. There is greater potential for migration of the pathogen aboveground than in it. Aboveground the pathogen contacts the plant; belowground the plant usually contacts the pathogen.

MICROORGANISMS ON AERIAL PARTS

A restricted characteristic flora is maintained on aerial plant parts in the same way that one is maintained in soil, despite the deposition of many diverse spores on each. Bacteria, yeasts, and yeastlike organisms usually outnumber Hyphomycetes. Unlike the bacteria in soil, pigmented forms sometimes predominate, and many are gram-negative, asporogenous, and slime-forming rods and cocci. The slimy characteristic favors adherence to leaves. Yeasts are generally common, predominate on pine needles and fruit, and may represent 1% of the wet weight of grass leaves in humid areas. *Aureobasidium (Pullularia) pullulans*, *Sporobolomyces* spp., *Tilletiopsis* spp., *Itersonilia* sp., and *Cladosporium* spp. are common fungi on leaves. Actinomycetes are rarely found on leaves and stems, although they are common in soil. The population of microorganisms on leaves is high, may reach a thickness of 22 μ on the surface of tropical plants (Ruinen, 1961–63), and has been estimated as millions of cells per g of tissue. Some of these organisms may be embedded in the wax and cuticle of the plant surface (Preece and Dickinson, 1971).

The subject of epiphytic microorganisms has recently been reviewed by Hudson (1968), Last and Deighton (1965), Last and Warren (1972), Leben (1965), Preece and Dickinson (1971), Ruinen (1961–63), Shigo (1967), and Van den Heuvel (1970).

In contrast to most plant surfaces, enclosed apical meristems have rarely been shown to harbor epiphytic microorganisms, a fact not commonly recognized. In commercial culturing of minute apical meristems to obtain plants free of viruses and microorganisms, it is usually unnecessary to sterilize the tissues before their transfer to nutrient solutions or agar. The meristem cells are so covered by clasping leaf primordia that microorganisms do not penetrate to this surface. Such a condition would appear to have evolutionary survival value for the plant; the vital growing points are little troubled by parasites or saprophytes that might produce metabolites or enzymes that could injure the delicate meristematic cells, or by growth substances that could affect their normal development. Perhaps the root cap serves the same function for root-tip meristems.

Resident Flora

Microorganisms that multiply on exposed healthy surfaces without noticeably affecting the plant are *residents*. They are generally nonspecific and occur on a range of plants, but a few are quite specific. They are widely distributed; fungi, in particular, are similar in the temperate zones of the northern and southern hemispheres. As leaves mature, microorganism numbers generally increase, perhaps because of accumulation of leakage through ectodesmata and the weathered cuticle. Residents may colonize pollen, flower parts, and other debris on the leaf surface. Leaves parasitized by rusts, powdery mildews, foliar nematodes, or mites, or having a mechanical injury, are apt to have higher populations of resident epiphytes than are healthy leaves.

Leaves in a tropical rain forest were reported by Ruinen (1961–63) to have nitrogen-fixing bacteria as primary colonizers, followed by fungi, algae, yeasts, and lichens. Trees here are also colonized by epiphytic higher plants, each of which has its own microflora. Young leaves leak more nutrients than aged ones under rain forest conditions, perhaps because the latter have been exposed to prolonged leaching by rain. In temperate areas the number and variety of epiphytic microorganisms is less than in the tropics, and they occur in succession. Because the climate is generally drier in temperate areas, and because the growing season for epiphytes is drastically shortened by the deciduous habit of the trees, the microflora of leaves is fairly inconspicuous. Algae occur on bark there, but rarely on leaves, whereas in the tropics they occur on both.

Microorganisms on the leaf surface may stimulate the formation of phytoalexins that will retard the activity of parasites there (Bailey, in Preece and Dickinson, 1971), and may also produce growth regulators, such as gibberellins, that may affect plant growth (Last and Warren, 1972).

Plant-pathogenic microorganisms also may have resident phases, as will be explained.

Transient Flora

Transient or casual microorganisms on aerial parts settle there from the air, but do not multiply except perhaps in foreign debris on the surface. The composition of this flora may, therefore, vary considerably with time and location.

Fokkema (1968, 1971) reported that increase in the number of pollen grains on rye leaves 1 month after flowering was correlated with a two- or threefold increase in the number of colonies of *Cladosporium herbarum,* and also with strongly increased infections by *Helminthosporium sativum* and *Septoria nodorum,* but there was no effect on the number of infections of *Puccinia recondita* f. sp. *recondita.* Warren (1972) obtained contradictory results with sugar beet on leaves of which its pollen had shed normally. Yeasts, *Cladosporium* spp., and *Aureobasidium pullulans* became abundant (280,500/cm^2 of leaf). Inoculations with spores of *Phoma betae* gave only 3–5% aggressive infections, and inoculation with rye pollen plus *P. betae* gave only 5% infection. However, on plants that had no flowers or pollen, and therefore low populations (9700/cm^2 of leaf) of antagonistic microorganisms, inoculation with rye pollen plus *P. betae* gave 88% aggressive infections. Normal accumulation of pollen apparently promotes an antagonistic microflora that inhibits pathogens, but pollen applied with the pathogen may provide a food base for infection, and in the absence of inhibitory microorganisms, cause severe disease.

Succession of Microorganisms

Plant pathogens commonly are transients on aerial parts before they penetrate the host. During this brief period they are exposed to interactions with the resident epiphytes. This certainly includes competition for nutrients, and probably exposure to antibiotics and enzymes of the resident flora as well. There is also interaction and succession among the epiphytes themselves. Bacteria may be the first to colonize rapidly growing tissue, and yeasts and other fungi become more plentiful as the host matures and becomes moribund (Leben, 1965).

Parasites that produce a latent infection or a lesion in the leaf on the plant usually maintain their dominance after leaf fall. Similarly, some saprophytes may occupy substomatal chambers of healthy leaves and thus be in position for a quick take-over after leaf fall (the possession principle).

Thereafter the first colonizers are saprophytic fungi (such as Mucorales) that digest sugars, pentosans, and possibly hemicelluloses. The cellulose-decomposing Ascomycetes and Fungi Imperfecti then dominate, followed by the lignin-decomposing Hymenomycetes and Gasteromycetes.

Microorganism succession in mechanical wounds and branch stubs of trees was summarized by Etheridge (in Nordin, 1972), Merrill (1970), Shigo (1967), and Shigo and Hillis (1973). Bacteria usually invade first and cause discoloration of the wood, followed by nonhymenomycetous fungi that do not cause decay, and finally by wood-rotting Basidiomycetes. The succession in *Populus* sp. starts with bacterial invasion of dead branches; about 4 years after branch death *Cytospora* spp. appear; 2 years later *Cytospora chrysosperma, Phoma* sp., and *Libertella* spp. take over; about 8–9 years after branch death *Corticium polyonium* and *Polyporus adustus* appear, and 9–10 years later *Fomes igniarius* var. *populinus* is found. A similar succession occurs in *Acer rubrum,* culminating in *Polyporus glomeratus, Fomes* spp., and *Daedalea unicolor.*

Some wood-rotting fungi, such as *Stereum* spp. and *Fomes annosus,* cannot compete with others, and must, therefore, be pioneer invaders. *Stereum* spp. are some of the most vigorous pioneer Hymenomycetes; they commonly follow *Hypoxylon* spp. and grow in close association with them in living trees, but cannot follow or accompany most other organisms. As others invade, *Stereum* declines (see below). *Stereum purpureum* is unable to invade wood occupied by *Diaporthe perniciosa* or *Sclerotinia cinerea.* As Shigo observed, "the pioneer organisms often determine the successional pattern. . . . pure cultures are rare in nature." *Stereum sanguinolentum* is prevented from infecting dying branches of *Abies balsamea* because nondecay fungi such as *Retinocyclus abietis* and *Kirschsteinella thujina* quickly colonize them.

Trees killed suddenly by fire or insect attack are commonly invaded by blue-stain fungi and sapwood-decay fungi. Competition is intense in dead wood, and the process is degradative in character.

Although forest pathology was founded on studies of wood decay caused by Basidiomycetes, it is now realized that this usually is the final phase of an ecological succession. Biological control has been successfully applied to a pioneer invader, *Fomes annosus,* which is sensitive to competitors. Whether it can be applied to wood-decay fungi further along in the successions, or whether it should be applied only to the pioneer invaders remains to be determined. However, the substitution of an antagonistic bacterium for the usual pioneer bacterial invader in order to alter the succession apparently has not been attempted.

Ricard and Bollen (1968) examined sound Douglas fir poles to determine why they escaped decay. The fungus *Scytalidium* sp. was found to be present; it penetrated wood readily without altering its strength. Invaded wood, even after it had been autoclaved, was undamaged by *Poria carbonica*, which decayed uninvaded wood. *Scytalidium* produces a thermostable antibiotic scytalidin (Stillwell et al., 1973), which is strongly inhibitory to *Poria* and other fungi. The possibility of commercial use of this antagonist is being studied in Oregon and Sweden. Six antagonistic fungi were also found to inhibit five different decay fungi introduced subsequently into wood blocks (Hulme and Shields, 1972).

Ricard (1970) originated the term "IC" (immunizing commensality) for microbial associations in which a noninjurious antagonist (*Trichoderma album, T. viride, Coryne sarcoides, Scytalidium* sp.) grows on the dead or living substrate and inhibits or prevents the development there of a harmful microorganism. Grosclaude et al. (1973) inoculated plum trees in the field with a *Trichoderma viride* spore suspension on the pruning shears at the time of pruning, and inoculated them with *Stereum purpureum* 2 days later. This completely prevented silver leaf, as compared with 25–75% infections in the checks.

There is also a succession in invasion of coniferous needles. Thus, *Hypodermella mirabilis* and *Bifusella faullii* invade needles of *Abies balsamea* and produce pycnidia there about a year later. *Stegopezizella balsamea, Lophodermium autumnale,* and *Leptosphaeria faullii* then follow and prevent formation of the ascigerous stage (Shigo, 1967). Once on the ground, saprophytic fungi that can grow at very low temperatures often are the most aggressive invaders.

PATHOGENS ON AERIAL PARTS

Plant pathogens invade aerial parts of the host through unbroken surfaces or natural openings, wounds, or senescent or dead tissues. When in the epiphytic phase before infection, they are vulnerable to biological control.

The stored nutrients in spores of pathogens are important in determining their ability to infect. Garrett (1970) pointed out that **specialized fungus parasites require fewer spores to infect than do low-grade pathogens.** Rusts, for example, will achieve 10–50% infection from single spores, whereas *Botrytis fabae* will require about four spores, and *B. cinerea* more than 100 times this number to achieve 10% effectiveness. The ability of spores to infect also declines with age, because of depletion of nutrient reserves. The energy available to spores determines their ability to overcome both the resistance of the host and the competition of the epiphytic flora. Furthermore, the poorer the energy level of a spore, the more susceptible the spore becomes to toxic materials. When infections of *B. fabae* on broad bean are scattered over the surface, host resistance keeps them small, but when they are crowded, host resistance is broken down and larger lesions develop.

Hudson (1968) observed that aerial plant parts are colonized by fungi while still standing—by specialized parasites while the plant is in prime condition, by facultative and weak parasites during the period of senescence, and by saprophytes after the shoot or leaf dies, but before abscission. Nutrients leaking from aerial parts of the host are as important in determining infection by pathogens there as are root exudates on subterranean parts.

Conditions are sufficiently different in the aerial and subterranean habitats, however, that no pathogen operates with equal effectiveness or in the same manner in both. It was suggested by Baker (in Parmeter, 1970) that the ubiquitous *Rhizoctonia solani* may have evolved from saprophytic types that developed near the soil surface. Some strains evolved parasitic capability in that zone, and from them were selected strains that were able to attack roots at depths of 30 cm or more. Other strains that evolved were able to attack aerial parts and to persist there.

Infection Through Unbroken Plant Surfaces or Natural Openings

Taphrina deformans persists as resident yeastlike cells on the bark of peach trees, as it does in culture media; mycelia develop only in infected leaves. Fitzpatrick (1934–35) and Mix (1935) showed that saprophytic "sprout conidia" multiply on the surface of the tree and are washed by spring rains onto opening leaf buds. The conidia can infect young leaves directly through the epidermis. The leaves grow so rapidly in summer that the susceptible period is too brief to permit successful establishment of the pathogen, although it can rapidly penetrate young leaves at 10–24° C. Fitzpatrick stated that it is "most probable that the spores of *Taphrina deformans* actually became part of the normal surface flora of peach trees in districts where leaf curl is present." This epiphytic phase should be subject to the effects of the saprophytic flora of aerial parts. Application of selective nutrients to stimulate antagonists, or perhaps of a combination of selective nutrients and antibiotics, might offer a reasonable chance of successful control of this pathogen.

Pseudomonas glycinea was found by Leben (1969) and associates to increase and spread to the tops of soybean plants under humid conditions from inoculated seeds planted in soil. The bacteria are perhaps carried up on the apical bud, and move locally by swimming. Insects and mites may also transfer this pathogen, as well as resident epiphytes, from plant to plant (Leben, 1965). Bacterial colonies could be seen on the surface of buds with the scanning electron microscope. Chakravarti et al. (1972) found that "buds" (the terminal 1.5 cm of stem) from field-grown soybean plants had 10^6–10^7 bacteria/g, wet weight. Of these, 25% were antagonistic *in vitro* to *P. glycinea.* The bacterial parasite of bacteria, *Bdellovibrio bacteriovorus,* was found (Scherff, 1973) to inhibit necrotic lesion development and systemic toxemia from *P. glycinea* when soybean leaves were inoculated (by wounding) with *P. glycinea* and *B. bacteriovorus* in ratios of 1:9 or more.

Is the interpretation correct that the unimportance of fire blight in colonial orchards in the northeastern United States for 150 years was caused by innate resistance resulting from semiwild growing conditions and lack of care? We do not know that blossoms of neglected trees are less susceptible to blight infection than those under proper cultivation. Apparently the epiphyic microflora of old abandoned orchards or of native stands of hawthorn, crabapple, serviceberry, and mountain ash in the northeastern United States has not been investigated as a possible source of antagonists.

According to Cunningham (1931), "When fireblight appeared in New Zealand [in 1919 in the northern North Island] it destroyed in a single season many acres of pear trees and threatened to become a menace to the fruit industry. But within two or three seasons . . . it had ceased to cause more damage than the death of occasional branches of pear trees, and the destruction of a proportion of blossom spurs on apples and pears. When the disease appeared in epiphytic form in . . . [the south of the North Island] in 1927 it likewise appeared at first in a virulent form, but after a season or so appeared to become relatively of slight economic importance. As a result, fireblight is to-day regarded as a serious disease only by growers in those localities in which it has not made its appearance. . . . in most commercial orchard areas the disease is seldom now troublesome, and rarely appears in epiphytotic form." Apparently the pathogen carries over in the abundant hawthorn hedges there. D. W. Dye (personal communication) reported that the disease is still of minor consequence even though the 1972–73 season was "considered by many pear growers . . . to be the worst fireblight season they have known, and for many the first time they have seen fireblight." New Zealand would appear to be an interesting place to seek antagonists for biological control of fire blight. Might not these situations be comparable to that of continuous monoculture on soilborne pathogens?

Inoculation of nectaries of pear with an effective antagonistic microorganism might reduce losses from fire blight in areas where transmission is largely by bees. This might be done by spraying with inoculum, as is now done with streptomycin, or an inoculum might be prepared that would be attractive to bees, so that the antagonist would be transferred by them with the pathogen. *Erwinia herbicola,* avirulent isolates of *E. amylovora,* or perhaps some other microorganisms might be used for this purpose. The yellow bacterium *E. herbicola* commonly associated with *E. amylovora* in fire-blight lesions, has been found (Chatterjee et al., 1969; Preece and Dickinson, 1971; Riggle and Klos, 1972) to inhibit development of the disease *in vivo,* and growth of the pathogen *in vitro.* Apparently *E. herbicola* produces a β-glucosidase that decomposes the arbutin of the host to hydroquinone and D-glucose. The hydroquinone inhibits the oxidative metabolism of D-glucose by the pathogen, and thus reduces its growth.

Keil and Wilson (1963) reported that mixing bacteriophage with suspensions of *Xanthomonas pruni* before spraying on plants in the glasshouse, or the spraying of phage on the plants before inoculation, gave control equal to that provided by application of zinc sulfate-hydrated lime. However, the use of bacteriophage for control of bacterial diseases of plants or animals generally has not proved successful.

Rust fungi infect directly through the epidermis or through stomata, and usually do not require exogenous nutrients for germination. They are, therefore, less subject to competition from epiphytes than are most fungi during infection. However, these obligate parasites are commonly subject to attack by antagonists, once the pustules (on leaves) or cankers (on stems) have ruptured for spore release.

Cronartium ribicola infects needles of white pine and advances into the stem, and after 2–3 years produces a perennating canker, which soon girdles the trunk or branch, killing the distal portion. *Tuberculina maxima* was known for many years to attack the fungus in these cankers, but was not considered to be very effective in control. J. L. Mielke concluded in 1933 that the fungus "is not considered an important factor in reducing aecial sporulation or in controlling *C. ribicola.*" The rust had spread into the Inland Empire of the Pacific Northwest by 1923 and had intensified by about 1931. R. T. Bingham and J. Ehrlich in 1941 estimated that 0–14% of the trees in this area had been saved by *T. maxima,* and that aecia had been reduced by 5-60% in the Pacific Northwest. Kimmey (1969) also reported for this area that "complete inactivation of cankers was rare during the period 1928–37, regardless of cause." However, *T. maxima* began to attract attention on the Pacific Coast with the finding that it, rather than the extensively applied phytoactin and cycloheximide, was the cause of the inactivity of many of the cankers observed. A 1965 survey by Kimmey (1969) of 48 young unsprayed stands revealed that 62% of all lethal-type trunk cankers were inactivated, and aecial production was reduced to 5.4% of potential on them. "*Tuberculina maxima* [is] believed to be the principal cause of the . . . inactivation. . . ." His figures were 2–3 times those of 1941 for percentage of trees saved. It seems clear that *T. maxima* has, for undetermined reasons, become increasingly effective in the past 30 years in this area. Because of the abundance of the rust in the extensive nearly pure stands of white pine (Mielke, 1943), conditions were favorable for this intensification. Could this situation be comparable to that of continuous monoculture for *Streptomyces scabies* on potato or *Gaeumannomyces graminis* on wheat (Chaper 4)? The effectiveness of *Tuberculina* against *Cronartium* is in accordance with the observation (Chapter 7) that hyperparasites are more effective against secondary than against primary infections, although it is not a hyperparasite.

E. F. Wicker and associates found in 1967–70 that *Tuberculina* conidia infect sporulating pycnia or aecia of *Cronartium,* and sporulate in them 3–5 weeks later. The antagonist invades only tissue penetrated by rust mycelia, and once it has killed the rust it dies. Pectin occurs in healthy pine tissue but not in rusted or callus tissues, which are invaded by *Tuberculina.* One year after infection by the antagonist, the cankers are hard, dry, and black. *Tuberculina* winters as mycelia in active rust cankers and as conidia, and is able to produce spores whenever cankers are erumpent and susceptible. The fungus is now abundant on western white pine, but is erratic on sugar pine, and is rare on eastern white pine. Wicker (1968) thought that "the direct canker treatment using these antibiotics may have been detrimental to blister rust control rather than just ineffective," because they are toxic to *Tuberculina.*

Nectria fuckeliana, Diplodia pinea, and *Gibberella lateritium* often kill substantial numbers of rust galls caused by *Peridermium harknessii* in California pine forests (Byler et al., 1972).

Puccinia antirrhini infects snapdragon leaves through stomata, and the uredospore pustules open in about 10 days, releasing new spores. Under conditions of low humidity, rusted leaves desiccate, but under humid conditions they remain turgid and show little damage although infected (Dimock and Baker, 1951). Under the latter conditions, however, the open pustules are commonly invaded by *Fusarium roseum* from soil. At 10° C the *Fusarium* invades only the center of the pustule; at 16° C it advances to the limits of the pustule; at 21–32° C it advances into healthy tissue, decays the leaves, and girdles the stems. Applications of Bordeaux mixture prevented infection by *Fusarium,* but failed to control rust. *Fusarium* produced no lesions when inoculated onto nonrusted leaves, but certain isolates in water drops were shown by W. Siefert and A. W. Dimock (unpublished) to cause collapse of the leaf mesophyll under the drops, but without invasion. Some isolates did not produce this effect and did not invade the tissue around the lesion at 21–30° C. Dimock commented,

"Highly efficient biological control of the rust, but a bit rough on the snapdragons!" However, by selecting among isolates from a large number of clones of the antagonist, particularly from lesions in which they did not advance beyond the rust lesion, an effective and noninjurious control agent might be obtained.

Snow mold, caused by *Typhula idahoensis* and *Fusarium nivale,* has long been known to destroy overwintering inoculum of stripe rust *(Puccinia striiformis)* in winter wheat in northern Washington. These cryophilic fungi cause indiscriminate destruction of the leaves under snow (Chapter 9). This hazardous type of biological control destroys any carryover of stripe rust in that area. The plants recover from crown tissue belowground, and have no stripe rust.

Crosse et al. (1968) found that application of a 5% urea spray to apple trees just before leaf fall caused rapid decomposition of leaves, and increased their bacterial flora 170-fold over that on unsprayed leaves by December, and 48-fold by January. After the spray the flora shifted from predominant gram-positive chromogenic leaf epiphytes to gram-negative nonchromogenic soil inhabitants. The combination of faster leaf decomposition, increased nitrogen content, toxicity of urea, altered type and number of bacteria, and increased ingestion by worms combined to reduce drastically the number of perithecia of apple scab *(Venturia inaequalis)* formed in the leaves, and of ascospores discharged by them. A *Pseudomonas* sp. was important in leaf decomposition, and a yellow, gram-negative, peritrichous rod inhibited ascospore release. A second spray of 2% urea on fallen leaves just before bud-bursting suppressed ascospore production and release. This general approach to pathogen control is promising, since the microflora on the tree can be altered by spraying with a fertilizer at little additional cost over regular fertilizer practices. Application of selective fungicides or antibiotics may also be useful. A captan spray on apple leaves increased the relative abundance of bacteria and yeasts, as compared with filamentous fungi (Hislop and Cox, 1969). However, some chemicals decrease palatability of fallen leaves to earthworms, and thus diminish scab control (Preece and Dickinson, 1971). Would application of various forms of nitrogen and cabohydrate, possibly combined with selective antibiotics, to aerial parts in the spring cause a different group of microorganisms antagonistic to the pathogen to occupy the plant surface?

Aureobasidium pullulans isolated from bean leaves reduced the number of infections of *Alternaria zinniae* on bean, a disease not known in the field (Van den Heuvel, 1970). Inoculation with *A. pullulans* 1 day before that with *Alternaria zinniae* reduced infections by 52%, on the same day by 47%, and 1 day after *Alternaria* by 17%, as compared with infections without inoculation with *A. pullulans*. Since few viable cells of *A. pullulans* remained on the leaves 7 days after inoculation, the antagonistic effect was on infection. The number of germ tubes per *Alternaria* spore was markedly reduced, and the percentage of germination slightly decreased. *Aureobasidium pullulans* inhibited *A. zinniae* only slightly in agar culture, and was considered to produce only low levels of an antibiotic. *Alternaria tenuissima* produced more antibiotic in culture and was more effective in reduction of leaf lesions by *A. zinniae.* It was concluded that the inhibitory effect in each instance was largely caused by phytoalexins or induced inhibitors formed by the leaves under stimulation of epiphytes. Antagonistic epiphytes, therefore, are best applied before inoculation with the pathogen.

Cicinnobolus spp. are common mycoparasites on powdery mildews in nature, but have seldom been observed to reduce the abundance of those pathogens or their injury to the plant.

Competition for nutrients probably is intense on aerial plant parts, and this aspect of antagonism is undoubtedly important in the prepenetration phase. Bacterial pathogens, which penetrate through natural openings and wounds, but not through epidermis, would be exposed to antagonism by resident epiphytes. Most fungus dispersal spores are partially to wholly nutrient-dependent in germination, an ecological advantage helping to insure that germination does not occur in locations lacking materials needed for continued growth. Thus, like bacteria, germ tubes of those spores are subjected to competition by resident epiphytes.

Emmett and Parbery (1974) pointed out that microorganisms growing on the surface of leaves may derive nutrients from: (a) aerosols produced by other plants, released into the atmosphere, and settled on all surfaces contacted; (b) metabolic by-products of other phyllosphere microorganisms; (c) chemical components of the host wax and cuticle; (d) metabolites from cells within the leaf that diffuse to the leaf surface. These materials may greatly affect the phyllosphere flora. Chemical pollutants of the atmosphere may also strongly affect this flora.

As A. Kerr and N. T. Flentje pointed out in 1957, parasites of aerial parts must either infect or die, unless blown or washed to a new site. There is a restricted time for infection, limited by favorable environment and leaf abscission. On the other hand, propagules of some pathogens may remain dormant in soil for long periods until stimulated to germinate by the presence of the host. There is thus less selective pressure for pathogens of aerial parts to require exogenous nutrients than there is for root pathogens.

Infection Through Wounds

Several *Pseudomonas* spp. may possess a resident phase in their life cycles on buds or leaves, and these organisms readily infect after the plant is wounded. Consequently, a disease outbreak may occur when such contaminated plants are injured by frost, hail, or blowing sand, but noncontaminated plants remain healthy. Freezing injury to pear flowers may in this way initiate an epidemic of flower blight. An outbreak of a bacterial disease may be preceded by an increase of a resident pathogen on the plant surface. Epiphytes may compete with the pathogens at that time (Preece and Dickinson, 1971).

Pseudomonas mors-prunorum was found by Crosse (1965) in England in 1955–65 to be resident in large numbers on healthy cherry leaves, along with *Aureobasidium pullulans* and saprophytic bacteria. The pathogenic bacteria were fixed to the leaves in dry weather, but were distributed by rain. The bacteria were unable to infect twigs during the growing season, but as twigs became dormant, infections occurred through leaf scars. Bacteria in the resulting cankers died by early summer, and those in leaf spots by early fall. Pathogen survival was thought to be reduced by epiphytic bacteria on leaf surfaces. The number of *P. mors-prunorum* on leaves of the susceptible variety Napoleon was three times that on the resistant Roundel, on which the threshold population required for infection was also higher than for Napoleon. This lower population, caused by the saprophytic epiphytes, combined with the required numbers of the pathogen for infection, was thought to explain resistance of the Roundel variety. The quantity of pathogen inoculum increased with age of the tree, perhaps because of increased shading and protection from rain, and slower drying of leaf surfaces. The populations were reduced by rain, desiccation, ultraviolet radiation, and perhaps bacteriophage. The pathogen outnumbered gram-negative epiphytic bacteria on leaves by 3 or 4 to 1. An *Erwinia* sp. from the leaves reduced infection when placed on the host *with* the pathogen, but disappeared in a few days when sprayed on trees in the field. This bacterium was therefore thought to be a transient rather than a resident on cherry leaves. It may prove as difficult to establish an antagonist on leaves as in soil, and the rules of the game are probably similar.

Fungi, actinomycetes, and bacteria on peach bark were found (Wensley, 1971) to be antagonistic to *Leucostoma cincta,* cause of perennial canker. The antagonists were prevalent in the spring and fall infection periods. Differences in varietal susceptibility were thought to be caused both by host resistance and by the population of antagonists on the plants.

A suspension of spores of *Fusarium lateritium* sprayed on apricot trees within 24 hours after pruning was found by Carter (1971) to provide some protection of wounds from infection by *Eutypa armeniacae.* The percentage of branches infected when inoculated by *Eutypa* was 55; inoculated with *Fusarium* and 24 hours later with *Eutypa,* 36; sprayed with thiobendazole and 24 hours later with *Eutypa,* 15; sprayed with benomyl and 24 hours later with *Eutypa,* 5.

It was suggested that this dieback disease of apricot intensified in South Australia after fall applications of a copper fungicide began to be used to control *Clasterosporium carpophilum,* perhaps as a result of the reduction of surface antagonists (Carter, 1971). Application of 0.3 ppm thiobendazole plus *F. lateritium* to pruning wounds was found by Carter (in Anonymous, 1973) to provide an integrated control of *Eutypa.* The fungicide apparently is effective against *Eutypa* but not *F. lateritium,* and therefore provides immediate protection before the antagonist becomes effective.

The silver-leaf disease of plum trees, caused by *Stereum purpureum,* was found by Grosclaude (1970) to be diminished by application of *Trichoderma viride* to fresh wounds. Painting the wounds with mercuric oxide, however, increased the percentage of infections from 18.6 to 79.1, presumably because of reduction of antagonists.

Bacillus subtilis isolated infrequently from leaf scars of apple in winter in Northern Ireland was found (Swinburne, 1973) to reduce the number of infections of such scars by *Nectria galligena* when sprayed on twigs immediately after leaf fall and prior to infection by *Nectria.*

Bier (1964) and associates in 1962–66 studied conditions affecting the production of cankers on poplar by *Hypoxylon pruinatum,* and on willow by *Cryptodiaporthe salicella.* In temperate climates the bark is somewhat dormant for 7 months each year, and during this time the cankers develop. The percentage of water present in bark tissue to the maximum that the tissue can hold in year-old stems drops below about 80 during that period. The saprophytes were actively growing on and in bark (lenticel tissue, nodes) of high water content, where they helped protect against infection by canker fungi. The saprophytes were less active in bark of low water content, in which canker fungi were able to develop. Several common saprophytic fungi and unspecified yeasts and bacteria were commonly involved. When a piece of healthy bark or leaf was shaken in sterile water for 5 days to increase the saprophyte population, and this decoction was applied to poplar cuttings or trees in the laboratory, attack by *Hypoxylon* was delayed. Autoclaving the bark, or passing the decoction through bacterial filters, destroyed the protective effect. This method perhaps has some effect in common with the prolonged soaking of cereal seeds in water to reduce transmission of pathogens, that is, an increase in inoculum of antagonistic microflora. Wood and French (1960) found that *Hypoxylon* ascospores discharged in the field were frequently contaminated with bacteria that reduced spore germinability. They showed that ascospore inoculum was discharged during the susceptible winter period when bark had a low water content.

Bier also found inhibitory effects of saprophytes on fir heartwood against *Stereum sanguinolentum.* Chemicals toxic to the pathogens would also inhibit the natural antagonistic flora.

A means of successfully preventing infection of pine stumps by *Fomes annosus* was devised by Rishbeth (1963) in 1950–63. Immediately after the tree is felled, the stump surface is inoculated with oidia of *Peniophora gigantea,* and the fungus occupies the cut surface as the dominant pioneer organism. The inoculum is produced on malt agar and washed off with a sucrose solution. To the suspension is added talc and Cellophos B600, and the mixture is poured into molds and desiccated to form tablets. Each tablet, dispersed in 100 ml water, gives a suspension of at least 1×10^6 viable oidia per ml, and will inoculate about 100 stumps. *Peniophora* spreads through the stump into the lateral roots, and checks the advance of *Fomes* in existing root infections. Sporophores of the antagonist are produced within a year, and the stump is rapidly decayed. Artman (1972) added the oidia to lubricating oil placed on the chain saw, eliminating an extra procedure in inoculation of the freshly cut stump. Apparently the effectiveness of Rishbeth's method is a result of successful competition and replacement of *Fomes* by *Peniophora,* made possible by mass inoculation (1×10^4 oidia per stump of 16 cm diameter) of a freshly exposed surface. *Stereum sanguinolentum* and *Polystictus abietinus* cannot replace *F. annosus* when inoculated at the same time. *Fomes annosus* can grow along roots occupied by only a few fungi (*Cylindrocarpon radicicola* and some blue-stain fungi). *Peniophora* appears to be an excellent choice for biological control of *F. annosus* on pine. However, it is ineffective on spruce in Norway, where *F. annosus* causes heartrot. Ricard (1970) found that *Trichoderma polysporum (T. album)* and *Ascocoryne (Coryne) sarcoides,* which occurred naturally in living spruce trees, prevented decay by *F. annosus* when inoculated into spruce logs. The possibility of inoculating spruce trees of various ages with these antagonists to diminish injury by *F. annosus* is under study.

Because of the more extensive leakage from injured host cells in wounds than from normal epidermis, competition with resident epiphytes for nutrients may be unimportant for many parasites infecting wounds. *Fomes annosus,* however, is apparently an exception. Because of the available substrate, however, organisms that produce antibiotics may be present and impede infection.

Infection Through Senescent or Dead Plant Parts

Old flowers, dead bracts, or leaves with salinity injury provide excellent infection courts for facultative parasites. Because epiphytic saprophytic antagonists may also infect such tissue, the potential for biological control is rather good for this type of pathogen.

Botrytis cinerea, the common gray mold of many kinds of plants, requires specialized conditions for infection (Baker et al., 1954):

1. Senescent or dead tissue in contact with living host tissue. Fallen petals, bracts, or pollen on the surface of leaves frequently serve as a food base for penetration of normal tissue. Many flowers have passed through the climacteric by the time they are fully open, and are very susceptible to *Botrytis* infection. The fungus infects senescent cells and forms toxins and enzymes that kill and digest the adjacent living tissue, and then spreads behind this advancing margin of dead tissue.
2. Free moisture permits development of the fungus, but unless well established in the tissue, it is killed if the lesion dries. If established, growth ceases until moist conditions return, but the fungus does not die. Lowering humidity to 80% will check an epidemic in the glasshouse.
3. Cool conditions (7–15° C) favor infection. Raising glasshouse temperatures to 21–27° C frequently will check an outbreak.

Newhook (1957) in 1951–57 used *Cladosporium herbarum* and *Penicillium* sp. to control *Botrytis* on lettuce leaves and on dead petals of tomato (which leads to fruit decay). Application of antagonists to recently dried tomato petals not yet colonized by *Botrytis* gave complete control of the pathogen, but was only 30% successful on petals that had dried several days before, and presumably were already colonized by the pathogen. *Cladosporium* and *Penicillium* were used because they colonized dead tissue under drier conditions than did most other microorganisms.

Field-grown lettuce in England was attacked by *B. cinerea* through injured or senescent basal leaves, especially on plants growing on ridges (Newhook, 1951). Soil that washed onto the bottom leaves in low areas inoculated the leaves with antagonists. Inoculation of lettuce and other plants with a soil suspension also reduced *Botrytis* infections. At 20° C after 8 days there were 21 *Botrytis* lesions out of 24 inoculations on lettuce leaves detached from check plants, and none on soil-inoculated leaves; at 4° C, where activity of antagonists was reduced, the figures were 20 and 7. [One wonders whether the occasional infection by soil organisms might have been prevented by treating the soil suspension at 60° C/30 minutes to kill pathogens.] Earthing up around plants promoted decay of petioles of senescent leaves and prevented *Botrytis* from entering the stems. Application of suspensions of *Cephalosporium, Fusarium,* or *Phoma* did not decrease *Botrytis* infection of lettuce. Inoculation with a mixture of soil bacteria was more effective than with individual organisms if applied before *Botrytis* infection. *Bacillus* sp. and *Escherichia coli* were effective only at 15–20° C, but a *Pseudomonas* sp. was effective at 4° C. Biological control required, in general, that the antagonist be inoculated before infection by the pathogen, and that moist conditions or high humidity and moderate temperatures (about 20° C) prevail.

The technique of using soil suspensions, or the washings from incubated plant parts, to inoculate plant surfaces with mixed antagonists may be germinal. Selection of the soils used, and possibly increasing the inoculum as outlined in Chapter 5, might enhance effectiveness, and fortifying with selective nutrients and antibiotics might intensify the effect. This approach is reminiscent of the suggestion of Le Berryais (1785) that pruning wounds of trees should be coated with fresh mud. A test by Grosclaude (1970) confirmed the efficacy of this method. Mud was applied to the fresh wounds for 24 hours before inoculating with a *Stereum purpureum* culture; this series gave 30% silver-leaf infection, compared with 100% in the checks.

Blakeman and Fraser (1971) showed that unidentified bacteria on the surface of chrysanthemum leaves were responsible for the failure of *Botrytis cinerea* to germinate in water drops there and to infect. Infection would occur, however, if dextrose was added to the drops, or if leaves were senescent. Lack of infection was not caused by host antifungal materials, and required the presence of the bacteria, which increased in number in association with *Botrytis* spores. Blakeman (1972) found that the total number of bacteria, and the relative proportion of a yellow *Pseudomonas* sp. that inhibited germination of *B. cinerea* spores, increased as the age of a beet leaf exceeded 9 weeks. The increased leakage of amino acids with leaf age was thought to selectively favor the *Pseudomonas,* which inhibited spore germination. There was no inhibition of germination of *Botrytis* spores by the sterile filtrate from bacteria-containing infection drops from the leaves. These studies reemphasize the value of seeking antagonists where a disease does not occur, rather than where it does (Chapter 3).

Latent Infections

Certain fungi penetrate the epidermis of actively growing tissue, but are prevented from further development by host resistance. They persist as thick-walled mycelia in substomatal chambers or under the cuticle and remain in a dormant condition until the host tissue matures and becomes senescent, when pathogen growth is resumed. The same type of situation exists in roots where saprophytic fungi inhabit outer tissues of healthy roots, and initiate decay when the root becomes senescent or dies. This is comparable to man's staking out a mining claim for later exploitation.

Gloeosporium musarum infects green Cavendish banana fruit, but remains dormant for up to 5 months, and produces decay only after the fruit begins to ripen. Simmonds (1963) found that a toxic material was present in green fruit, and that the fungus produced little pectinase until the fruit ripened. Many of the latent infections did not resume growth. Similar infections occur with *Colletotrichum phomoides* on tomato fruit, *Guignardia citricarpa* on orange fruit, and *Glomerella cingulata* on mango. Fumigation of green papaya fruit with methyl bromide to destroy the fruit fly breaks latency of *G. cingulata* and increases the number of lesions.

Spores of *Heterosporium (Acroconidiella) tropaeoli* infect green fruits of *Tropaeolum majus*, but lesions do not develop until the fruits turn yellow and become senescent. Invasion then continues until the seed is penetrated (Baker and Davis, 1950). Infection of gladiolus leaves and stems by *Stemphylium botryosum* follows almost the same course.

Latent infections of the above types have survival value for the pathogen, but may have a negative effect on biological control. Host resistance to invasion is breached by the pathogen just enough to establish a bridgehead, but prevents other organisms from infecting. As the tissue matures and becomes susceptible, the resident grows and occupies it (the possession principle). Since epiphytic residents usually increase in numbers with maturation of host tissue, latent infections probably enable a parasite to evade a good deal of competition.

Perennial Mycelia

Some rusts develop perennial mycelia, which persist for several years in woody tissue. *Cronartium ribicola*, already mentioned, does this in pine stems, and *Gymnoconia peckiana* survives in *Rubus* stems and systemically invades leaves as they are formed. This is an effective means of avoiding the inhibitory effect of resident epiphytic antagonists, since a single systemic infection is protected within host tissue, and can produce a maximum number of spores. However, this may be offset by providing a larger target for infection by antagonists for a longer period of time. *Tuberculina maxima* is possibly a successful antagonist of *C. ribicola* because it attacks the perennial stage of the rust on pine instead of leaf infections on *Ribes*, thus minimizing the effect of competitive residents.

Viruses and Mycoplasmas

Since most plant viruses infect through aerial parts, they fall within the purview of this chapter. Many viruses are placed within the plant by insect vectors during feeding or are seedborne. For these reasons, and because they multiply only in living cells, there is little opportunity for direct effect by antagonists in the usual sense. Biological control of insect vectors has recently been reviewed (DeBach, 1964; Huffaker, 1971; Rabb and Guthrie, 1970), and is not, therefore, discussed here. However, it offers a valuable means of biologically controlling plant viruses.

Control of plant viruses by cross-protection with attenuated or related strains of the viruses shows some possibilities. Hazards of the method include expense, the increased risk of infection of plants by a highly infectious virus such as tobacco mosaic during inoculation, the virus reservoir provided by infected plants, and the possibility that protection might not extend to some strains of the virus. Attempts to protect fruit trees from virulent viruses by prior inoculation with an attenuated strain have not often come into commercial use because mild strains may themselves cause some reduction in yield, and because of the risk that the virus may mutate to a more virulent form, or that subsequent infection by a different virus might produce greater loss than would have resulted from infection of healthy plants (Matthews, 1970). Such intensification of effect by a mixture of viruses is well known. For example, the combination of potato X virus and tobacco mosaic virus in tomato produces a severely damaging tomato streak disease.

Grapefruit infected with mild stem-pitting strains of the tristeza virus have been protected for 14 years from more severe strains of the virus in New South Wales. Since they are growing on rootstocks that minimize the effects of the virus, little injury results (Fraser et al., 1968). The tristeza virus of citrus produces a serious disease when orange trees are grown on sour-orange rootstocks, because of severe vascular injury at the graft union. Planting oranges grafted on trifoliate orange, mandarin, or certain other rootstocks has minimized losses from tristeza virus and is now a standard method of control in many areas of the world (Wallace, 1956). The use of hybrid rootstock of passion fruit infected with a mild strain of the woodiness virus has also commercially protected the plants in Queensland against severe strains. Because the rootstock on which the vine is grafted is resistant to fusarium wilt, control of the woodiness virus entails no additional expense (Simmonds, 1959; Greber, 1966).

An early-inoculation procedure was suggested by Broadbent (1964) to minimize the injury from tobacco mosaic virus on glasshouse tomatoes. This highly infectious virus usually spreads to all glasshouse plants before they mature; if infection occurs after flowering commences, fruit yield and quality are severely reduced. By intentionally inoculating the tomato seedlings before planting in the beds, crop losses are greatly reduced, although yield does not equal that of healthy plants. Mild-strain protection was not effective for this disease.

Control of insect-borne virus diseases by isolation of new fields from sources of viruliferous insects is another type of biological control because it involves both the host and an insect. Shepherd and Hills (1970) showed, for example, that aphid-borne beet mosaic and beet yellows were effectively controlled in California by planting new sugar beet fields 19–24 or 24–32 km, respectively, from old infected fields.

The use of virus-free seed or vegetative planting material is another very effective means of biological control involving the host. The source of primary inoculum is thus eliminated, and the pathogen cycle disrupted. Planting certified virus-free seed potatoes is now the standard method for prevention of potato mosaic and leaf roll.

Infection of a plant by a virus may greatly intensify plant loss in the field from root-infecting organisms. For example, Beute and Lockwood (1968) showed that peas infected by bean yellow mosaic virus or common pea mosaic virus released more amino acids and carbohydrates from the roots 4–16 days later than did noninfected plants. This increased the inoculum of *Fusarium solani* f. sp. *pisi* and *Aphanomyces euteiches* in the rhizosphere, and the number of infections, but not the severity of individual lesions. Hourly leaching of the soil removed the exudates and prevented intensification of fusarium root rot, and addition of the leachate to pots of plants noninfected by virus increased root-rot lesions. The number of root-rot lesions was increased only when invasion followed virus infection by less than 12 days, that is, during the period of shock effect. A similar intensification of root rot from *Rhizoctonia solani, Pythium* sp., and *Thielaviopsis basicola* is commonly observed on sweet peas in the Lompoc Valley, California, following early infection by pea enation mosaic virus (K. F. Baker and W. C. Snyder, unpublished). Gold and Faccioli (1972) demonstrated severe effects on tomato roots from infection of the plants with the curly-top virus. Similar symptoms on roots and tops were produced by exposure to ethylene and by infection with the virus. Since viruses are known to induce ethylene production in plants, and ethylene is a factor in dormancy and fungistasis (Chapter 6), it is probable that infection of plants by viruses may thus influence the invasion of roots by microorganisms.

Other methods of biological control of virus diseases have proved more useful in practice. The pear decline mycoplasma, severe in commercial pear varieties grafted on the rootstocks *Pyrus serotina* or *P. ussuriensis,* has been commercially restricted by using *P. calleryana* or Old Home instead (Griggs et al., 1968).

Where curly top virus causes frequent losses to sugar beet or tomato in the semiarid western states, it is commonly controlled by planting earlier than usual (Bennett, 1971; Chapter 2). By the time the beet leafhoppers migrate, the crop plants present an unfavorable habitat. Replacement control of disturbed desert vegetation to restore it to plants unfavorable to the beet leafhopper (Bennett, 1971) is another type of biological control. Spotted wilt of tomato is rare in hot interior valleys of California, but it is prevalent on the coast. The thrips vectors are favored by the cool moist conditions of the coast, and are rare in the dry interior valleys. Tomatoes grown in the latter areas thus escape infection by spotted wilt. These are examples of biological control operating through the insect vector.

EXUDATION TO EXTERNAL SURFACES

Aerial parts of plants, as well as roots, leak and absorb materials. There is also a good deal of leaching by rain from the exposed parts. Morgan and Tukey (1964) stated, "All of the common inorganic nutrients found in plants can be leached, including both the macro- and microelements. In addition, large quantities of organic materials are also leached. . . ." The materials lost are not entirely metabolic waste products. Exudation from leaves may take place through ectodesmata, stomata, and hydathodes, and from stems through lenticels and wounds.

There is a good deal of exudation through the cuticle via ectodesmata or microcapillaries, and on drying this may leave a residue on the leaf surface or in the surface wax (Preece and Dickinson, 1971). Water drops may form in stomata under humid conditions, and may evaporate or be retracted as humidity decreases. *Philodendron "hastatum"* exudes concentrated sugar solutions from secretory cells beneath stomata of leaves (Munnecke and Chandler, 1957). When this flows out on the leaf surface, sooty molds and saprophytic bacteria grow in it. Necrotic spots are common on this plant and may be caused by toxins formed by the bacteria.

Exudation of nutrients through hydathodes is common. These exudation drops may be blown or shaken off the leaf, evaporate, or be retracted when humidity falls. If they evaporate, they may leave a toxic salt accumulation that will kill the cells. These necrotic, nutrient-charged cells have provided excellent infection sites for *Botrytis cinerea* on young leaves of *Matthiola incana* (Baker et al., 1954).

Some leakage occurs through lenticels, at least into spongy tissue under the opening. Bleeding from wounds in stems provides a good substrate for growth of microorganisms. Plant surfaces therefore provide unique ecological niches for microorganisms. Nutrient leaching is often greater from nutrient-deficient than from adequately fertilized plants, from mechanically injured or frosted than from normal leaves, from leaves kept in darkness than from those illuminated, and from old than from young leaves (Last and Deighton, 1965). The supply of nutrients exuded by leaves of healthy vigorous plants may be limited, and those that are supplied may be lost by leaching. Microorganisms on aerial parts must compete for this meager energy source.

Leaf saprophytes undoubtedly sometimes act as scavengers, reducing the nutrients available for pathogens that may fall on the plant. If the pathogen requires nutrients for germination, the presence of the resident epiphytes will be detrimental to its successful invasion. As expressed by Leben (1965), resident epiphytes may reduce the inoculum potential of pathogens, reduce their epiphytic resident phase, affect disease development and propagule production, and alter host resistance.

NATURAL DISSEMINATION OF EPIPHYTES

Air currents disseminate both transient and resident epiphytes to aerial plant surfaces, where the transients remain inactive or die, and the residents may multiply. Of the wide variety of soil microorganisms that may be airborne to the plant surface, only a portion survive as epiphytes because of the selective environment there. For example, *Cladosporium,* usually the most common airborne fungus, is uncommon on young leaves, but accumulates and becomes abundant on leaves as they reach maturity.

Leben (1965) showed that bacteria on planted seeds migrated to the shoots of cucumber and tomato under humid conditions. However, fungi, yeasts, and actinomycetes from the soil with which the seeds were treated rarely appeared on the seedlings. Simmonds (1947) found that the normal bacterial flora of wheat seeds reduced *Helminthosporium sativum* seedling blight. Ledingham et al. (1949) gathered seed from "weathered" wheat plants that had a high population of epiphytic bacteria, and found that they developed less seedling blight from *H. sativum* than did plants from nonweathered seed (Chapter 8). The seed-soak method of treatment, which apparently grew out of these studies, increases the population of the resident flora on seed, and probably is effective through antagonistic action (Baker, 1972). Two explanations of the effectiveness of the method are that antagonists are increased on the seed surface, or that microorganisms kill the pathogens during the soak by toxins or altered physical conditions in the water.

COMMENTARY

Biological control has been used less for pathogens attacking aerial parts of plants than for those in the subterranean habitat, perhaps because aboveground the method is in direct competition with control by chemical sprays. However, the commercial success of the application of *Peniophora gigantea* on pine stumps for control of *Fomes annosus* has exceeded that obtained by chemical methods. This calls into question the assumption that chemical control is necessarily always the right approach. There undoubtedly are many other pathogens against which biological control could be equally effective.

Since the pathogen and the normal epiphytic microflora occur on the host surface together, there must be considerable interaction between them. As J. E. Bier commented in 1963, "healthy foliage and bark should not be considered as entities but rather as complex biological communities." It seems probable in such instances, as has been mentioned for soilborne pathogens, that antagonists are already reducing disease incidence of aerial parts. Only three serious attempts [those of Blakeman and Fraser (1971) on *Botrytis cinerea,* Ricard and Bollen (1968) on *Poria carbonica,* and Ricard (1970) on *Fomes annosus* on spruce, discussed above] have been made to find instances in which a pathogen is present but not causing disease on susceptible plants or plant products under conditions favorable for infection. A wider search for plants that escape infection might provide more effective resident epiphytic antagonists than has the present practice of studying the flora on plants that contract the disease. Perhaps pathologists mistake the incidence of disease in check plots for normal, when it may actually represent a substantial disease reduction from antagonists. Is this base line illusory? Are there any checks that are free of biological control? Are epidemiologists justified in attributing disease outbreaks solely to a favorable juxtaposition of host, abiotic environment, and pathogen?

Is it not probable that application of fungicides to aboveground parts may have changed, reduced, or eliminated the antagonistic flora of the plant surface (Dickinson, 1973; Hislop and Cox, 1969), as application of insecticides is known to have reduced populations of predators and parasites of insect pests?

It should be emphasized that antagonists of pathogens aboveground, as in the subterranean sphere, should occupy the site in advance of the pathogen's arrival.

In an otherwise pessimistic assessment of the possibilities of biological control of plant disease, Wood and Tveit (1955) commented, "since pathogens become established, there should be no reason why a particular antagonist should not behave similarly once the right conditions are found." This is a reasonable prognosis on which to close this evaluation of biological control of aerial plant pathogens.

11

WHITHER BIOLOGICAL CONTROL?

In practical matters the end is not mere speculative knowledge of what is to be done, but rather the doing of it.
—ARISTOTLE, Fourth Century B.C.

Only those ideas in which we have faith ever reach complete expression.
—CHARLES FILLMORE

Man now realizes that he cannot with impunity destroy the balance evolved in nature, and that he cannot continue to use the world as his refuse dump. He is finding that he lacks the wisdom to determine which living things may safely be sacrificed to his dominion. In the words of Ishi, last of the Yahi Indians, man is clever but not wise. Awareness is also slowly developing that man cannot suppress one organism without initiating a resurgence of another. As the American naturalist John Muir commented, "When we try to pick out anything by itself, we find it hitched to everything else in the Universe."

In matters of disease and pest control in food crops, these comprehensions are leading to practices that restore biological balance by slow, subtle, specific, and permanent adjustment, rather than by rapid, drastic, wide-spectrum, and temporary destruction. The thesis of this book is that a degree of biological control of diseases in crop plants is being achieved, even under present methods, and that by specific and judicious manipulation of environmental factors, perhaps without destroying the pathogen or even preventing it from infecting, satisfactory control of many additional diseases may be achieved. This requires a good deal of information on pathogen ecology and biology. "Biological control is . . . an endless series of individual problems, in applied microbial ecology" (Garrett, in Baker and Snyder, 1965). The evidence is clear, however, that it is unnecessary to await a time of total understanding to achieve success.

Rishbeth (1963) showed admirable tenacity over nearly 15 years in his studies on *Fomes annosus* on pines in England, continuing these until commercial application of his technique for inoculation with the antagonist *Peniophora gigantea* was successful. Many early investigators, however, not fully apprehending the complexities of their subject, became discouraged. Conditioned by assurances that there is insufficient essential knowledge, overly aware of past failures, and perhaps lulled into indifference by the controls provided by chemicals, plant pathologists have infrequently and briefly ventured into biological control. When an investigator has approached biological control with determination to make it work, or at least to find out why it did not (which often leads to a better approach), as in instances where alternative controls were unavailable (cereal and field crops), the results have been encouraging. Although "nothing succeeds like success," nothing discourages like failure. We think there has been overemphasis on the difficulties and failures of biological control of plant pathogens. We have, therefore, emphasized the positive, cited many examples of success, and presented explanations for successes and failures.

> Oh, let us never, never doubt
> What nobody is sure about.
>
> —HILAIRE BELLOC

Interacting elements of established microbiological associations have developed interlocking systems of ecological niches that are flexibly stable or buffered. It is, however, possible to change a pattern by nudging with slight environmental changes, by swamping it with large numbers of some microorganism, or by application of some "shock" such as chemical or heat treatment, or addition of organic amendments to soil (Chapters 3, 8, and 9). All biological systems have tolerance limits, and man has long upset biological balance by exceeding them. It is time that he began using these potent practices for positive rather than for negative ends.

STAGE IN PATHOGEN CYCLE TO APPLY BIOLOGICAL CONTROL

Based on the general information on the pathogen, antagonists, host, and environment presented in Chapters 6–9, where in the cycle of the pathogen may one best apply biological control? Should one try to eliminate some developmental stage (dissemination structures, mycelia in debris, resting structures, active pathogen in host tissue), or to prevent some process (sporulation, germination or hatching, infection, development of disease, saprophytic colonization)? Since pathogens are versatile, variable, and adaptable to changed circumstances, it is axiomatic that there is no single procedure that will be effective against all pathogens or even against a given one in different localities or environmental conditions, any more than there is a fungicide that will be universally effective. There are, however, some generalizations that may prove helpful in specific situations.

It is a long-standing principle of plant pathology that **control procedures should be directed at the weakest point in the cycle of the pathogen.** We should now add that the antagonist should attack or influence the pathogen at a weak point, preferably the most vulnerable, when the physical environment is favorable to the antagonist, unfavorable to the pathogen, or both. Environmental conditions favorable to the active phase of the pathogen should also favor the active phase of the antagonist(s) of that pathogen. This implies a series of compromises to select a suitable stage of the pathogen, an active phase of the antagonist, and a physical environment notably lacking in limiting factors for the antagonist.

Examples are known of effective biological control at nearly every stage of the pathogen's development, as shown in Figure 11.1, in which the applicable methods are given by number:

1. Formation of dissemination structures is prevented or inhibited.

The completion of a disease cycle is generally characterized by sporulation, which subsequently contributes to dissemination of the pathogen by wind or water. Burial of colonized tissues generally prevents sporulation through action of the soil microflora. The basidiocarps of *Fomes annosus,* which produce basidospores for several years, are vulnerable to destruction by hyperparasites because of their perennial habit and their exposed position on the tree. Where dissemination structures are produced from resistant structures in soil, as with conidia from microsclerotia of *Verticillium albo-atrum,* or zoosporangia and zoospores from chlamydospores in *Phytophthora cinnamomi,* the use of proper crop residues and cultivation practices to increase the level of soil fungistasis might prevent such sporulation (Chapter 7). In all instances, the object is to prevent sporulation or dissemination of the pathogen, whether as secondary spread to another plant, or as introduction of primary inoculum into the soil.

1a. Dissemination structures are prevented from forming on the diseased plant.

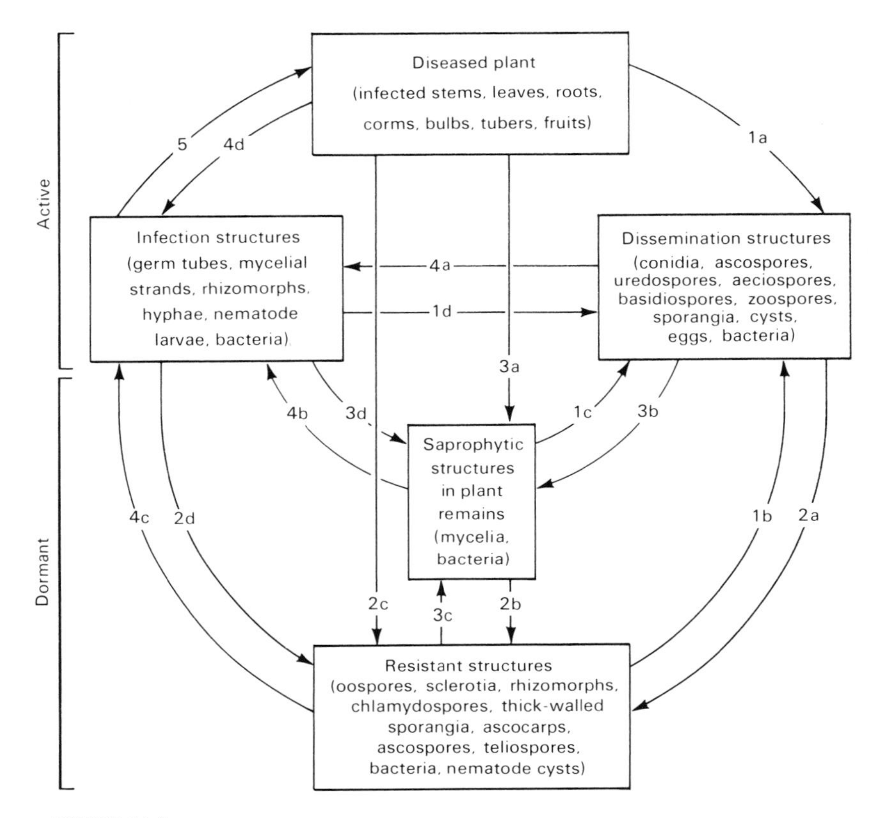

FIGURE 11.1.
Stages in the life history of plant pathogens in which biological control may be effective. Numbers refer to methods of biological control discussed in the text; each indicates prevention or inhibition of the given process.

Production of aeciospores by white pine blister rust (*Cronartium ribicola*) is diminished by destruction of the perennial mycelia in pine stems by the parasitic *Tuberculina maxima;* uredospore production by snapdragon rust (*Puccinia antirrhini*) in snapdragon leaves is decreased by *Fusarium roseum* (Chapter 10). The soil population of *Streptomyces scabies,* cause of common scab of potato, is prevented from increasing by plowing under a green soybean manure crop on which *Bacillus subtilis* grows and an antibiotic is formed. Populations of *S. scabies* also decline after a few years of potato monoculture, possibly because an increase of antagonists prevents inoculum formation, eliminates that formed, or both (Chapters 4 and 7).

1b. Dissemination structures are prevented from forming on resistant structures.

Prevention of apothecial formation from sclerotia of *Sclerotinia sclerotiorum* by antagonists is stimulated by exudation from sclerotia as a result of rapid drying and later rewetting (Chapters 4 and 9). Sclerotia of *Colletotrichum coccodes* (*C. atramentarium*), cause of the black dot disease of potato and tomato, survive 1–2 years in nonsterile soil, but acervular conidia produced on the sclerotia are destroyed within 3 weeks in nonsterile soil. Sclerotial germination occurs with wetting of the soil in the absence of the tomato host. The production and subsequent elimination of conidial inoculum between tomato crops perhaps explains why "tomato anthracnose is not significant" in glasshouse tomatoes in England (Blakeman and Hornby, 1966).

1c. Dissemination structures are prevented from forming on saprophytic structures in plant remains.

Formation of zoospores by *Phytophthora cinnamomi* from mycelia in host refuse is somehow prevented by bacteria in some soils; since zoospores are the principal agents of infection, root invasion is reduced (Chapters 4 and 7). Some soils are apparently less favorable than others to sporulation on *Verticillium* microsclerotia, suggesting that antagonists may inhibit this stage of the pathogen (Chapter 9). Sporulation of *Cephalosporium gramineum* in wheat straw incubated at -150 to -210 bars water potential, or buried in alkaline soil, is suppressed by *Penicillium* spp. and certain other fungi that invade the straw under dry or alkaline conditions (Chapter 9). Spraying apple leaves with urea before leaf fall causes increased microbial attack and faster decomposition of the leaves on the ground; mycelia of *Venturia inaequalis,* cause of apple scab, are destroyed, and few perithecia form (Chapter 10).

1d. Dissemination structures are not formed from infection structures.

The prevention of infection of pine stumps by *Fomes annosus* through prior inoculation with *Peniophora gigantea* reduces the number of *Fomes* sporophores and thus the production of basidiospores, the principal means of dissemination (Chapter 10). *Crotalaria spectabilis,* planted either before a susceptible crop or as an intercrop in an orchard, becomes infected by the root-knot nematode, *Meloidogyne* spp., but prevents maturation and egg-laying by the nematode and thus diminishes its population (Chapter 4). A breeding program with various *Crotalaria* spp.

might considerably improve their trapping effectiveness. A short picking-cycle of mushrooms, and prevention of cap maturation on weekends, prevent spore dissemination in the growing houses; viruses, which are spread by anastomosis of germ tubes from infected basidiospores with uninfected mycelia, are thus prevented (Chapter 4).

2. Formation of resistant structures suited to long-term survival on or in soil or plant remains is prevented or inhibited, or those formed are destroyed.

Resting structures are generally produced in host tissue, are slowly released by its decomposition, and remain in the soil. They are more resistant to unfavorable conditions while embedded in host tissue than when free in soil (Chapter 9). Since application of soil fumigants is less effective against resting structures in plant tissue or in clods than against those in soil pores, acceleration of tissue decomposition by nutrient addition, and breakdown of clods by intensive working of the soil before treatment, would be helpful. With greater knowledge, we might find antagonists that would penetrate partially decomposed tissue and hasten pathogen destruction even more than would soil fumigation. These might be grown on favorable organic amendments and added to the soil to reinforce the antagonistic effect (see 3 below). Resident antagonists may also be assisted by cultivation practices: *Verticillium albo-atrum* on potato is reduced in Washington and Idaho by plowing under a stand of clover, followed by flooding to induce anaerobic fermentation (Chapter 7). Application of manure to areas infested with *Phymatotrichum omnivorum* leads to decomposition of sclerotia (Chapter 6).

2a. Resistant structures are not formed by dissemination structures.

The conversion of macroconidia of *Fusarium solani* f. sp. *phaseoli* to chlamydospores in soil may be prevented by bacterial activity that causes rapid lysis of macroconidia (Chapter 6). Alternatively, absence of the proper quantity or type of microorganisms needed to induce chlamydospore formation by this pathogen will also result in its reduced survival. This may be the basis for soils "resistant" to this fungus in Washington (Burke, 1965a,b). Conversion of zoosporangia of *Pythium ultimum* to thick-walled sporangia may be inhibited by actinomycetes (Chapter 7).

2b. Resistant structures are not formed by saprophytic structures in plant remains.

Formation of sclerotia by saprophytic mycelia of *Rhizoctonia solani* may be decreased by bacterial action, and sclerotia that are formed may be destroyed (Baker et al., 1967).

2c. Resistant structures are not formed by the pathogen, or they are destroyed in the diseased plant.

Phymatotrichum omnivorum, cause of cotton root rot, may be controlled by plowing under manure or organic matter, which increases the number of soil microorganisms and hastens the decay of mycelia and sclerotia (Chapter 6). With papago pea as a green-manure crop, supplemented with inorganic nitrogen, it is possible to grow cotton annually with little disease caused by *P. omnivorum* (Chapter 6).

2d. Resistant structures are not formed by infection structures.

Root rot of bean, caused by *Fusarium solani* f. sp. *phaseoli,* is reduced in severity by plowing under a mature crop of barley. The nitrogen deficiency thus produced decreases infection of beans; germ tubes that survive form chlamydospores (Chapter 4). Bean exudates rich in amino acids stimulate chlamydospore germination, but also favor rapid lysis, some of which occurs rapidly enough to preempt further chlamydospore formation (Chapter 7).

3. Saprophytic survival in plant remains is prevented, either by precluding saprophytic colonization or by elimination of the pathogen from tissues after its establishment.

This widely applicable method has far-reaching potential in biological control, since many plant pathogens have a saprophytic stage in their life cycle. Even those not dependent on dead host tissue for survival (*Verticillium, Pythium,* chlamydospore-forming fusaria) may pass through this stage before the resistant stage is released by decay of host tissue.

The many plant pathogens (*Cephalosporium gramineum, Gaeumannomyces graminis, Cercosporella herpotrichoides, Armillaria mellea, Fusarium nivale, F. roseum* f. sp. *cerealis* 'Graminearum,' and bacterial pathogens) that depend almost entirely on host remains for survival are especially vulnerable to attack in this phase of their life cycles.

There are several advantages in attacking the pathogen at this stage. There is a longer exposure time of the pathogen to antagonists than during actual infection of the host. Nutrients from decomposing plant remains are available. The environment around plant remains is more easily controlled than around a growing plant. At this stage, hyperparasites, predators, antibiotic-producing organisms, or competitors may be effective, whereas only the last two are generally useful at the infection site (Chapter 7).

The environment can be manipulated to favor antagonists by altering fertilizer practices and timing of irrigation (Chapter 9). The addition to nontreated soil of massive quantities of antagonists grown on an organic amendment may, in some instances overwhelm the resident microflora (Chapters 4 and 7). A similar effect may be attained by selective soil treatment to eliminate the pathogens and leave the resident antagonists; aerated steam (60° C/30 minutes) is useful for this purpose on soil that has resident antagonists (Baker, 1962; Bollen, 1969; Moore, 1972; Olsen and Baker, 1968). Application of fungicides such as thiram or PCNB to soil may have prolonged effects on the microflora (Farley and Lockwood, 1969); chemicals may thus be used selectively to alter the antagonistic flora when more information on specificity is available (Chapters 5 and 9).

3a. Saprophytic mycelia of the pathogen in infected tissue are eliminated or replaced by saprophytes.

Mycelia of *Gaeummanomyces graminis* are quantitatively decreased in wheat-plant remains through application of nitrogen to hasten straw decomposition (Chapter 8). *Cephalosporium gramineum* in the vascular system of growing wheat is able to retain possession of the tissue in dead straw in acid or wet soil, because of production of an antibiotic that inhibits fungi. In alkaline soil or in straw at −150 to −210 bars water potential the fungus is, however, suppressed by *Penicillium* (Chapter 9).

If the depletion or suppression of a pathogen by a 3–4 year rotation could be shortened, the advantage would be economically significant. If we knew why *Cercosporella herpotrichoides* is able to delay decomposition of straw it occupies (Chapter 6 and 8), we might be able to counteract the mechanism and decrease the survival time of the pathogen.

3b. Colonization of plant refuse by dissemination structures is prevented.

Cephalosporium gramineum can colonize fresh wheat straw on the soil surface by water splashing of conidia produced on nearby straws already colonized. Clean straw buried directly beneath infected straw is not invaded, possibly because of its more rapid colonization by soil fungi (Bruehl and Lai, 1968a). Straw colonized by airborne or soil saprophytes before sporulation by *C. gramineum* probably is resistant to saprophytic colonization by *C. gramineum*. Living plant tissue plowed under will resist colonization by saprophytes because of its general resistance (Chapter 8), but will be vigorously attacked by pathogens such as *Pythium, Phytophthora,* and *Rhizoctonia.* Such amendments may thus increase the population of some pathogens. It is therefore better to kill plant tops by flame or perhaps application of an herbicide before plowing under. They might even be spray-inoculated with selected saprophytic antagonists several days before plowing under, to preestablish an inhibitory population.

Pythium aphanidermatum and *Phytophthora parasitica* are effective colonists of papaya residue in Hawaii, thus increasing their disease potential; they subsequently form more zoosporangia and zoospores. Perhaps if the papaya residue was contaminated with saprophytic fungi, these would retain possession of it against *P. aphanidermatum* and *P. parasitica* when subsequently buried, as in prevention of wood colonization by *P. mamillatum* (Barton, in Parkinson and Waid, 1960).

3c. Colonization of plant refuse from resistant structures is prevented.

Fusarium roseum f. sp. *cerealis* 'Culmorum' is an effective colonizer of clean bright wheat straw from chlamydospores in the soil, and the inoculum density consequently increases when such straw is buried. Leaving wheat stubble exposed so that it becomes colonized with saprophytes and antagonists before being plowed under reduces the development of this fungus on it.

3d. Colonization of plant refuse from infection structures is prevented.

The inoculum potential of *Armillaria mellea* is increased when rhizomorphs from infected trees colonize branch prunings of tea buried in soil. Leach (1939) showed that this could be prevented by leaving the branches on the soil surface for a few days to permit their colonization by saprophytes.

4. Prevent formation of infection structures or stages.

Even where dissemination of the pathogen has successfully occurred, resistant structures have survived and are located near the host, or the pathogen has survived saprophytically in refuse of a previous host, it may still be possible to prevent formation of infection structures by the pathogen, or to prevent host penetration. The disadvantage of this approach is the speed with which most pathogens can germinate and infect (Chapters 6 and 7). An advantage is that the pathogen has broken dormancy and probably becomes vulnerable to competition and antibiosis.

4a. Dissemination structures prevented from forming infection structures.

The roots of *Tagetes* spp. exude terthienyl compounds that are nematocidal to *Pratylenchus, Haplolaimus,* and *Tylenchorhynchus* spp.; asparagus roots exude a glycoside that protects intermingled tomato roots from the nematode *Trichodorus chris-*

tiei (Chapter 4). A breeding program with these plants might considerably improve production of inhibitory materials. Nematode-trapping fungi, as well as parasitic sporozoans, bacteria, and predacious tardigrades, mites, and nematodes, attack larvae and eggs of plant-parasitic nematodes, decreasing their infective populations. Addition of organic matter to soil increases this activity (Chapter 7). Nematode-transmitted plant viruses may be controlled by clean cultivation for a period sufficiently long for infective nematodes to die, and by control of weeds, the seeds of which commonly carry the viruses (Chapter 8).

Spread of *Fomes annosus* as basidiospores from old stumps to newly planted pines may be decreased by stump inoculation with *Peniophora gigantea,* which occupies the roots and prevents spread of *Fomes* through them (Chapter 10).

4b. Saprophytic structures prevented from forming infection structures.

Armillaria mellea is controlled in orchards by soil treatment with carbon disulfide, which kills exposed mycelia, and so weakens it in dead roots that it is destroyed by *Trichoderma viride* (Chapter 4). *Fomes lignosus, Ganoderma pseudoferreum,* and *F. noxious,* causes of root diseases of rubber, are controlled by planting creeping-legume cover crops between tree rows; these "decoy crops" are infected and provide favorable conditions for antagonists that destroy the pathogen (Chapter 4). Damping-off of seedlings, caused by *Rhizoctonia solani,* may be controlled in the glasshouse by inoculating treated soil of near-sterility with selected antagonists, or if the soil is naturally suppressive, by treating with aerated steam at 60° C/30 minutes (Chapter 5). Rotation of winter wheat with a single crop of spring wheat or peas will reduce speckled snow mold (*Typhula idahoensis*), pink snow mold (*Fusarium nivale*), and cephalosporium stripe disease (*C. gramineum*) in a subsequent wheat crop, because spring wheat avoids the winter environment needed by these pathogens for pathogenesis (Chapter 4). *Gaeumannomyces graminis,* cause of take-all of wheat, may be diminished in plant refuse in the soil by crop rotation, addition of organic amendments, or continuous cropping to wheat (Chapter 4), or by treating soil to induce near-sterility and inoculating it with small quantities of a suppressive soil (Chapter 7).

4c. Resistant structures prevented from forming infection structures.

There are numerous examples of soils suppressive to a pathogen: *Fusarium roseum* f. sp. *cerealis* 'Culmorum' is unable to establish and maintain itself in the Palouse wheat area of eastern Washington; *F. oxysporum* f. sp. *pisi* has been unable to establish in pea-growing areas of California; establishment of *F. oxysporum* f. sp. *cubense* in some banana-producing areas of Central America has been extremely slow; *F. oxysporum* f. sp. *melonis* and f. sp. *batatas* do not establish in some areas in California; *F. oxysporum* does not establish in pine duff in many areas, and dies if introduced (Chapter 4 and 5). *Phymatotrichum omnivorum*, cause of cotton root rot, may be controlled by plowing under manure or organic matter, thus stimulating soil microbiota to destroy the sclerotia (Chapter 6). Plowing under the surface plant refuse before sowing peanuts will reduce losses from *Sclerotium rolfsii* by moving sclerotia into a less favorable soil environment (Chapters 4 and 6); rapid drying of these sclerotia causes them to leak nutrients when rewetted, stimulating growth of microorganisms, which destroy the sclerotia (Chapter 6). Ethylene produced in the biological breakdown of organic matter may prevent germination of sclerotia of *S. rolfsii* (Chapter 6). Sclerotia of *Typhula idahoensis* are decayed more rapidly if alfalfa, rather than bare fallow, is used in rotation with wheat (Chapter 4). Germinated chlamydospores of *Fusarium roseum* f. sp. *cerealis* 'Culmorum' are destroyed by bacteria in moist soil, but the fungus is able to develop and infect under dry conditions (Chapter 6). Mustard plants give off isothiocyanates from their roots which prevent hatching of the cysts of the golden nematode of potato, *Heterodera rostochiensis* (Chapter 4).

4d. Infection structures are prevented from forming on the diseased plant.

Planting a creeping-legume cover crop between rows of rubber trees provides favorable conditions for antagonists, decreases the food base of *Fomes lignosus*, *Ganoderma pseudoferreum*, and *F. noxious*, and reduces rhizomorph formation (Chapter 4). *Rhizoctonia solani* mycelia may spread through a disease-suppressive soil without infecting seedlings growing there (Chapter 5).

5. Infection or disease development is prevented.

The infection process appears to be inhibited by antagonists in the soil, possibly by preventing formation of infection cushions or pegs. Weinhold and Bowman (1971) showed that 3-0-methyl glucose prevented infection of cotton seedlings by suppression of infection cushions of *R. solani*, even though the mycelia grew over the stems, and pectolytic enzymes were produced. *Streptomyces scabies*, cause of common scab of potato, may be prevented from infecting through lenticels of tubers under moist soil conditions, possibly by bacteria developing there (Chapter 9). Infection of seedlings by *Rhizoctonia solani* may be prevented by *Bacillus subtilis,* although growth of mycelia through the soil is not affected (Chapters 4 and 6). Application of competitive bacteria to the casing soil protects mushroom caps from infection by *Pseudomonas tolaasii,* cause of brown blotch (Chapter 4). Infection of roots by *Phytophthora cinnamomi* may be prevented by ectomycorrhizal fungi on them (Chapter 7). Selection or breeding of mycorrhizal fungi might improve this protective function. Infection of corn seedlings by *Fusarium roseum* f. sp. *cerealis* 'Graminearum' has been controlled almost as well by inoculating seed with *Chaetomium globosum* or *Bacillus subtilis* as with chemical treatments. *Helminthosporium victoriae* on oats is naturally controlled in Brazil by *C. globosum* and *C. cochlioides* on the seeds. *Fusarium nivale* on oats was controlled by seed inoculation with *C. cochlioides* (Chapter 4). Infection of corn by *Phymatotrichum omnivorum* is prevented by antagonists occurring naturally on the surface of the roots (Chapter 8). The extension of lesions caused by *Gaeumannomyces graminis* on wheat roots may be restricted in suppressive soils by antagonists such as *Phialophora radicicola* growing saprophytically on the roots (Chapter 7). Papaya seedlings planted in cores of pathogen-free virgin soil inserted in field soil infested with *Phytophthora palmivora* in Hawaii were said to be protected until they developed resistance to infection (Chapter 4). Root-knot nematodes (*Meloidogyne* spp.) may be controlled by nematocides which are applied to the tops, translocated to the roots, and there exuded (Chapter 8). Infection of senescent or damaged plant parts by the gray mold, *Botrytis cinerea,* has been diminished by application of a dilute slurry of soil microflora to plant tops (Chapter 10).

It has been shown by Cruickshank (in Baker and Snyder, 1965) and others that phytoalexin production by the host plant may be stimulated by microorganisms, nonpathogenic as well as pathogenic to it, and by numerous antibiotics of microbial origin. It is intriguing that it may be possible to select or breed microorganisms especially suited to stimulate phytoalexin production on root surfaces as a means of intensifying host resistance to certain pathogens. Such organisms would not need to be fully active or in possession of the rhizosphere to be effective.

Facultative microorganisms such as *Fusarium roseum, F. solani,* and *Rhizoctonia solani* apparently may exist in root cortical tissues with relatively little economic damage until the host is placed under stress (Chapter 9). Development by plant breeders of crop varieties less subject to environmental stress, or prevention of stress by cultivation practices, may enable the host to keep the pathogen in check.

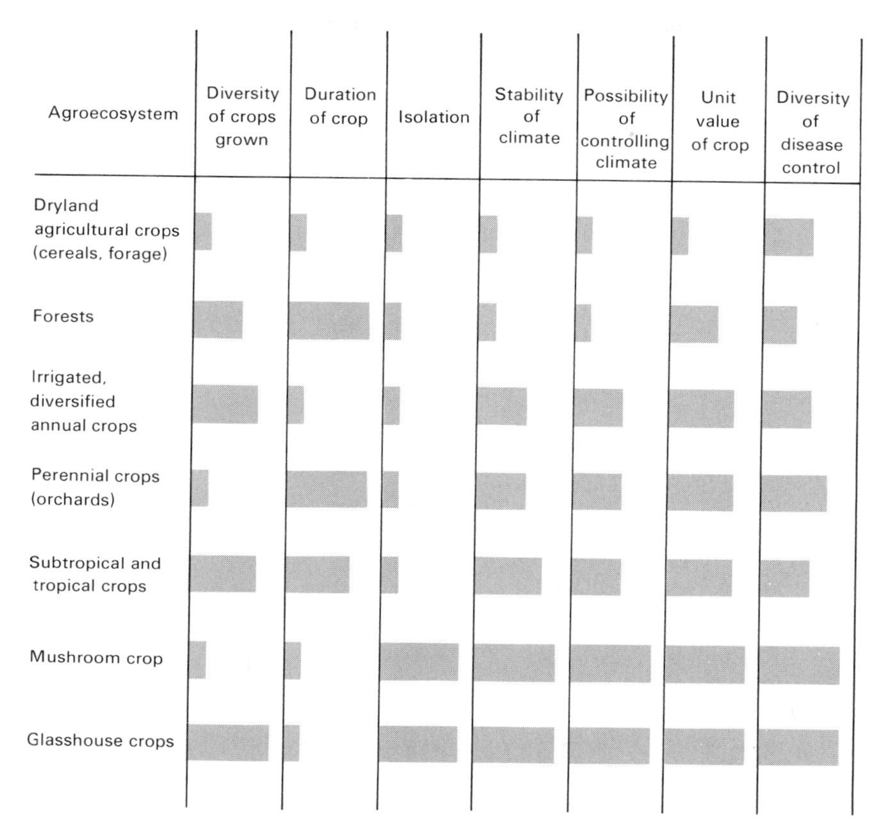

FIGURE 11.2.
Features of representative agroecosystems related to disease control. The areas in each column range from low (left) to high (right) for each given feature.

AGROECOSYSTEMS IN RELATION TO BIOLOGICAL CONTROL

Biological control of plant pathogens will not fit into all types of agriculture with equal facility and effectiveness. Are there distinctive features of different types of agriculture that determine whether biological control is apt to be economic in a given situation? Seven representative agroecosystems are analyzed in Figure 11.2 with regard to seven features of importance in biological control; the systems are arranged approximately in ascending order of the average degree or intensity of these features.

The greater the diversity of crops, the better would appear to be the chance for their rotation. However, this is not necessarily true. Cereal crops in a dryland area cannot be rotated with many other types of crops, but to a pathogen, different varieties may be as significant as a change of species; a change from a winter- to a spring-sown crop may also be very effective. Although orchards afford little opportunity for diversity, some is possible through use of different rootstocks or understory groundcovers. A mushroom crop presents almost no opportunity for diversification, yet the industry is based on complex and highly effective biological and ecological control of weed molds and pathogens. Glasshouse crops present almost the ultimate in diversification and have a high potential use of biological control. There is clearly no relationship between diversity of crops and effectiveness of biological control. Application of biological control may be achieved more rapidly, however, in an 8–10 week crop, such as mushrooms, than in an orchard.

There is also no relationship between the degree of isolation of the crop and effectiveness of biological control, although isolation is important in reducing risk of contamination and increasing inoculum density, as in mushroom growing (Chapter 4).

The seven agroecosystems show progressively increasing stability of climate and possibility of its control, but almost no relationship to effectiveness of biological control. For example, although little can be done to modify the climate of cereal crops, biological control is very effectively used on them.

The unit value of the crops increases progressively in Figure 11.2, but again this bears no relationship to effectiveness of biological control. True, the more valuable the crop, the greater the economic possibility of doing something to control disease in it. However, this potential has not really been used.

The diversity of possible disease-control methods has shown only a loose relationship to the success of biological control, but such a connection may develop with greater study.

Biological control probably will initially prove most effective and be most used in situations where alternative control measures are unavailable, difficult to apply, or economically impractical, without regard to the agroecosystem in which they appear. Suitability of the method will depend on features of the life cycle of the pathogen and antagonists involved, and on flexibility of crop management practices. Each application of biological control is a case unto itself.

It may become significant to determine whether a pathogen is indigenous in the given area, as explained for *Phytophthora cinnamomi* (Chapter 4).

As biological control of plant pathogens gains momentum, and essential facts are developed by purposeful research, it will be applied in most types of agriculture. There is no reason to believe that its use will largely be confined to the high-value crops. Indeed, present evidence would suggest that it will prove as useful in low-value crops such as cereals as it will in glasshouse crops. It is a safe prediction that biological control of plant pathogens will take its place in the pathologists' arsenal alongside present methods, and will be used in conjunction with them in an effective integrated control program.

APPLYING, ACTIVATING, OR ASSISTING ANTAGONISTS

Although there appears to be no useful relationship between types of agroecosystems and the potential for biological control, there are specific agricultural situations in which the chance of success is enhanced. Of many, a few are considered here. Biological control, as would be expected, has proved easier for some pathogens than for others.

Application of Antagonists to Soil

Addition of antagonists to nontreated soil has the best chance of initiating successful biological control when applied in such quantity as to swamp the resident saprophytic microbiota, or when the population of resident microorganisms has been severely diminished by treatment. In either case the population balance is drastically swung to favor the antagonists; this is illustrated by comparing the studies of Menzies (1959) with *Streptomyces scabies* in nontreated soil, and Shipton et al. (1973) with *Gaeumannomyces graminis* in treated soil, inoculated with suppressive soil in each instance (Chapter 7). Addition of large amounts of a sterilized, favorable, finely ground organic amendment inoculated with the antagonists might shift the balance so completely as to have a prolonged desirable effect. This is comparable to the introduction of a pathogen with the host as the selective medium. So few rational attempts have been made to disturb the biological balance by swamping untreated soil or by inoculating treated soil that it is difficult to evaluate future prospects of the method. The time of application of inoculum in relation to the life cycle of the pathogen may be very important in the degree of success attained, as explained in the discussion of Figure 9.6.

It is usually better and more practical to use a mixture of antagonists rather than a large quantity of one alone (Chapter 5). Since pathogens are part of a network of interacting biota in a variable environment and interact spatially and temporally with many other organisms, most of them will be affected more by groups of antagonists acting collectively than by a single species acting alone. If this interpretation proves generally true, as present evidence indicates, it will profoundly affect the approach to biological control. *The very phenomenon sought, however, may be missed if refinement of research techniques is begun too early in the investigation.* It is better here to work from the general to the specific, rather than from the specific to the general.

There are several conditions necessary for successfully modifying a microbial association:

1. The introduced organism(s), or those residents that are to be favored, must be sufficiently adapted to maintain their position in the microbiota under the prevailing environment (Chapter 7).
2. Some shock to the resident microbiota may be necessary in order successfully to introduce an organism into a balanced ecosystem (Chapter 3). A new balance is created by such a shock every time a virgin field is brought under cultivation, and it is maintained by subsequent culture (Chapter 8). Treatment of soil by fumigation or steam eliminates most or all of the microbiota and makes it possible to establish alien microorganisms. It has often been pointed out, on tenuous evidence, that the microflora tends eventually to return after a shock to approximately the previous balance. It does not follow, however, that biological control is necessarily temporary.
3. Maintenance of an antagonist in a dominant position in the association is possible by continuation of the conditions that favor it more than the other microbiota.

Inoculation of the Host with Antagonists

The genetic resistance of corn to *Phymatotrichum omnivorum,* peas to *Gaeumannomyces graminis,* and of certain wheat varieties to *Helminthosporium sativum,* is mediated by the rhizosphere flora, the nature of which apparently is determined by the host exudates (Chapter 8). The antagonistic effect may thus be an extension of genetic resistance. The resistance of Brazilian varieties of oats to *Helminthosporium victoriae* (Chapter 4) appears to result from the natural occurrence of *Chaetomium* spp. on the seed, and from their establishment in the rhizosphere on planting. The commonplace successful establishment of *Rhizobium* spp. in legume roots by inoculation of seed before planting and the inoculation of cereal seed with *Bacillus* and *Streptomyces* spp. as a means of increasing growth (Chapter 4) show the potential usefulness of the technique. Studies in England and Australia have shown that *Azotobacter chroococcum* inoculated on several kinds of seeds successfully established in the rhizosphere. Because of the rapid sorption and decomposition of antibiotics in soil (Brian, 1957), as well as their cost, it is not feasible to add these materials to soil for disease control. It is considered more promising to use living microorganisms to produce the antibiotic(s) continuously in low concentration *in situ*. The dominance of the antagonists on the seedling may persist only a short time, or may be retained for the life of an annual crop. If the susceptibility of the host to the pathogen is brief, even a short protection may be enough to avoid the disease (Chang and Kommedahl, 1968).

The seed should be disinfected with a nonresidual material such as sodium hypochlorite before inoculation to insure the dominance of the antagonist. Addition of a suitable amendment to the soil, or addition of an antagonist-inoculated amendment to the seed or soil, will help the microorganisms maintain their dominance.

Inoculation of freshly cut pine stumps with *Peniophora gigantea* protects against *Fomes annosus* (Chapter 10).

Since antagonists active against pathogens in the presence of the host will usually operate in the rhizophere, those used should be resident there, and even obtained from that zone (Chapter 5).

Activation of Resident Antagonists

Because of their adaptation to biotic and abiotic factors of the soil, including the presence of the pathogen itself, as well as to the climatic environment, it is desirable in biological control to activate resident antagonists whenever possible. For the same reasons, resident antagonists on aboveground parts should be stimulated for control of foliage pathogens.

Antagonists in soil may be stimulated by addition of organic amendments selectively favorable to them (Chapter 8), by continued monoculture (Chapter 4), by reinforcing the antagonists by addition of more of the specifically active ones, or by selective treatment with chemicals or steam to reduce microorganisms that inhibit the antagonists (Chapter 5) or to weaken the pathogen and make it more vulnerable to antagonists (Chapter 9).

Since a good deal of inhibition of pathogens results from weakening of survival structures and mycelia by an active general soil microflora (Chapters 6, 8, and 9), factors such as addition of plant remains, manure, inorganic fertilizers, and alteration of pH, which increase the total number of microorganisms, are useful in biological control. There has been insufficient emphasis on biological control by an active microflora that weakens the pathogen, pehaps by a phenomenon akin to fungistasis.

Starvation of the Pathogen

Pathogens may be deprived of nutrients by crop rotation, organic amendments, or other treatments that promote microbial competition. Their populations may diminish through attrition when denied a host or organic matter as an energy source (Chapters 4 and 6). This starvation aspect is intensified by microorganisms that surround the resting stage of the pathogen and hasten its exhaustion (Chapter 6). Several root rots of rubber trees are controlled by planting creeping susceptible legumes between the tree rows; the pathogenic fungi grow on them and deplete the food reserves (Chapter 4). Organic amendments may stimulate the pathogen to germinate, but endolysis may quickly follow because of the intensified microbial activity around the growing thallus (Chapter 8).

It is possible to deplete the nitrogen supply of soil in the area of pathogen attack (in the upper 15 cm of soil) by adding high-carbon substrates (Chapter 4). Thus, barley straw worked into the surface soil, with nitrogen for the plant applied below that level, will decrease fusarium root rot of bean by immobilizing nitrogen needed by the pathogen for growth (Chapter 4).

Suppressive Soil

Antagonists resident in a soil may prevent establishment of an introduced pathogen or prevent disease production by it after it is present in the soil, or they may cause progressive reduction of disease under continued monoculture (Chapter 4). These soils, eventually if not initially, may be pathogen-suppressive. Conducive soils may become suppressive with continued monoculture, but this is not always true.

Suppressive soils are recognized only when a susceptible crop and a pathogen are brought together in it, and it is therefore necessary to distinguish suppression by antagonists from that induced by the physical environment. For example, *Spongospora subterranea* does not cause powdery scab on potatoes in Florida, since soil temperatures are too high, although the fungus may survive in soil there (Melhus et al., 1916).

Field Inoculation with Suppressive Soil

The evidence suggests that it is quite possible to convert a conducive into a suppressive soil by transfer of the antagonistic flora *en masse,* and that this conversion may last for several years. This approach to biological control has such great potential that it is surprising it has not been exploited.

Perhaps plant pathologists have developed a one-to-one syndrome from exposure to the idea that a single pathogen was the cause of a given disease, that successful biological control of a given insect was by a single parasite or predator, and that successful control of a pathogen was by a single fungicide or by single-factor resistance; even the gene-for-gene concept of resistance emphasizes this feature. Perhaps the early discovery that many individual antagonists provided biological control when applied to sterile soil led pathologists off the track by emphasizing the soil treatment or shock aspect instead of seeking soils in which biological control was naturally working and attempting to make use of and understand such examples.

Menzies (Chapter 7) showed that addition of a soil that naturally suppressed *Streptomyces scabies* to an untreated conductive soil at a dilution of 1:9 plus 1% alfalfa meal greatly reduced the amount of potato scab. Shipton et al. (1973) also transferred the antagonistic microflora from a suppressive to a conducive soil in the field by a 1:199 dilution after chemical fumigation, with significant control of *Gaeumannomyces graminis* (Chapter 7). It is not yet known how long this conversion will persist, but it has now lasted for 3 years.

The prevalent idea that it is impossible to introduce alien organisms into untreated soil is refuted by the success of *Rhizobium* inoculation on legume seeds and by other bacteria on various seeds (Chapter 4), and also by the commonplace introduction of root pathogens into new areas. In all these instances the organism introduced is strongly and selectively favored by the presence of the host.

It has long been a practice in planting forest trees to place the seedling in the hole with a small quantity of soil from an area forested with the same trees, to establish the mycorrhizal fungi (Harley, 1969). Australian nurserymen place a small bit of root from an older plant of waratah (*Telopea speciosissima*) free of root rot, on roots of seedlings of that plant to insure the presence of the necessary fungus (J. Pike, unpublished).

Starting Plant Propagative Material in Suppressive Soil

One of the best methods of instituting biological control is to start seedlings and cuttings in propagative beds inoculated with suitable antagonists. Because the antagonists dominate the flora in the medium where the roots of the seedlings or cuttings are started, the roots and stems are infested by them. The fact that the environment can be readily controlled in these beds enables one to develop optimal conditions for establishment of the antagonists in the rhizosphere.

The fairly uniform rooting media of propagative beds is an advantage in inoculation with suppressive soil, as antagonists cannot be established in some soils. Seed beds are more variable, but the medium can be standardized with one of the modern porous mixes (Baker, 1957).

When the propagules are transplanted to pathogen-infested soil, they are protected by the antagonistic rhizophere flora. This dominance may well protect an annual crop through the critical period of susceptibility or primary injury without the antagonistic microflora being perfectly adapted to the crop (the possession principle).

On biennial or perennial plants the adaptation of the antagonists to the rhizosphere would have to be more perfectly matched to retain dominance for a longer period of time. **The longer the life of the host, the more perfectly must antagonists be adapted to the rhizosphere, and the more difficult biological control becomes.**

Olsen and Baker (1968) showed that seedlings started in soil infested with antagonistic *Bacillus subtilis* and transplanted to soil infested with *Rhizoctonia solani* were somewhat protected from damping-off. The inoculation of a carnation propagative bed with *B. subtilis* or a pseudomonad, so that cuttings became infested while rooting, has given some protection against *F. roseum* f. sp. *cerealis* (Chapter 4) when planted in infested soil.

Selective Treatment of Soil

Selective treatment of soil with aerated steam or chemicals usually will increase the total number of active antagonists, besides eliminating the pathogen and leaving residual antagonists. This is because the treatments may selectively kill some microorganisms and may activate dormant spores of some antagonists to grow, and because reduced antagonism enables the survivors to develop more rapidly (Chapter 5). Treatment of soil with aerated steam at 60° C/30 minutes has often left the soil as suppressive as before, while eliminating much of the total microbiota. The method is in widespread commercial use. Since antagonists responsible for the suppressiveness of soils to *Gaeumannomyces graminis* are destroyed at 48–54° C (Chapter 4), effective heat-sensitive antagonists should also be expected. The 60° C-intolerant antagonists are likely to be fungi, nonspore-forming bacteria, or sensitive actinomycetes. Do they, perhaps, represent the nonspecific competitors that weaken pathogens? The 60° C-tolerant antagonists are apt to be spore-forming bacteria and resistant actinomycetes.

Trap and Inhibitory Plants

Planting *Crotalaria* in peach orchards will reduce the population of root-knot nematode by trapping, and diminish injury to the trees (Chapter 4).

Planting perennial ryegrass between cruciferous crops reduces injury from club root (Chapter 8). *Tagetes* spp. produce terthienyls from their roots, and asparagus produces a glycoside, both of which are inhibitory to nematodes (Chapter 4). African savannah grass produces materials bactericidal to *Nitrosomonas* and *Nitrobacter* spp., and inhibitory to rhizomorph formation by *Armillaria mellea* (Chapter 4). *Datura stramonium* rotated with potato causes resting spores of *Spongospora subterranea* to germinate and die (Chapter 8). Growing trap or inhibitory plants with the crop, or in rotation with it, provides an effective means of biological control that has not been sufficiently exploited.

Translocation of Chemicals Applied to Foliage

Several fungicidal materials, such as benomyl, may be applied to soil and be translocated to the tops, but relatively few, when applied to the tops, are carried to and exuded from roots. This type of chemotherapy is quite new. Some nematocides of this nature have, however, been developed (Chapter 8). It is now an accepted fact that many organic and inorganic compounds are absorbed through the leaves, and that a few are translocated downward to the roots and may be exuded from the roots in the original form or in a modified state. Some of these cause a change in the rhizosphere flora (Agnihotri, 1964; Audus, 1970), as would be expected. This important and promising research area should receive a great deal of future attention from the standpoint of manipulation of rhizosphere flora for the control of root pathogens.

LARGE-SCALE PRODUCTION OF ANTAGONISTS

The need will eventually develop in biological control for large quantities of an antagonist or a group of antagonists to inoculate seed, soil, or other material. One method is to increase the level of suppression of a soil by amendment or selective treatment with chemicals or heat, coupled with mass antagonist inoculum, either as giant cultures or as a quantity of a suppressive soil. As explained in Chapter 5, such an increase by inoculation amounts to a "starter soil" which can then be used to amend nonsuppressive soils, added to propagative beds, or applied to seeds or cuttings. It is not yet known how long such inoculated soils will remain suppressive, but the time will undoubtedly vary directly with the suitability of the soil and the environment for the antagonists.

Individual specific antagonists may also be increased for use, as is being done in the commercial production of antibiotics, *Rhizobium* cultures, and bacterial cultures against insects.

The commercial production of mass cultures of *Bacillus thuringiensis* (a variety of *B. cereus*) for use in controlling larvae of insects, is a directly comparable operation. The cultures are grown in huge tanks of nutrient media for 28–32 hours, and the spores and the parasporal crystal-toxin are sold as dry powder (Briggs, in Steinhaus, 1963; Dulmage and Rhodes, in Burges and Hussey, 1971). There are four companies in the United States and four in Europe that produce and market this bacterium (Steinhaus, 1963). *Bacillus popilliae* and *B. lentiniorbus* are grown on Japanese beetle larvae, and nuclear polyhedrosis virus of cabbage looper is grown on larvae of that insect in a necessarily more complicated procedure.

When it becomes desirable to produce mass cultures of microorganisms antagonistic to plant pathogens, there should be little difficulty in doing so. Many of the active microorganisms are bacteria that grow very rapidly, and readily form long-lived spores resistant to drying.

INTEGRATED CONTROL

Biological control must work within the context of biological balance. A method of biological control that flouts this principle will probably achieve no better ultimate success than will single-minded emphasis on control by application of pesticides.

It is becoming clear that all biological processes are affected by many biotic and abiotic factors, and that each probably has more than one pathway along which it can proceed. Such a situation has had obvious survival value in evolutionary development. It is overlooked by man at his peril, as shown by his mistaken assumption that proprietary insecticides and fungicides, each effective against a highly specific metabolic process, would be as secure against development of insect or pathogen resistance as were the old nonspecific oil sprays and lime sulfur. The widespread development in insects of resistance to insecticides, and the recent appearance of resistance to fungicides in fungi (Chapter 12), has shattered this idea.

Nature, like a successful criminal lawyer, is well supplied with alternatives to each main plan of action. For this reason, man cannot rely on a single method in the control of plant diseases. Multiple control procedures usually are necessary for consistent success. A measure of host resistance, reduction of inoculum or infection by proper cultivation practices, a mild fungicidal spray, soil or seed treatment, use of pathogen-free propagules, and biological control may be required for effective control. Man tends to emphasize one method and thus places unreasonable demands on its effectiveness. As J. E. Van der Plank noted in 1963, "Nature seldom draws lines without smudging them. Biologists waste time seeking the perfect classification and clean lines that often, in fact, do not exist."

A modest level of resistance is not to be scorned, and if combined with other partially effective control measures, may be quite valuable. There is a cumulative effect, each control measure reinforcing the other. No single measure, then, has to be pushed to injurious limits. As stated by Strickland (in Rabb and Guthrie, 1970), "the complexities of crop production require an integrated multi-disciplinary approach if current problems are to be effectively solved."

The methods for producing pathogen-free seed and vegetative propagative material have been reviewed in detail (Baker, 1957, 1972). Sanitation procedures, which vary greatly with the crop and location, have also been reviewed (Baker, 1957; Stevens, in Horsfall and Dimond, 1959–60). Soil treatments have been reviewed (Munnecke, 1972), and fungicides for application to aerial parts are reviewed annually (Zehr, 1972). To the time-tested triad for effective plant-disease control—pathogen-free planting stock (or fungicidal application on trees and perennials already infected), pathogen-free soil, and sanitation—may now be added biological control. This rational approach has the merit of emulating the way nature controls populations. The control of the white, red, and brown root diseases of plantation rubber in Malaya (Chapter 4) is an excellent example of integrated multiple control procedures: poisoning of old trees, creosoting of stump surfaces, ditching around old stumps that provide persistent sources of infection, planting a mixed cover of creeping legumes between tree rows, quarterly foliage inspections and removal of infected trees, and treatment of the exposed collars of neighboring trees with PCNB or drazoxolon formulation to inhibit pathogenesis of *Fomes lignosus* and *Ganoderma pseudoferreum*, respectively.

An integrated program to reduce losses from fusarium foot rot of wheat is being used in the semiarid Pacific Northwest. Some tillage practices (a dust-mulch immediately after rains during the summer fallow; chisel plowing in the fall to increase water infiltration) conserve water in the soil profile and delay onset of soil-water depletion and plant-water stress in the subsequent wheat crop. Maximum straw residue is left on the soil surface for as long as possible after harvest to reduce wind erosion and also to permit development of molds, with preemption of the substrate against saprophytic colonization by *Fusarium roseum* f. sp. *cerealis 'Culmorum.'* No more nitrogen is applied than the crop can use in a single growing season for optimum growth and yield; excess nitrogen accumulates in the profile, and when combined with new application of nitrogen, contributes to excessive growth in the subsequent crop, increased rate of water use, and earlier plant-water stress. Seeding should be delayed enough to keep plants small during fall and early spring, decrease the rate of water use, and delay plant-water stress. This also places the plant in cooler soil during periods when infection commonly occurs. Oats should be avoided in the rotation because, when they are infected by Culmorum, massive sporulation by the pathogen (and hence population increase) is favored. The varieties Luke or Paha are grown, rather than Nugaines; although neither is fully resistant, both show less disease than Nugaines because of greater water-use efficiency, physiological resistance, or both. None of these practices alone can control fusarium foot rot of wheat, but combined over a long period, they provide an effective and stable control.

Biological control should be regarded, therefore, as one facet of the whole control program (Garrett, 1970), rather than as a method to be used alone and to be judged on its solo performance. In the words of R. L. Rudd, "Integrated control is biological common sense." Biological control is a logical extension of good cultivation practices that the best growers are already using.

Addendum

Recent unpublished evidence of A. M. Smith and R. J. Cook that ethylene, and hence fungistasis, is produced by anaerobes (page 169) opens many avenues for biological control. Fusarium-conducive soils (pages 62–66) are predictably low in ethylene production (sandy, acid, low in organic matter). Perhaps amendment with organic matter properly conditioned and fortified with *Clostridium* spores would suppress *Fusarium*. Pathogens more tolerant of ethylene (probably those associated with wet or waterlogged soils) may be suppressed by higher than normal ethylene levels. Thus the *Phytophthora cinnamomi*-suppressive Queensland soils (page 67) produced 10 ppm or more of ethylene under air without additional substrate (A. M. Smith, unpublished), and their suppressive nature remained following treatment at 80° C/30 minutes. Although pathogens may be controlled by judicious management of ethylene concentrations, caution is necessary if the host is ethylene-sensitive.

The success of aerated steam treatment of nursery soils (page 122) may relate to residual *Clostridium* spp. as ethylene producers, and *Bacillus* spp. as oxygen consumers, thereby providing a balanced, biologically buffered system. Much of the general pathogen suppression by soils, and its increase by introduced organic matter probably relates to rate of oxygen consumption and volume of ethylene-producing anaerobic microsites. Pathogen-conducive virgin desert soils (page 72) produce very little ethylene. Pasture legumes or organic amendments increase ethylene production; cultivation lowers, whereas minimal or no tillage increases it (A. M. Smith and R. J. Cook, unpublished). Control of the ethylene level also has potential for management of energy turnover in soil, nitrification, and improved soil structure.

12

WHY BIOLOGICAL CONTROL?

The greater is the circle of light, the greater is the darkness by which it is confined.
—J. PRIESTLEY, 1781

In the colonization of any new territory, the unfounded optimism of ignorance is invariably followed by disappointment, which is succeeded in its turn by solid achievement.
—S. D. GARRETT, 1956

THE ROLE OF BIOLOGICAL CONTROL IN PLANT PATHOLOGY

Man slipped gradually into the direct control of plant diseases after he began to use toxic chemicals in the mid-eighteenth century. This seemed the rational thing to do, because fungicides could often be applied after the disease appeared and rapid effective control was provided. The concept of overkill slowly evolved, particularly in soil treatments. However, unexpected and disturbing results increasingly appeared, indicating that there was more to soil treatment than merely killing microorganisms (Baker, in Toussoun et al., 1970). About the same time, insects became increasingly resistant to proprietary insecticides, and soon fungi resistant to fungicides, and bacteria to antibiotics, began to appear. In the 1960s the public became aware of the pollution problem, and of the increasing concentrations of toxic chemicals in nature's food chains. The pendulum of public opinion, once so favorable to this cheap and rapid control of pests and diseases, now threatens to swing too far in the direction of forbidding use of any chemicals, even to providing organic gardening with a new talking point. Paralleling the public's rediscovery of its environment, plant pathologists are becoming increasingly aware that upsetting the biological balance can lead to severe outbreaks of disease. Thus, there is evident a renewed and increased interest in biological control of plant pathogens. Pathologists are beginning to purposefully involve biological control in their integrated control programs for plant diseases.

Until the 1940s, insecticides (oil sprays, pyrethrum, nicotine sulfate, arsenate of lead) and fungicides (lime sulfur, Bordeaux mixture, mercuric chloride, formaldehyde) were of a nonspecific or broad-spectrum character. Development of compounds that affected specific metabolic processes increased effectiveness, and by the 1960s these had largely replaced the earlier materials. More than 200 species of insects, mites, and ticks have developed resistance to one or more pesticides. "In fact, only a very few species regularly treated with pesticides have not developed significant levels of resistance" (Smith, in Rabb and Guthrie, 1970). Under field conditions the wheat smut, *Tilletia foetida,* has developed resistance to hexachlorobenzene in Australia, *Helminthosporium avenae* to organomercurials in Scotland, Ireland, Netherlands, New Zealand, New York, and perhaps Austria (Baker, 1972), and powdery mildews and other fungi to benomyl (Schroeder and Provvidenti, 1969). It is probable that many more examples of fungicide-resistant plant pathogens will be found.

The hazards of insecticides to human health have been widely publicized. Most of the proprietary fungicides, fortunately, have low toxicity to man, and have not received as much publicity as the insecticides. Mercurials, an exception to this, are being removed from the market because of toxicity to humans and wildlife. For a similar reason, the treatment of soil with live steam (100° C) is being replaced by aerated steam (60° C) because of lessened discomfort and hazard to the operators. Growers who have used the lower temperature have indicated that labor relations alone would prevent a return to live steam.

Most controls by fungicide application are temporary in effect, requiring repeated applications. Biological control, when effective, usually is more enduring. Certain fungicides may actually increase the disease by reducing the number or effectiveness of antagonists, as is thought to have been the situation when applications of phytoactin and cycloheximide for control of white pine blister rust apparently adversely affected the hyperparasite *Tuberculina maxima* (Chapter 10), or when PCNB or Dexon were used for control of *Rhizoctonia* or *Pythium* damping-off (Chapter 9). A further example was the application of Bordeaux mixture for control of snapdragon rust, which in fact reduced the antagonist *Fusarium roseum* (Chapter 10). The biota of the ecosystem may be markedly reduced and simplified by the use of fungicides, leading to aggravated disease losses. The well-known flourishing of a pathogen after its introduction into soil of near-sterility is a case in point. Biological control has not similarly upset the biological balance, although theoretically it could do so.

Man is learning that the best answer to the severe fluctuations in pathogen attack resulting from his use of "dynamite" or overkill treatments is not ever-more-potent treatments that produce successively stronger and more extensive oscillations. Biological balance is achieved in nature by gentler, more subtle methods, to which there is little resistance (Chapter 1). The end result may provide just as effective, and much more stable, control in a new biological balance. Man must learn to achieve control of plant diseases by more imaginative, subtle, and sophisticated means than he previously thought necessary.

A plant disease should never be viewed as an isolated phenomenon, unrelated to other phases or practices of growing. It is possible to control cercosporella foot rot of wheat in the Palouse area of Washington by seeding about October 15 rather than September 15, but the small size of the plants produced before winter favors increased soil erosion. This control is used, nevertheless, with the result that the erosion problem in eastern Washington is one of the most acute in the United States.

A disease-control program either must fit into current cultivation practices, or these must be modified before it can be adopted. Both often develop together. If an advance is made in disease control, its successful adoption may necessitate changes in other practices, and its benefits may extend far beyond the initial objective. For example, in the 1930s only chrysanthemum varieties resistant to verticillium wilt were grown in the fall months as cut flowers. With the development of methods of securing pathogen-free propagules through the cultured-cutting technique of A. W. Dimock in 1943, and the apical-meristem-culture technique of several workers in 1954-57, coupled with soil treatment, the disease situation was reduced to unimportance. When this advance was combined with control of flowering by manipulating day length, the year-round production of pot chrysanthemums became a reality. After the public became conditioned to having chrysanthemums every month in the year, and attractive varieties suitable for rapid production in pots were developed, the "year-round pot mum" became the leading florist crop. Thus, new developments in disease control and scheduled day length have revolutionized the culture of the crop, and profoundly affected the economics of the whole florist industry.

Biological control should be viewed as a part of the whole disease-control program, assuming a role of varying importance with different diseases—dominant in some instances, of minor importance in others. In every case, the methods of biological control used must be integrated with the other methods of disease control and with the culture of the crop. At least one large fungicide manufacturer is studying the possibility of combining use of fungicides with biological and tillage control of plant diseases.

With soilborne pathogens, the opportunities for biological control are perhaps greater than for aboveground parts. As stated by J. E. Weaver and F. E. Clements in 1929, "the part of the plant environment beneath the surface of the soil is more under the control of the plant grower than is the part which lies above. He can do relatively little toward changing the composition, temperature, or humidity of the air or the amount of light. But much may be done by proper cultivating, fertilizing, irrigating, draining, etc., to influence the structure, fertility, aeration, and temperature of the soil."

Biological control fits particularly well into the increasingly prevalent large-scale agriculture, which can afford the expense of developing a program that may take several years to become profitable. For example, a program aimed at reducing potato scab by continued monoculture, or by growing a soybean green-manure crop, may not be practicable for the small grower. The preconditioning of virgin land before it is planted to the principal crops (Chapter 8) is an expense a small-scale farmer may be unable to bear. The cost of biological control will probably prove to be less in the long run than traditional controls, but this may not be true at first when disease potential is greatest; this again favors the large operator.

The customer, particularly in America, has been conditioned to agricultural produce that is of high quality and free of spoilage and such blemishes as apple scab and potato common scab, which affect appearance but not palatability. Some suggest that the part of the control of plant diseases that is primarily "cosmetic" may, because it is unnecessary, be eliminated to reduce the quantity of fungicide that must be applied. It is improbable that a lowering of appearance of produce will be tolerated by the consuming public, and alternative methods of reducing the amount of fungicides used must be found. Biological control may be that method.

EPILOGUE

The soil microflora and fauna are now recognized to be in a state of dynamic equilibrium in response to the biotic and abiotic environment, rather than in a relatively static system, as was once thought. Within limits, this balance is quite stable and has, like the spider's web, remarkable resiliency because it is biologically buffered. This subterranean world is relentlessly competitive, an organism only briefly gaining numerical ascendency, soon to be pushed back into a dormant or inferior condition. Recognition of this led Hepting (in Baker and Snyder, 1965) to comment that "the old terra firma is a very uncongenial habitat for our pathogens unless they are asleep." The study of soil microorganisms is, for these reasons, one of the most complex, difficult, exciting, and rewarding problems facing man today. As in most of man's contests with nature, the game is to understand the processes involved, so as to be better able to control them—the ultimate objective is thus biological control. In this game, though, there is relatively little to exploit beyond the control of plant pathogens.

One of the cornerstones of civilization is man's confidence that he lives in an orderly universe where cause and effect are knowable, and in which he has some control over his destiny. That there existed a biological balance in the world before man, and even today in areas undisturbed by him, is clear. That it exists today in areas disturbed by man is more obscure, because the biota there is in a continuous state of flux. This has led to such captious comments as "The balance of nature is not an achievable ideal, if it is an ideal at all." If, however, there were no balance, the biological world would be so unpredictable that crops would not be sown. Agriculture is possible because there is limited biological balance and biological control.

Plant pathologists have tended to approach biological control by accumulating ecological facts, hoping that by some form of alchemical transmutation, these will produce biological control. There have been few direct or empirical trials in the field, yet some, such as those on root rots of rubber in Malaya, have yielded significant benefits (Fox, in Baker and Snyder, 1965). It has become the "in thing" to question the value of biological control, and this has discouraged many from attempting it, and undermined the confidence and tenacity of others. It is easier to concentrate on more traditional projects than to convince skeptical administrators or selection panels of granting agencies. Biological control is not inherently spectacular, and its successes tend to be overlooked or attributed to other factors.

Man must learn to visualize the pathogen on his crops as a partner at the feast, there before himself and, in the overview, as much a part of the scene as he, and more likely to hold a residence permit. Each organism is as much the center of its own universe as man believes himself to be. If there is one conclusion from the analyses in this book, it is that **there is no one system by which biological control works.** Each relationship is absolutely unique, and understanding it requires imaginative ingenuity and a worm's-eye view of events. In this milieu, it is easy to become so academically absorbed in unraveling details that the objective—reduction of damage from disease in man's crops—is never attained. Instances where biological control is working should be studied with the objective of extending the benefit to other areas. This implies mass transfer of all or a selected part of the microbiota from a suppressive to a conducive soil under conditions favorable for establishment. *Analysis of the microorganisms involved, and the biochemistry of their relationships, then becomes a means of perfecting the result obtained, not a necessary precursor to attempting biological control.*

Nature has achieved astoundingly successful biological control, which man only dimly perceives and does not understand. It is, therefore, not necessary, and usually not desirable, to await complete knowledge before making the attempt. Biological control should be confidently approached with the idea of putting it to work, not solely as a field for academic discoveries. Isolate the phenomenon, make it work, then study it carefully to perfect its accomplishment. Much of human progress has traced that route!

Man's attempt to feed his teeming multitudes frequently disturbs the delicate balance belowground, as well as above. His naïve assaults on the subterranean biological network often result in his entrapment; in freeing himself from one strand he becomes entangled in others. Numerous fortuitous demonstrations have shown that agriculture can function within the ecological limits of the soil flora and fauna, and these examples man must study deeply and reflectively. Unable to win by annihilating all competitors belowground, as he can the animals and plants above, man must come at length to understand the effects of what he does, and work with, rather than ignore, the established order of the earth.

LITERATURE CITED

Abeles, F. B. 1973. Ethylene in Biology. Academic Press, New York. 302 pp.

Adams, P. B., J. A. Lewis, and G. C. Papavizas. 1968a. Survival of root-infecting fungi in soil. IX. Mechanism of control of Fusarium root rot of bean with spent coffee grounds. Phytopathology **58:**1603–1608.

Adams, P. B., G. C. Papavizas, and J. A. Lewis. 1968b. Survival of root-infecting fungi in soil. III. The effect of cellulose amendment on chlamydospore germination of *Fusarium solani* f. sp. *phaseoli* in soil. Phytopathology **58:**373–377.

Agnihotri, V. P. 1964. Studies on aspergilli. XIV. Effect of foliar spray of urea on the aspergilli of the rhizosphere of *Triticum vulgare* L. Plant Soil **20:**364–370.

Agnihotri, V. P. 1970. Solubilization of insoluble phosphates by some soil fungi isolated from nursery beds. Can. J. Microbiol. **16:**877–880.

Ajl, S. J., A. Ciegler, S. Kadis, T. C. Montie, and G. Weinbaum. 1970–72. Microbial Toxins. I–VIII. Academic Press, New York. 4057 pp.

Aldrich, J., and R. Baker. 1970. Biological control of *Fusarium roseum* f. sp. *dianthi* by *Bacillus subtilis*. Plant Dis. Rep. **54:**446–448.

Alexander, M. 1961. Introduction to Soil Microbiology. John Wiley, New York. 472 pp.

Amin, K. S., and L. Sequeira. 1966. Phytotoxic substances from decomposing lettuce residues in relation to the etiology of corky root rot of lettuce. Phytopathology **56**:1054–1061.

Anderson, E. J. 1962–64. Indirect effects of agricultural chemicals in soil. Long-term effects of soil fungicides. Proc. Ann. Conf. Control Soil Fungi, San Francisco and San Diego, Calif. **9**:17; **10**:13–14.

Anonymous. 1971. Root rot disease of native forests. Rural Res. CSIRO (Australia) No. 74:2–8.

Anonymous. 1973. Perspectives de lutte biologiques les champignons parasites des plantes cultivees et les pourritures des tissus ligneux. Colloque scientifique international. 22 pp. Mimeographed paper. Sta. Fed. Rech. Agron. Lausanne, Switzerland.

Artman, J. D. 1972. Further tests in Virginia using chain saw-applied *Peniophora gigantea* in loblolly pine stump inoculation. Plant Dis. Rep. **56**:958–960.

Ashworth, L. J., Jr., J. E. Waters, A. G. George, and O. D. McCutcheon. 1972. Assessment of microsclerotia of *Verticillium albo-atrum* in field soils. Phytopathology **62**:715–719.

Atanasoff, D. 1920. Fusarium-blight (scab) of wheat and other cereals. J. Agr. Res. **20**:1–32.

Atkinson, T. G., J. L. Neal, Jr., and R. I. Larson. 1974. Rootrot reaction in wheat: resistance not mediated by rhizosphere or laimosphere antagonists. Phytopathology **64**:(in press).

Audus, L. J. 1970. The action of herbicides and pesticides on the microflora. Meded. Fac. Landbouw. Wetenschappen Rijksuniv. Gent **35**:465–492.

Avizohar-Hershenzon, Z., and P. Shacked. 1969. Studies on the mode of action of inorganic nitrogenous amendments on *Sclerotium rolfsii* in soil. Phytopathology **59**:288–292.

Ayers, W. A. 1971. Induction of sporangia in *Phytophthora cinnamomi* by a substance from bacteria and soil. Can. J. Microbiol. **17**:1517–1523.

Ayers, W. A., and G. A. Zentmyer. 1971. Effect of soil solution and two soil pseudomonads on sporangium production by *Phytophthora cinnamomi*. Phytopathology **61**:1188–1193.

Baker, K. F. 1948. Fusarium wilt of garden stock (*Mathiola incana*). Phytopathology **38**:399–403.

Baker, K. F. (Ed.) 1957. The U. C. system for producing healthy container-grown plants. Calif. Agr. Exp. Sta. Manual 23:1–332.

Baker, K. F. 1961. Control of root-rot diseases. In Recent Advances in Botany **1**:486–490. Univ. Toronto Press, Toronto. 947 pp.

Baker, K. F. 1962. Principles of heat treatment of soil and planting material. J. Aust. Inst. Agr. Sci. **28**:118–126.

Baker, K. F. 1969. Aerated-steam treatment of seed for disease control. Hort. Res. **9**:59–73.

Baker, K. F. 1972. Seed pathology. In T. T. Kozlowski, Ed. Seed Biology **2**:317–416. Academic Press, New York. 447 pp.

Baker, K. F., and L. H. Davis. 1950. Heterosporium disease of nasturtium and its control. Phytopathology **40**:553–566.

Baker, K. F., N. T. Flentje, C. M. Olsen, and H. M. Stretton. 1967. Effect of antagonists on growth and survival of *Rhizoctonia solani* in soil. Phytopathology **57**:591–597.

Baker, K. F., O. A. Matkin, and L. H. Davis. 1954. Interaction of salinity injury, leaf age, fungicide application, climate, and *Botrytis cinerea* in a disease complex of column stock. Phytopathology **44**:39–42.

Baker, K. F., and R. H. Sciaroni. 1952. Diseases of major floricultural crops in California. California State Florists' Association, Los Angeles. 57 pp.

Baker, K. F., and W. C. Snyder, Eds. 1965. Ecology of Soil-borne Plant Pathogens. Prelude to Biological Control. Univ. Calif. Press, Berkeley. 571 pp.

Baker, R. 1968. Mechanisms of biological control of soil-borne pathogens. Annu. Rev. Phytopathol. **6**:263–294.

Balis, C. 1970. A comparative study of *Phialophora radicicola,* an avirulent fungal root parasite of grasses and cereals. Ann. Appl. Biol. **66**:59–73.

Ballesta, J.-P. G., and M. Alexander. 1972. Susceptibility of several Basidiomycetes to microbial lysis. Trans. Brit. Mycol. Soc. **58**:481–487.

Baltruschat, H., and F. Schönbeck. 1972. Untersuchungen über den Einfluss der endotrophen Mycorrhiza auf die Chlamydosporenbildung von *Thielaviopsis basicola* in Tabakwurzeln. Phytopathol. Z. **74**:358–361.

Barker, W. G., and O. T. Page. 1954. The induction of scab lesions on aseptic potato tubers cultured *in vitro.* Science **119**:286–287.

Barnett, H. L., and F. L. Binder. 1973. The fungal host-parasite relationship. Annu. Rev. Phytopathol. **11**:273–292.

Barron, G. L. 1968. The Genera of Hyphomycetes from Soil. Williams and Wilkins, Baltimore. 364 pp.

Barron, G. L., and J. T. Fletcher. 1970. *Verticillium albo-atrum* and *V. dahliae* as mycoparasites. Can. J. Bot. **48**:1137–1139.

Barrons, K. C. 1939. Studies of the nature of root knot resistance. J. Agr. Res. **58**:263–272.

Barton, R. 1957. Germination of oospores of *Pythium mamillatum* in response to exudates from living seedlings. Nature **180**:613–614.

Beckman, C. H. 1966. Cell irritability and localization of vascular infections in plants. Phytopathology **56**:821–824.

Beckman, C. H. 1968. An evaluation of possible resistance mechanisms in broccoli, cotton, and tomato to vascular infections by *Fusarium oxysporum.* Phytopathology **58**:429–433.

Beckman, C. H., and S. Halmos. 1962. Relation of vascular occluding reactions in banana roots to pathogenicity of root-invading fungi. Phytopathology **52**:893–897.

Beckman, C. H., and G. E. Zaroogian. 1967. Origin and composition of vascular gel in infected banana roots. Phytopathology **57**:11–13.

Beckman, C. H., D. M. Elgersma, and W. E. MacHardy. 1972. The localization of fusarial infections in the vascular tissue of single-dominant-gene resistant tomatoes. Phytopathology **62**:1256–1260.

Beckman, C. H., S. Halmos, and M. E. Mace. 1962. The interaction of host, pathogen, and soil temperature in relation to susceptibility to Fusarium wilt of bananas. Phytopathology **52**:134–140.

Beirne, B. P. 1967. Biological control and its potential. World Rev. Pest Control **6**(1):7–20.

Bennett, C. W. 1971. The curly top disease of sugarbeet and other plants. Amer. Phytopathol. Soc. Monogr. 7:1–81.

Bent, K. J. 1969. Fungicides in perspective. Endeavour **28**:129–134.

Bergeson, G. B., S. D. Van Gundy, and I. J. Thomason. 1970. Effect of *Meloidogyne javanica* on rhizosphere microflora and Fusarium wilt of tomato. Phytopathology **60**:1245–1249.

Beute, M. K., and J. L. Lockwood. 1968. Mechanisms of increased root rot in virus-infected peas. Phytopathology **58**:1643–1651.

Bier, J. E. 1964. The relation of some bark factors to canker susceptibility. Phytopathology **54**:250–253.

Black, H. S. 1968. Influence of the microflora on monocot resistance to *Phymatotrichum omnivorum*. Bios **39**:157–161.

Blakeman, J. P. 1972. Effect of plant age on inhibition of *Botrytis cinerea* spores by bacteria on beetroot leaves. Physiol. Plant Pathol. **2**:143–152.

Blakeman, J. P., and A. K. Fraser. 1971. Inhibition of *Botrytis cinerea* spores by bacteria on the surface of chrysanthemum leaves. Physiol. Plant Pathol. **1**:45–54.

Blakeman, J. P., and D. Hornby. 1966. The persistence of *Colletotrichum coccodes* and *Mycosphaerella ligulicola* in soil, with special reference to sclerotia and conidia. Trans. Brit. Mycol. Soc. **49**:227–240.

Bliss, D. E. 1946. The relation of soil temperature to the development of Armillaria root rot. Phytopathology **36**:302–318.

Bliss, D. E. 1951. The destruction of *Armillaria mellea* in citrus soils. Phytopathology **41**:665–683.

Bloomfield, B. J., and M. Alexander. 1967. Melanins and resistance of fungi to lysis. J. Bact. **93**:1276–1280.

Bloss, H. E., and G. A. Gries. 1967. Physiological responses of resistant and susceptible root tissues infected with *Phymatotrichum omnivorum*. Phytopathology **57**:380–384.

Bollard, E. G. 1960. Transport in the xylem. Annu. Rev. Plant Physiol. **11**:141–166.

Bollen, G. J. 1969. The selective effect of heat treatment on the microflora of a greenhouse soil. Neth. J. Plant Pathol. **75**:157–163.

Boosalis, M. G., and A. L. Scharen. 1959. Methods for microscopic detection of *Aphanomyces euteiches* and *Rhizoctonia solani* and for isolation of *Rhizoctonia solani* associated with plant debris. Phytopathology **49**:192–198.

Boughey, A. S., P. E. Munroe, J. Meiklejohn, R. M. Strang, and M. J. Swift. 1964. Antibiotic reactions between African savanna species. Nature **203**:1302–1303.

Bowen, G. D., and A. D. Rovira. 1973. Are modelling approaches useful in rhizosphere biology? Bull. Econ. Res. Comm. (Stockholm) **17**:443–450.

Bowman, P., and J. R. Bloom. 1966. Breaking the resistance of tomato varieties to *Fusarium* wilt by *Meloidogyne incognita*. Phytopathology **56**:871.

Boyce, J. S. 1938. Forest Pathology. McGraw-Hill, New York. 600 pp.

Brian, P. W. 1957. The ecological significance of antibiotic production. pp. 168–188. In R. E. O. Williams and C. C. Spicer, Microbial Ecology. Cambridge Univ. Press, London. 388 pp.

Broadbent, L. 1964. The epidemiology of tomato mosaic. VII. The effect of TMV on tomato fruit yield and quality under glass. Ann. Appl. Biol. **54**:209–224.

Broadbent, P. 1969. Observations on the mode of infection of *Phytophthora citrophthora* in resistant and susceptible citrus roots. Proc. 1st Internat'l. Citrus Symp. **3**:1207–1210.

Broadbent, P., and K. F. Baker. 1974a. Behaviour of *Phytophthora cinnamomi* in soils suppressive and conducive to root rot. Aust. J. Agr. Res. (in press).

Broadbent, P., and K. F. Baker. 1974b. Association of bacteria with sporangium formation and breakdown of sporangia in *Phytophthora* spp. Aust. J. Agr. Res. (in press).

Broadbent, P., K. F. Baker, and N. Johnston. 1974. Effect of *Bacillus* spp. on growth of seedlings in steamed soil. Aust. J. Agr. Res. (in press).

Broadbent, P., K. F. Baker, and Y. Waterworth. 1971. Bacteria and actinomycetes antagonistic to fungal root pathogens in Australian soils. Aust. J. Biol. Sci. **24**:925–944.

Brooks, D. H. 1965. Root infection by ascospores of *Ophiobolus graminis* as a factor in epidemiology of the take-all disease. Trans. Brit. Mycol. Soc. **48**:237–248.

Brown, M. E. 1961. Stimulation of streptomycin-resistant bacteria in the rhizosphere of leguminous plants. J. Gen. Microbiol. **24**:369–377.

Brown, M. E. 1971. Bacteriosis versus rhizosphere stimulation. Rothamsted Exp. Sta. Annu. Rep. **1970**(1):90.

Brown, M. E. 1972. Plant growth substances produced by micro-organisms of soil and rhizosphere. J. Appl. Bact. **35**:443–451.

Brown, M. E. 1973. Soil bacteriostasis limitation in growth of soil and rhizosphere bacteria. Can. J. Microbiol. **19**:195–199.

Brown, M. E. 1974. Seed bacterization. Annu. Rev. Phytopathol. **12**: (in press).

Brown, M. E., S. K. Burlingham, and R. M. Jackson. 1964. Studies on *Azotobacter* species in soil. III. Effects of artificial inoculation on crop yields. Plant Soil **20**:194–214.

Brown, R. W. 1970. Measurement of water potential with thermocouple psychrometers: construction and applications. U. S. Forest Serv. Res. Paper INT-80:1–27.

Bruehl, G. W., Ed. 1974. Biology and Control of Soilborne Plant Pathogens. American Phytopathological Society, St. Paul, Minn. (in press).

Bruehl, G. W., and B. Cunfer. 1971. Physiologic and environmental factors that affect the severity of snow mold of wheat. Phytopathology **61**:792–799.

Bruehl, G. W., and P. Lai. 1968a. The probable significance of saprophytic colonization of wheat straw in the field by *Cephalosporium gramineum*. Phytopathology **58**:464–466.

Bruehl, G. W., and P. Lai. 1968b. Influence of soil pH and humidity on survival of *Cephalosporium gramineum* in infested wheat straw. Can. J. Plant Sci. **48**:245–252.

Bruehl, G. W., B. Cunfer, and M. Toiviainen. 1972. Influence of water potential on growth, antibiotic production, and survival of *Cephalosporium gramineum*. Can. J. Plant Sci. **52**:417–423.

Bruehl, G. W., R. L. Millar, and B. Cunfer. 1969. Significance of antibiotic production by *Cephalosporium gramineum* to its saprophytic survival. Can. J. Plant Sci. **49**:235–246.

Bruehl, G. W., W. L. Nelson, F. Koehler, and O. A. Vogel. 1968. Experiments with *Cercosporella* foot rot (straw breaker) disease of winter wheat. Wash. Agr. Exp. Sta. Bull. 694:1–14.

Bruehl, G. W., R. Sprague, W. R. Fisher, M. Nagamitsu, W. L. Nelson, and O. A. Vogel. 1966. Snow molds of winter wheat in Washington. Wash. Agr. Exp. Sta. Bull. 677:1–21.

Bull, A. T. 1970a. Inhibition of polysaccharases by melanin: enzyme inhibition in relation to mycolysis. Arch. Biochem. Biophys. **137**:345–356.

Bull, A. T. 1970b. Chemical composition of wild-type and mutant *Aspergillus nidulans* cell walls. The nature of polysaccharide and melanin constituents. J. Gen. Microbiol. **63**:75–94.

Burges, A., and F. Raw, Eds. 1967. Soil Biology. Academic Press, New York. 532 pp.

Burges, H. D., and N. W. Hussey. 1971. Microbial Control of Insects and Mites. Academic Press, New York. 861 pp.

Burgess, L. W., and D. M. Griffin. 1967. Competitive saprophytic colonization of wheat straw. Ann. Appl. Biol. **60**:137–142.

Burke, D. W. 1965a. Plant spacing and Fusarium root rot of beans. Phytopathology **55**:757–759.

Burke, D. W. 1965b. Fusarium root rot of beans and behavior of the pathogen in different soils. Phytopathology **55**:1122–1126.

Burke, D. W., D. E. Miller, L. D. Holmes, and A. W. Barker. 1972. Counteracting bean root rot by loosening the soil. Phytopathology **62**:306–309.

Butler, E. E. 1957. *Rhizoctonia solani* as a parasite of fungi. Mycologia **49**:354–373.

Byler, J. W., F. W. Cobb, Jr., and J. R. Parmeter, Jr. 1972. Occurrence and significance of fungi inhabiting galls caused by *Peridermium harknessii*. Can. J. Bot. **50**:1275–1282.

Cailloux, M. 1972. Metabolism and the absorption of water by root hairs. Can. J. Bot. **50**:557–573.

Carter, M. V. 1971. Biological control of *Eutypa armeniacae*. Aust. J. Exp. Agr. Anim. Husb. **11**:687–692.

Chakravarti, B. P., C. Leben, and G. C. Daft. 1972. Numbers and antagonistic properties of bacteria from buds of field-grown soybean plants. Can. J. Microbiol. **18**:696–698.

Chang, I-pin, and T. Kommedahl. 1968. Biological control of seedling blight of corn by coating kernels with antagonistic microorganisms. Phytopathology **58**:1395–1401.

Chatterjee, A. K., L. N. Gibbins, and J. A. Carpenter. 1969. Some observations on the physiology of *Erwinia herbicola* and its possible implication as a factor antagonistic to *Erwinia amylovora* in the "fire-blight" syndrome. Can. J. Microbiol. **15**:640–642.

Chilvers, G. A., and E. G. Brittain. 1972. Plant competition mediated by host-specific parasites—a simple model. Aust. J. Biol. Sci. **25**:749–756.

Chinn, S. H. F. 1971. Biological effect of Panogen PX in soil on common root rot and growth response of wheat seedlings. Phytopathology **61**:98–101.

Chinn, S. H. F., and R. J. Ledingham. 1957. Studies on the influence of various substances on the germination of *Helminthosporium sativum* spores in soil. Can. J. Bot. **35**:697–701.

Chinn, S. H. F., R. J. Ledingham, B. J. Sallans, and P. M. Simmonds. 1953. A mechanism for the control of the common root rot of wheat. Phytopathology **43**:701.

Christiansen, J. A. 1971. Effect of pisatin on clones of *Fusarium solani* pathogenic and nonpathogenic to peas. Ph.D. Thesis. Wash. State Univ., Pullman. 63 pp.

Christias, C., and K. F. Baker. 1967. Chitinase as a factor in the germination of chlamydospores of *Thielaviopsis basicola.* Phytopathology **57**:1363–1367.

Clark, F. E. 1942. Experiments towards the control of the take-all disease of wheat and the Phymatotrichum root rot of cotton. U.S. Dep. Agr. Tech. Bull. 835:1–27.

Clark, F. E., and E. A. Paul. 1970. The microflora of grassland. Advances Agron. **22**:375–435.

Clark, F. E., W. E. Beard, and D. H. Smith. 1960. Dissimilar nitrifying capacities of soils in relation to losses of applied nitrogen. Proc. Soil Sci. Soc. Amer. **24**:50–54.

Clough, K. S., and Z. A. Patrick. 1972. Naturally occurring perforations in chlamydospores of *Thielaviopsis basicola* in soil. Can. J. Bot. **50**:2251–2253.

Cochrane, V. W. 1958. Physiology of Fungi. John Wiley, New York. 524 pp.

Cohen, E., S. B. Helgason, and W. C. McDonald. 1969. A study of factors influencing the genetics of reaction of barley to root rot caused by *Helminthosporium sativum.* Can. J. Bot. **47**:429–443.

Colbaugh, P. F. 1973. Environmental and microbiological factors associated with parasitic and saprophytic activities of *Helminthosporium sativum* Pam. King and Bakke on *Poa pratensis* L. Ph.D. Thesis. Univ. of Calif. Riverside. 92 pp.

Coley-Smith, J. R., and R. C. Cooke. 1971. Survival and germination of fungal sclerotia. Annu. Rev. Phytopathol. **9**:65–92.

Colhoun, J. 1957. A technique for examining soil for the presence of *Plasmodiophora brassicae* Woron. Ann. Appl. Biol. **45**:559–565.

Colhoun, J., and D. Park. 1964. Fusarium diseases of cereals. I. Infection of wheat plants, with particular reference to the effects of soil moisture and temperature on seedling infection. Trans. Brit. Mycol. Soc. **47**:559–572.

Conroy, J. J., R. J. Green, Jr., and J. M. Ferris. 1972. Interaction of *Verticillium albo-atrum* and the root lesion nematode, *Pratylenchus penetrans,* in tomato roots at controlled inoculum densities. Phytopathology **62**:362–366.

Cook, R. J. 1962. Influence of barley straw on the early stages of pathogenesis in Fusarium root rot of bean. Phytopathology **52**:728.

Cook, R. J. 1964. Influence of the nutritional and biotic environments of soil on the bean root rot *Fusarium.* Ph.D. Thesis. Univ. Calif., Berkeley. 81 pp.

Cook, R. J. 1968. Fusarium root and foot rot of cereals in the Pacific Northwest. Phytopathology **58:**127–131.

Cook, R. J. 1969. International Workshop on *Fusarium,* 11th–13th July 1968. A report of discussions and outcome. Bull. Brit. Mycol. Soc. **3:**15–18, 55–58.

Cook, R. J. 1970. Factors affecting colonization of wheat straw by *Fusarium roseum* f. sp. *cerealis* 'Culmorum.' Phytopathology **60:**1672–1676.

Cook, R. J. 1973. Influence of low plant and soil water potentials on diseases caused by soilborne fungi. Phytopathology **63:**451–458.

Cook, R. J., and G. W. Bruehl. 1968. Relative significance of parasitism versus saprophytism in colonization of wheat straw by *Fusarium roseum* 'Culmorum' in the field. Phytopathology **58:**306–308.

Cook, R. J., and R. I. Papendick. 1970. Soil water potential as a factor in the ecology of *Fusarium roseum* f. sp. *cerealis* 'Culmorum.' Plant Soil **32:**131–145.

Cook, R. J., and R. I. Papendick. 1972. Influence of water potential of soils and plants on root diseases. Annu. Rev. Phytopathol. **10:**349–374.

Cook, R. J., and M. N. Schroth. 1965. Carbon and nitrogen compounds and germination of chlamydospores of *Fusarium solani* f. *phaseoli.* Phytopathology **55:**254–256.

Cook, R. J., and W. C. Snyder. 1965. Influence of host exudates on growth and survival of germlings of *Fusarium solani* f. *phaseoli* in soil. Phytopathology **55:**1021–1025.

Cook, R. J., and R. D. Watson, Eds. 1969. Nature of the influence of crop residues on fungus-induced root diseases. Wash. Agr. Exp. Sta. Bull. 716:1–32.

Cook, R. J., R. I. Papendick, and D. M. Griffin. 1972. Growth of two root rot fungi as affected by osmotic and matric water potentials. Proc. Soil Sci. Soc. Amer. **36:**78–82.

Cooley, J. S. 1944. Some host-parasite relations in the black root rot of apple trees. J. Agr. Res. **69:**449–458.

Crosse, J. E. 1965. Bacterial canker of stonefruits. VI. Inhibition of leaf-scar infection of cherry by a saprophytic bacterium from the leaf surfaces. Ann. Appl. Biol. **56:**149–160.

Crosse, J. E., C. M. E. Garrett, and R. T. Burchill. 1968. Changes in the microbial population of apple leaves associated with the inhibition of the perfect stage of *Venturia inaequalis* after urea treatment. Ann. Appl. Biol. **61:**203–216.

Crozier, A., and D. M. Reid. 1971. Do roots synthesize gibberellins? Can. J. Bot. **49:**967–975.

Cunningham, G. H. 1931. Fireblight and its control. N. Z. J. Agr. **43:**111–118.

Curl, E. A. 1963. Control of plant diseases by crop rotation. Bot. Rev. **29:**413–479.

Darley, E. F., and W. D. Wilbur. 1954. Some relationships of carbon disulfide and *Trichoderma viride* in the control of *Armillaria mellea.* Phytopathology **44:**485.

Daubenmire, R. 1959. Plants and Environment. A Textbook of Autecology. 2nd ed. John Wiley, New York. 422 pp.

Daubenmire, R. 1968. Plant Communities. A Textbook of Plant Synecology. Harper & Row, New York. 300 pp.

Daulton, R. A. C., and R. F. Curtis. 1963. The effects of *Tagetes* spp. on *Meloidogyne javanica* in Southern Rhodesia. Nematologica **9**:357–362.

Davis, R. J. 1925. Studies on *Ophiobolus graminis* Sacc. and the take-all disease of wheat. J. Agr. Res. **31**:801–825.

Day, P. R. 1973. Genetic variability of crops. Annu. Rev. Phytopathol. **11**:293–312.

Deacon, J. W. 1973. Control of the take-all fungus by grass leys in intensive cereal cropping. Plant Pathol. **22**:88–94.

Dean, A. C. R., and C. Hinshelwood. 1966. Growth, Function and Regulation in Bacterial Cells. Oxford Univ. Press, London. 439 pp.

DeBach, P., Ed. 1964. Biological Control of Insect Pests and Weeds. Reinhold, New York. 844 pp.

DeBach, P. 1971. The use of imported natural enemies in insect pest management ecology. Proc. Tall Timbers Conf. Ecol. Animal Control by Habitat Management No. 3:211–233.

Dennis, C., and J. Webster. 1971. Antagonistic properties of species-groups of *Trichoderma*. I–II. Trans. Brit. Mycol. Soc. **57**:25–39, 41–48.

Dickinson, C. H. 1973. Effects of ethirimol and zineb on phylloplane microflora of barley. Trans. Brit. Mycol. Soc. **60**:423–431.

Dickson, J. G. 1923. Influence of soil temperature and moisture on the development of the seedling-blight of wheat and corn caused by *Gibberella saubinetti*. J. Agr. Res. **23**:837–870.

Dimock, A. W., and K. F. Baker. 1951. Effect of climate on disease development, injuriousness, and fungicidal control, as exemplified by snapdragon rust. Phytopathology **41**:536–552.

Dodd, A. P. 1940. The biological campaign against prickly-pear. Commonwealth Prickly Pear Board (Australia), Brisbane. 177 pp.

Domsch, K. H. 1959. Untersuchungen zur Wirkung einiger Bodenfungizide. Mitt. Biol. Bundesanst. Land- Forstwirt. Berlin-Dahlem Heft 97: 100–106.

Domsch, K. 1963. Einflüsse von Pflanzenschutzmitteln auf die Bodenmikroflora. Mitt. Biol. Bundesanst. Land- Forstwirt. Berlin-Dahlem Heft 107:1–52.

Domsch, K. H., and W. Gams. 1972. Fungi in agricultural soils. (transl. by P. S. Hudson). John Wiley, New York. 290 pp.

Drayton, F. L. 1929. Bulb growing in Holland and its relation to disease control. Sci. Agr. **9**:494–509.

Duddington, C. L., and C. H. E. Wyborn. 1972. Recent research on the nematophagous Hyphomycetes. Bot. Rev. **38**:545–565.

Duffus, J. E. 1971. Role of weeds in the incidence of virus diseases. Annu. Rev. Phytopathol. **9**:319–340.

Durbin, R. D. 1959a. Factors affecting the vertical distribution of *Rhizoctonia solani* with reference to CO_2 concentration. Amer. J. Bot. **46**:22–25.

Durbin, R. D. 1959b. The possible relationship between *Aspergillus flavus* and albinism in citrus. Plant Dis. Rep. **43**:922–923.

Durbin, R. D. 1961. Techniques for the observation and isolation of soil microorganisms. Bot. Rev. **27**:522–560.

Easton, G. D. 1964. The results of fumigating *Verticillium* and *Rhizoctonia* infested potato soils in Washington. Amer. Potato J. **41**:296.

Eaton, F. M., and N. F. Rigler. 1946. Influence of carbohydrate levels and root-surface microfloras on Phymatotrichum root rot in cotton and maize plants. J. Agr. Res. **72:**137–161.

Eaton, F. M., E. W. Lyle, and D. R. Ergle. 1947. Relations between carbohydrate accumulation and resistance of cotton plants to Phymatotrichum root rot in dry summers. Plant Physiol. **22:**181–192.

Edmunds, L. K. 1964. Combined relation of plant maturity, temperature, and soil moisture to charcoal stalk rot development in grain sorghum. Phytopathology **54:**514–517.

Ellenby, C. 1945. Control of the potato-root eelworm, *Heterodera rostochiensis* Wollenweber, by allyl isothiocyanate, the mustard oil of *Brassica nigra* L. Ann. Appl. Biol. **32:**237–239.

Emmett, R. W., and D. G. Parbery. 1974. Appressoria. Annu. Rev. Phytopathol. **12:** (in press).

Erwin, D. C., and H. Katznelson. 1961. Suppression and stimulation of mycelial growth of *Phytophthora cryptogea* by certain thiamine-requiring and thiamine-synthesizing bacteria. Can. J. Microbiol. **7:**945–950.

Evans, G., S. Wilhelm, and W. C. Snyder. 1967. Quantitative studies by plate counts of propagules of the Verticillium wilt fungus in cotton field soils. Phytopathology **57:**1250–1255.

Farley, J. D., and J. L. Lockwood. 1969. Reduced nutrient competition by soil microorganisms as a possible mechanism for pentachloronitrobenzene-induced disease accentuation. Phytopathology **59:**718–724.

Farley, J. D., S. Wilhelm, and W. C. Snyder. 1971. Repeated germination and sporulation of microsclerotia of *Verticillium albo-atrum* in soil. Phytopathology **61:**260–264.

Faulkner, L. R., W. J. Bolander, and C. B. Skotland. 1970. Interactions of *Verticillium dahliae* and *Pratylenchus minyus* in Verticillium wilt of peppermint: Influence of the nematode as determined by a double root technique. Phytopathology **60:**100–103.

Fellows, H., and C. H. Ficke. 1934. Wheat take-all. Kansas Agr. Exp. Sta. Annu. Rep. **1932–34:**95–96.

Fellows, H., and C. H. Ficke. 1939. Soil infestation by *Ophiobolus graminis* and its spread. J. Agr. Res. **58:**505–519.

Ferguson, J. 1953. Factors in colonization of sclerotia by soil organisms. Phytopathology **43:**471.

Ferguson, J. 1958. Reducing plant disease with fungicidal soil treatment, pathogen-free stock, and controlled microbial colonization. Ph.D. Thesis. Univ. Calif., Berkeley. 169 pp.

Fitzpatrick, R. E. 1934–35. The life history and parasitism of *Taphrina deformans*. Sci. Agr. **14:**305–326; **15:**341–344.

Flentje, N. T., and H. K. Saksena. 1964. Pre-emergence rotting of peas in South Australia. III. Host-pathogen interaction. Aust. J. Biol. Sci. **17:**665–675.

Fokkema, N. J. 1968. The influence of pollen on the development of *Cladosporium herbarum* in the phyllosphere of rye. Neth. J. Plant Pathol. **74:**159–165.

Fokkema, N. J. 1971. The effect of pollen in the phyllosphere of rye on colonization by saprophytic fungi and on infection by *Helminthosporium sativum* and other leaf pathogens. Neth. J. Plant Pathol. **77** (Suppl. 1):1–60.

Ford, E. J., A. H. Gold, and W. C. Snyder. 1970. Soil substances inducing chlamydospore formation by *Fusarium*. Induction of chlamydospore formation in *Fusarium solani* by soil bacteria. Interaction of carbon nutrition and soil substances in chlamydospore formation by *Fusarium*. Phytopathology **60:**124–128, 479–484, 1732–1737.

Ford, H. W. 1968. Burrowing nematode resistant rootstocks as biological barriers in citrus groves. Citrus Indust. **49**(9):18–19.

Fox, R. A. 1964. The principles of root disease control. Planters' Bull. (Malaya) **75:**210–217.

Fox, R. A. 1966. White root disease of *Hevea brasiliensis:* collar protectant dressings. J. Rubber Res. Inst. Malaya **19:**231–241.

Franklin, H. J. 1940. Cranberry growing in Massachusetts. Mass. Agr. Exp. Sta. Bull. 371:1–44.

Fraser, L. R., K. Long, and J. Cox. 1968. Stem pitting of grapefruit—field protection by the use of mild virus strains. Proc. 4th Conf. Internat'l. Organ. Citrus Virologists, pp. 27–31. Univ. Fla. Press, Gainesville. 404 pp.

French, E. R., and L. W. Nielsen. 1966. Production of macroconidia of *Fusarium oxysporum* f. *batatas* and their conversion to chlamydospores. Phytopathology **56:**1322–1323.

Fries, N. 1973. Effects of volatile organic compounds on the growth and development of fungi. Trans. Brit. Mycol. Soc. **60:**1–21.

Frobisher, M. 1968. Fundamentals of Microbiology. 8th ed. W. B. Saunders, Philadelphia. 629 pp.

Gardner, J. M., I. S. Mansour, and R. P. Scheffer. 1972. Effects of the host-specific toxin of *Periconia circinata* on some properties of sorghum plasma membranes. Physiol. Plant Pathol. **2:**197–206.

Garren, K. H. 1963. Evidence for two different pathogens of peanut pod rot. Phytopathology **53:**746.

Garrett, S. D. 1936. Soil conditions and the take-all disease of wheat. Ann. Appl. Biol. **23:**667–699.

Garrett, S. D. 1940. Soil conditions and the take-all disease of wheat. V. Further experiments on the survival of *Ophiobolus graminis* in infected wheat stubble buried in the soil. Ann. Appl. Biol. **27:**199–204.

Garrett, S. D. 1944. Root Disease Fungi. Chronica Botanica, Waltham, Mass. 177 pp.

Garrett, S. D. 1956. Biology of Root-infecting Fungi. Cambridge Univ. Press, London. 293 pp.

Garrett, S. D. 1970. Pathogenic Root-infecting Fungi. Cambridge Univ. Press, London. 294 pp.

Gerlagh, M. 1968. Introduction of *Ophiobolus graminis* into new polders and its decline. Neth. J. Plant Pathol. **74** (Suppl. 2):1–97.

Ghaffar, A., and D. C. Erwin. 1969. Effect of soil water stress on root rot of cotton caused by *Macrophomina phaseoli.* Phytopathology **59:**795–797.

Gibson, I. A. S., M. Ledger, and E. Boehm. 1961. An anomalous effect of pentachloronitrobenzene on the incidence of damping-off caused by a *Pythium* sp. Phytopathology **51:**531–533.

Gilbert, R. G., and R. G. Linderman. 1971. Increased activity of soil microorganisms near sclerotia of *Sclerotium rolfsii* in soil. Can. J. Microbiol. **17**:557–562.

Gilman, J. C. 1957. A Manual of Soil Fungi. 2nd ed. Iowa State Univ. Press, Ames. 450 pp.

Giuma, A. Y., A. M. Hackett, and R. C. Cooke. 1973. Thermostable nematotoxins produced by germinating conidia of some endozoic fungi. Trans. Brit. Mycol. Soc. **60**:49–56.

Gold, A. H., and G. Faccioli. 1972. Comparative effect of curly top infection and ethylene treatment of tomatoes and other plants. Phytopathol. Mediterr. **11**:145–153.

Goldberg, H. S., Ed. 1959. Antibiotics, Their Chemistry and Non-medical Uses. D. Van Nostrand, Princeton, N. J. 608 pp.

Goode, M. J., and J. M. McGuire. 1967. Relationship of root knot nematodes to pathogenic variability in *Fusarium oxysporum* f. sp. *lycopersici*. Phytopathology **57**:812.

Goring, C. A. I. 1962. Control of nitrification of ammonium fertilizers and urea by 2-chloro-6-(trichloromethyl) pyridine. Soil Sci. **93**:431–439.

Goss, R. W., and M. M. Afanasiev. 1938. Influence of rotations under irrigation on potato scab, Rhizoctonia, and Fusarium wilt. Nebr. Agr. Exp. Sta. Bull. 317:1–18.

Gottlieb, D., and P. D. Shaw. 1970. Mechanism of action of antifungal antibiotics. Annu. Rev. Phytopathol. **8**:371–402.

Gould, C. J., R. L. Goss, and V. L. Miller. 1966. Effects of fungicides and other materials on control of Ophiobolus patch disease on bent grass. J. Sports Turf Res. Inst. **42**:41–48.

Gray, T. R. G., and D. Parkinson, Eds. 1968. The Ecology of Soil Bacteria. Univ. Toronto Press, Toronto. 681 pp.

Greber, R. S. 1966. Passion-fruit woodiness virus as the cause of passion vine tip blight disease. Queensland J. Agr. Anim. Sci. **23**:533–538.

Greenwood, D. J. 1967. Studies on the transport of oxygen through the stems and roots of vegetable seedlings. New Phytol. **66**:337–347.

Greenwood, D. J., and D. Goodman. 1964. Oxygen diffusion and aerobic respiration in soil spheres. J. Sci. Food Agr. **15**:579–588.

Griffin, D. M. 1968. A theoretical study relating the concentration and diffusion of oxygen to the biology of organisms in soil. New Phytol. **67**:561–577.

Griffin, D. M. 1972. Ecology of Soil Fungi. Chapman and Hall, London. 193 pp.

Griffin, D. M., and N. G. Nair. 1968. Growth of *Sclerotium rolfsii* at different concentrations of oxygen and carbon dioxide. J. Exp. Bot. **19**:812–816.

Griffin, D. M., and G. Quail. 1968. Movement of bacteria in moist, particulate systems. Aust. J. Biol. Sci. **21**:579–582.

Griffin, G. J., and T. Pass. 1969. Behavior of *Fusarium roseum* 'Sambucinum' under carbon starvation conditions in relation to survival in soil. Can. J. Microbiol. **15**:117–126.

Griffiths, D. A., and W. Campbell. 1970. Interaction between hyphae of *Verticillium dahliae* Kleb. during microsclerotial development. Can. J. Microbiol. **16**:1132–1133.

Griggs, W. H., D. D. Jensen, and B. T. Iwakiri. 1968. Development of young pear trees with different rootstocks in relation to *Psylla* infestation, pear decline, and leaf curl. Hilgardia **39:**153–204.

Grosclaude, C. 1970. Premiers essais de protection biologique des blessures de taille vis-a-vis du *Stereum purpureum* Pers. Ann. Phytopathol. **2:**507–516.

Grosclaude, C., J. Ricard, and B. Dubos. 1973. Inoculation of *Trichoderma viride* spores via pruning shears for biological control of *Stereum purpureum* on plum tree wounds. Plant Dis. Rep. **57:**25–28.

Hacskaylo, E., Ed. 1971. Mycorrhizae. U. S. Dep. Agr. Misc. Publ. 1189:1–255.

Hadwiger, L. A., and M. R. Schwochau. 1969. Host resistance responses—an induction hypothesis. Phytopathology **59:**223–227.

Hameed, K. M., and H. B. Couch. 1972. Effects of *Penicillium simplicissimum* on growth, chemical composition, and root exudation of axenically grown marigolds. Phytopathology **62:**669.

Hankin, L., and J. E. Puhalla. 1971. Nature of a factor causing interstrain lethality in *Ustilago maydis*. Phytopathology **61:**50–53.

Harding, H. 1971. Effect of *Bipolaris sorokiniana* on germination and seedling survival of varieties or lines of 14 *Triticum* species. Can. J. Bot. **49:**281–287.

Hare, R. 1970. The Birth of Penicillin and the Disarming of Microbes. George Allen and Unwin, London. 236 pp.

Harley, J. L. 1969. The Biology of Mycorrhiza. 2nd ed. Leonard Hill, London. 334 pp.

Harper, J. L. 1950. Studies in the resistance of certain varieties of banana to Panama disease. II. The rhizosphere. Plant Soil **2:**383–394.

Harris, R. F., and L. E. Sommers. 1968. Plate-dilution frequency technique for assay of microbial ecology. Appl. Microbiol. **16:**330–334.

Hawker, L. E. 1957. The Physiology of Reproduction in Fungi. Cambridge Univ. Press, London. 128 pp.

Hayes, W. A., P. E. Randle, and F. T. Last. 1969. The nature of the microbial stimulus affecting sporophore formation in *Agaricus bisporus* (Lange) Sing. Ann. Appl. Biol. **64:**177–187.

Hendrix, F. F., Jr., and E. G. Kuhlman. 1965. Factors affecting direct recovery of *Phytophthora cinnamomi* from soil. Phytopathology **55:**1183–1187.

Henis, Y., and I. Chet. 1968. The effect of nitrogenous amendments on the germinability of sclerotia of *Sclerotium rolfsii* and on their accompanying microflora. Phytopathology **58:**209–211.

Henry, A. W. 1932. Influence of soil temperature and soil sterilization on the reaction of wheat seedlings to *Ophiobolus graminis* Sacc. Can. J. Res. **7:**198–203.

Hijink, M. T., and R. Winoto Suatmadji. 1967. Influence of different Compositae on population density of *Pratylenchus penetrans* and some other root-infesting nematodes. Neth. J. Plant Pathol. **73:**71–82.

Hislop, E. C., and T. W. Cox. 1969. Effects of captan on the non-parasitic microflora of apple leaves. Trans. Brit. Mycol. Soc. **52:**223–235.

Hodges, C. S., Jr., J. W. Koenigs, E. G. Kuhlman, and E. W. Ross. 1971. *Fomes annosus:* A bibliography with subject index—1960–1970. U. S. Forest Serv. Res. Paper SE-84:1–75.

Holland, A. A., and C. A. Parker. 1966. Studies on microbial antagonism in the establishment of clover pasture. II. The effect of saprophytic soil fungi upon *Rhizobium trifolii* and the growth of subterranean clover. Plant Soil **25**:329–340.

Hollings, M., and O. M. Stone. 1971. Viruses that infect fungi. Annu. Rev. Phytopathol. **9**:93–118.

Hollis, J. P. 1951. Bacteria in healthy potato tissue. Phytopathology **41**:350–366.

Holton, C. S., et al., Eds. 1959. Plant Pathology Problems and Progress, 1908–1958. Univ. Wisc. Press, Madison. 588 pp.

Hopp, A. D. 1972. The influence of antibiotic production and soil pH on survival of *Cephalosporium gramineum* in infested wheat straw. Ph.D. Thesis. Wash. State Univ., Pullman. 28 pp.

Hoppe, P. E. 1949. Differences in Pythium injury to corn seedlings at high and low temperatures. Phytopathology **39**:77–84.

Hoppe, P. E. 1957. An improved technique for "cold testing" maize seed. Proc. 4th Internat'l. Congr. Crop Prot., Hamburg **2**:1489–1490.

Hora, T. S., and R. Baker. 1972. Soil fungistasis: microflora producing a volatile inhibitor. Trans. Brit. Mycol. Soc. **59**:491–500.

Hornby, D. 1969. Quantitative estimation of soil-borne inoculum of the take-all fungus (*Ophiobolus graminis* (Sacc.) Sacc.). Proc. 5th Brit. Insectic. Fungic. Conf. **2**:65–70.

Hornby, D., and C. A. I. Goring. 1972. Effects of ammonium and nitrate nutrition on take-all disease of wheat in pots. Ann. Appl. Biol. **70**:225–231.

Horsfall, J. G., and A. E. Dimond. 1959–60. Plant Pathology, an Advanced Treatise. 3 vols. Academic Press, New York. 2064 pp.

Hsu, S. C., and J. L. Lockwood. 1971. Responses of fungal hyphae to soil fungistasis. Phytopathology **61**:1355–1362.

Huber, D. M., and H. C. McKay. 1968. Effect of temperature, crop, and depth of burial on the survival of *Typhula idahoensis* sclerotia. Phytopathology **68**:961–962.

Huber, D. M., and R. D. Watson. 1970. Effect of organic amendment on soil-borne plant pathogens. Phytopathology **60**:22–26.

Huber, D. M., A. L. Andersen, and A. M. Finley. 1966. Mechanisms of biological control in a bean root rot soil. Phytopathology **56**:953–956.

Huber, D. M., C. G. Painter, H. C. McKay, and D. L. Peterson. 1968. Effect of nitrogen fertilization on take-all of winter wheat. Phytopathology **58**:1470–1472.

Huber, D. M., R. D. Watson, and G. W. Steiner. 1965. Crop residues, nitrogen, and plant disease. Soil Sci. **100**:302–308.

Hudak, J., and P. Singh. 1970. Incidence of Armillaria root rot in balsam fir infested by balsam woolly aphid. Can. Plant Dis. Surv. **50**:99–101.

Hudson, H. J. 1968. The ecology of fungi on plant remains above the soil. New Phytol. **67**:837–874.

Huffaker, C. B., Ed. 1971. Biological Control. Plenum Press, New York. 511 pp.

Hulme, M. A., and J. K. Shields. 1972. Interaction between fungi in wood blocks. Can. J. Bot. **50**:1421–1427.

Ingham, J. L. 1972. Phytoalexins and other natural products as factors in plant disease resistance. Bot. Rev. **38**:343–424.

Isaac, I., P. Fletcher, and J. A. C. Harrison. 1971. Quantitative isolation of *Verticillium* spp. from soil and moribund potato haulm. Ann. Appl. Biol. **67**:177–183.

Jackson, R. S. 1972. Environmental factors regulating the production of conidia by sclerotia of *Botrytis convoluta*. Can. J. Bot. **50**:869–875.

Javed, Z. U. R., and J. R. Coley-Smith. 1973. Studies on germination of sclerotia of *Sclerotium delphinii*. Trans. Brit. Mycol. Soc. **60**:441–451.

Jenkins, W. R., and B. W. Coursen. 1957. The effect of root-knot nematodes, *Meloidogyne incognita acrita* and *M. hapla*, on *Fusarium* wilt of tomato. Plant Dis. Rep. **41**:182–186.

Johann, H., Holbert, J. R., and J. G. Dickson. 1931. Further studies on Penicillium injury to corn. J. Agr. Res. **43**:757–790.

Johnson, J. E. 1971. Safety in the development of herbicides. Down to Earth **27**(1):1–7.

Johnson, L. F. 1954. Antibiosis in relation to Pythium root rot of sugarcane and corn. Phytopathology **44**:69–73.

Johnson, L. F. 1962. Effect of the addition of organic amendments to soil on root knot of tomatoes. II. Relation of soil temperature, moisture, and pH. Phytopathology **52**:410–413.

Johnson, L. F., and E. A. Curl. 1972. Methods for Research on the Ecology of Soil-borne Plant Pathogens. Burgess, Minneapolis. 247 pp.

Johnson, S. P. 1953. Some factors in the control of the southern blight organism, *Sclerotium rolfsii*. Phytopathology **43**:363–368.

Jones, D., and E. Griffiths, 1964. The use of thin soil sections for the study of soil micro-organisms. Plant Soil **20**:232–240.

Keil, H. L., and R. A. Wilson. 1963. Control of peach bacterial spot with *Xanthomonas pruni* bacteriophage. Phytopathology **53**:746–747.

Kendrick, E. L., and C. S. Holton. 1961. Racial population dynamics in *Tilletia caries* and *T. foetida* as influenced by wheat varietal populations in the Pacific Northwest. Plant Dis. Rep. **45**:5–9.

Kerr, A. 1963. The root rot-Fusarium wilt complex of peas. Aust. J. Biol. Sci. **16**:55–69.

Kerr, A. 1972. Biological control of crown gall: seed inoculation. J. Appl. Bact. **35**:493–497.

Kerr, A., and M. Bumbieris. 1969. Effects of amendments on numbers of soil micro-organisms and on the root rot-*Fusarium* wilt complex of peas. Waite Agr. Res. Inst. (South Australia) Annu. Rep. **1968–69**:77–78.

Kevan, D. K. McE. 1962. Soil Animals. Philosophical Library, New York. 237 pp.

Kimmey, J. W. 1969. Inactivation of lethal-type blister rust cankers on western white pine. J. Forest. **67**:296–299.

King, C. J. 1923. Habits of the cotton rootrot fungus. J. Agr. Res. **26**:405–418.

King, C. J., and C. Hope. 1932. Distribution of the cotton root-rot fungus in soil and in plant tissues in relation to control by disinfectants. J. Agr. Res. **45**:725–740.

King, J. E., and J. R. Coley-Smith. 1968. Effects of volatile products of *Allium* species and their extracts on germination of sclerotia of *Sclerotium cepivorum* Berk. Ann. Appl. Biol. **61**:407–414.

Ko, W.-h. 1971. Biological control of seedling root rot of papaya caused by *Phytophthora palmivora.* Phytopathology **61**:780–782.

Ko, W.-h., and F. K. Hora. 1971. A selective medium for the quantitative determination of *Rhizoctonia solani* in soil. Phytopathology **61**:707–710.

Ko, W.-h., and J. L. Lockwood. 1967. Soil fungistasis: relation to fungal spore nutrition. Phytopathology **57**:894–901.

Koehler, B., and C. M. Woodworth. 1938. Corn-seedling virescence caused by *Aspergillus flavus* and *A. tamarii.* Phytopathology **28**:811–823.

Koenigs, J. W. 1960. *Fomes annosus:* A bibliography with subject review. U. S. Forest Serv. Exp. Sta. Occas. Paper 181:1–35.

Kraft, J. M., and D. D. Roberts. 1970. Resistance in peas to Fusarium and Pythium root rot. Phytopathology **60**:1814–1817.

Kramer, P. J. 1969. Plant and Soil Water Relationships. A Modern Synthesis. McGraw-Hill, New York. 482 pp.

Krupa, S., and N. Fries. 1971. Studies on ectomycorrhizae of pine. I. Production of volatile organic compounds. Can. J. Bot. **49**:1425–1431.

Kuo, M. J., and M. Alexander. 1967. Inhibition of the lysis of fungi by melanins. J. Bact. **94**:624–629.

Labruyère, R. E. 1971. Common scab and its control in seed-potato crops. Inst. Plantenziekt. Onderzoek, Wageningen Med. 575:1–71.

Lacy, M. L., and C. E. Horner. 1965. Verticillium wilt of mint: Interactions of inoculum density and host resistance. Phytopathology **55**:1176–1178.

Lamanna, C., and M. F. Malette. 1965. Basic Bacteriology; its Biological and Chemical Background. 3rd ed. Williams and Wilkins, Baltimore. 1001 pp.

Lambert, E. B., and T. T. Ayers. 1952. An improved system of mushroom culture for better control of diseases. Plant Dis. Rep. **36**:261–268.

Langton, F. A. 1969. Interactions of the tomato with two formae speciales of *Fusarium oxysporum.* Ann. Appl. Biol. **62**:413–427.

Lapierre, H., J.-M. Lemaire, B. Jouan, and G. Molin. 1970. Mise en évidence de particules virales associées à une perte de pathogenicité chez le Piétin-échaudage des céréales, *Ophiobolus graminis* Sacc. C. R. Acad. Sci. (Paris), Sér. D, **271**:1833–1836.

Lapwood, D. H., and T. F. Hering. 1970. Soil moisture and the infection of young potato tubers by *Streptomyces scabies* (common scab). Potato Res. **13**:296–304.

Larson, R. I., and T. G. Atkinson. 1970. A cytogenetic analysis of reaction to common root rot in some hard red spring wheats. Can. J. Bot. **48**:2059–2067.

Last, F. T. 1971. The role of the host in the epidemiology of some nonfoliar pathogens. Annu. Rev. Phytopathol. **9**:341–362.

Last, F. T., and F. C. Deighton. 1965. The non-parasitic microflora on the surfaces of living leaves. Trans. Brit. Mycol. Soc. **48**:83–99.

Last, F. T., and R. C. Warren. 1972. Non-parasitic microbes colonizing green leaves: their form and functions. Endeavour **31**:143–150.

Leach, L. D. 1947. Growth rates of host and pathogen as factors determining the severity of preemergence damping-off. J. Agr. Res. **75**:161–179.

Leach, L. D., and A. E. Davey. 1938. Determining the sclerotial population of

Sclerotium rolfsii by soil analysis and predicting losses of sugar beets on the basis of these analyses. J. Agr. Res. **56**:619–631.
Leach, L. D., and A. E. Davey. 1942. Reducing southern sclerotium rot of sugar beets with nitrogenous fertilizers. J. Agr. Res. **64**:1–18.
Leach, R. 1939. Biological control and ecology of *Armillaria mellea* (Vahl.) Fr. Trans. Brit. Mycol. Soc. **23**:320–329.
Leath, K. T., F. L. Lukezic, H. W. Crittenden, E. S. Elliott, P. M. Halisky, F. L. Howard, and S. A. Ostazeski. 1971. The Fusarium root rot complex of selected forage legumes in the northeast. Penn. Agr. Exp. Sta. Bull. 777:1–64.
Leben, C. 1965. Epiphytic microorganisms in relation to plant disease. Annu. Rev. Phytopathol. **3**:209–230.
Leben, C. 1969. Colonization of soybean buds by bacteria: observations with the scanning electron microscope. Can. J. Microbiol. **15**:319–320.
Le Berryais, L. R. 1785. Traité des jardins, ou le nouveau de la Quintinye, etc. 2nd ed., p. 98. P. F. Didot, Paris.
Lederberg, J., and E. M. Lederberg. 1952. Replica plating and indirect selection of bacterial mutants. J. Bact. **63**:399–406.
Ledingham, R. J., B. J. Sallans, and P. M. Simmonds. 1949. The significance of the bacterial flora on wheat seed in inoculation studies with *Helminthosporium sativum*. Sci. Agr. **29**:253–262.
Lemaire, J. M., B. Jouan, B. Perraton, and M. Sailly. 1971. Perspectives de lutte biologique contre les parasites des céréales d'origine tellurique en particulier *Ophiobolus graminis* Sacc. Sci. agron., Rennes **1971**, 8 pp.
Lester, E., and P. J. Shipton. 1967. A technique for studying inhibition of the parasitic activity of *Ophiobolus graminis* (Sacc.) Sacc. in field soils. Plant Pathol. **16**:121–123.
Leukel, R. W. 1948. *Periconia circinata* and its relation to milo disease. J. Agr. Res. **77**:201–222.
Levitt, J. 1972. Responses of Plants to Environmental Stresses. Academic Press, New York. 697 pp.
Lewis, B. G. 1970. Effects of water potential on the infection of potato tubers by *Streptomyces scabies* in soil. Ann. Appl. Biol. **66**:83–88.
Lewis, J. A., and G. C. Papavizas. 1971. Effect of sulfur-containing volatile compounds and vapors from cabbage decomposition on *Aphanomyces euteiches*. Phytopathology **61**:208–214.
Lilly, D. M., and R. H. Stillwell. 1965. Probiotics: growth-promoting factors produced by microorganisms. Science **147**:747–748.
Linderman, R. G., and T. A. Toussoun. 1968a. Breakdown in *Thielaviopsis basicola* root rot resistance in cotton by hydrocinnamic (3-phenylpropionic) acid. Phytopathology **58**:1431–1432.
Linderman, R. G., and T. A. Toussoun. 1968b. Pathogenesis of *Thielaviopsis basicola* in nonsterile soil. Phytopathology **58**:1578–1583.
Linford, M. B. 1937. The feeding of some hollow-stylet nematodes. Proc. Helminth. Soc. Wash. **4**:41–46.
Linford, M. B., F. Yap, and J. M. Oliveira. 1938. Reduction of soil populations of

the root-knot nematode during decomposition of organic matter. Soil Sci. **45**:127–140.

Lingappa, B. T., and J. L. Lockwood. 1964. Activation of soil microflora by fungus spores in relation to soil fungistasis. J. Gen. Microbiol. **35**:215–227.

Littlefield, L. J. 1969. Flax rust resistance induced by prior inoculation with an avirulent race of *Melampsora lini*. Phytopathology **59**:1323–1328.

Lloyd, A. B. 1969. Dispersal of streptomycetes in air. J. Gen. Microbiol. **57**:35–40.

Lockwood, J. L. 1964. Soil fungistasis. Annu. Rev. Phytopathol. **2**:341–362.

Louis, L., and G. T. Nightingale. 1937. The growth responses of planting material of different status and carbohydrate content. Pineapple Quarterly **7**(1):1–6.

Lyda, S. D., and E. Burnett. 1971. Changes in carbon dioxide levels during sclerotial formation by *Phymatotrichum omnivorum*. Phytopathology **61**:858–861.

Macer, R. C. F. 1961. The survival of *Cercosporella herpotrichoides* Fron. in wheat straw. Ann. Appl. Biol. **49**:165–172.

MacFarlane, I. 1952. Factors affecting the survival of *Plasmodiophora brassicae* Wor. in the soil and its assessment by a host test. Ann. Appl. Biol. **39**:239–256.

Malalasekera, R. A. P., and J. Colhoun. 1968. *Fusarium* diseases of cereals. III. Water relations and infection of wheat seedlings by *Fusarium culmorum*. Trans. Brit. Mycol. Soc. **51**:711–720.

Manandhar, J. B., and G. W. Bruehl. 1973. In vitro interactions of Fusarium and Verticillium wilt fungi with water, pH, and temperature. Phytopathology **63**:413–419.

Manning, W. J., and D. F. Crossan. 1966. Effects of a particular soil bacterium on sporangial production in *Phytophthora cinnamomi* in liquid culture. Phytopathology **56**:235–237.

Marshall, K. C. 1969. Studies by microelectrophoretic and microscopic techniques of the sorption of illite and montmorillonite to Rhizobia. J. Gen. Microbiol. **56**:301–306.

Marshall, K. C. 1971. Sorptive interactions between soil particles and microorganisms. In A. D. McLaren and J. Skujins, Eds. Soil Biochemistry **2**:409–445. Marcel Dekker, New York. 527 pp.

Marshall, K. C., and R. H. Cruickshank. 1973. Cell surface hydrophobicity and the orientation of certain bacteria at interfaces. Arch. Microbiol. **91**:29–40.

Marshall, K. C., M. J. Mulcahy, and M. S. Chowdhury. 1961. Second-year clover mortality in Western Australia—a microbiological problem. J. Aust. Inst. Agr. Sci. **29**:160–164.

Martin, M. M. 1970. The biochemical basis of the fungus-attine ant symbiosis. Science **169**:16–20.

Martin, N. E., and J. W. Hendrix. 1967. Comparison of root systems produced by healthy and stripe rust-inoculated wheat in mist-, water-, and sand-culture. Plant Dis. Rep. **51**:1074–1076.

Martin, W. J., L. D. Newsom, and J. E. Jones. 1956. Relationship of nematodes to the development of Fusarium wilt in cotton. Phytopathology **46**:285–289.

Martinson, C. A. 1963. Inoculum potential relationships of *Rhizoctonia solani* measured with soil microbiological sampling tubes. Phytopathology **53:** 634–638.

Marx, D. H. 1972. Ectomycorrhizae as biological deterrents to pathogenic root infections. Annu. Rev. Phytopathol. **10:**429–454.

Marx, D. H. 1973. Growth of ectomycorrhizal and nonmycorrhizal shortleaf pine seedlings in soil with *Phytophthora cinnamomi.* Phytopathology **63:**18–23.

Maschwitz, U., K. Koob, and H. Schildknecht. 1970. Ein Beitrag zur Funktion der Metathoracaldrüse der Ameisen. J. Insect Physiol. **16:**387–404.

Matta, A. 1971. Microbial penetration and immunization of uncongenial host plants. Annu. Rev. Phytopathol. **9:**387–410.

Matthews, R. E. F. 1970. Plant Virology. Academic Press, New York. 778 pp.

Maurer, C. L., and R. Baker. 1965. Ecology of plant pathogens in soil. II. Influence of glucose, cellulose, and inorganic nitrogen amendments on development of bean root rot. Phytopathology **55:**69–72.

McBeth, C. W., and A. L. Taylor. 1944. Immune and resistant cover crops valuable in root-knot-infested peach orchards. Proc. Amer. Soc. Hort. Sci. **45:**158–166.

McCain, A. H., O. V. Holtzmann, and E. E. Trujillo. 1967. Concentration of *Phytophthora cinnamomi* chlamydospores by soil sieving. Phytopathology **57:**1134–1135.

McClure, T. T. 1951. Fusarium foot rot of sweet-potato sprouts. Phytopathology **41:**72–77.

McKeen, W. E. 1949. A study of sugar beet rootrot in southern Ontario. Can. J. Res. **27:**284–311.

McKinney, H. H., and R. J. Davis. 1925. Influence of soil temperature and moisture on infection of young wheat plants by *Ophiobolus graminis.* J. Agr. Res. **31:**827–840.

McNamara, H. C., D. R. Hooton, and D. D. Porter. 1931. Cycles of growth in cotton root rot at Greenville, Tex. U.S. Dep. Agr. Circ. 173:1–18.

Meiklejohn, J. 1962. Microbiology of the nitrogen cycle in some Ghana soils. Emp. J. Exp. Agr. **30:**115–126.

Meiler, D., and A. Taylor. 1971. The effect of cochliodinol, a metabolite of *Chaetomium cochliodes,* on the respiration of microspores of *Fusarium oxysporum.* Can. J. Microbiol. **17:**83–86.

Melendéz, P. L., and N. T. Powell. 1967. Histological aspects of the Fusarium wilt-root knot complex in flue-cured tobacco. Phytopathology **57:**286–292.

Melhus, I. E., J. Rosenbaum, and E. S. Schultz. 1916. *Spongospora subterranea* and *Phoma tuberosa* on the Irish potato. J. Agr. Res. **7:**213–254.

Menzies, J. D. 1959. Occurrence and transfer of a biological factor in soil that suppresses potato scab. Phytopathology **49:**648–652.

Menzies, J. D. 1963. Survival of microbial plant pathogens in soil. Bot. Rev. **29:**79–122.

Menzies, J. D., and R. G. Gilbert. 1967. Responses of the soil microflora to volatile components in plant residues. Proc. Soil Sci. Soc. Amer. **31:**495–496.

Menzies, J. D., and G. E. Griebel. 1967. Survival and saprophytic growth of *Verticillium dahliae* in uncropped soil. Phytopathology **57:**703–709.

Merrill, W. 1970. Spore germination and host penetration by heartrotting Hymenomycetes. Annu. Rev. Phytopathol. **8:**281–300.

Merriman, P. R., R. D. Price, J. F. Kollmorgen, T. Piggott, and E. H. Ridge. 1974. Effect of seed inoculation with *Bacillus subtilis* and *Streptomyces griseus* on the growth of cereals and carrots. Aust. J. Agr. Res. (in press).

Meyer, J. A., and H. Maraite. 1971. Multiple infection and symptom mitigation in vascular wilt diseases. Trans. Brit. Mycol. Soc. **57:**371–377.

Meyers, J. A., and R. J. Cook. 1972. Induction of chlamydospore formation in *Fusarium solani* by abrupt removal of the organic carbon substrate. Phytopathology **62:**1148–1153.

Michael, A. H., and P. E. Nelson. 1972. Antagonistic effect of soil bacteria on *Fusarium roseum* 'Culmorum' from carnation. Phytopathology **62:**1052–1056.

Mielke, J. L. 1943. White pine blister rust in western North America. Yale School Forestry Bull. 52:1–155.

Miller, T. D., and M. N. Schroth. 1972. Monitoring the epiphytic population of *Erwinia amylovora* on pear with a selective medium. Phytopathology **62:**1175–1182.

Mircetich, S. M. 1970. Inhibition of germination of chlamydospores of *Phytophthora cinnamomi* by some antimicrobial agents in *Phytophthora* selective media. Can. J. Microbiol. **16:**1227–1230.

Misaghi, I., and R. G. Grogan. 1969. Nutritional and biochemical comparisons of plant-pathogenic and saprophytic fluorescent pseudomonads. Phytopathology **59:**1436–1450.

Mitchell, D. J., and J. E. Mitchell. 1973. Oxygen and carbon dioxide concentration effects on the growth and reproduction of *Aphanomyces euteiches* and certain other soil-borne plant pathogens. Phytopathology **63**: 1053–1059.

Mitchell, J. E., H. S. Bhalla, and G. H. Yang. 1969. An approach to the study of the population dynamics of *Aphanomyces euteiches* in soil. Phytopathology **59:**206–212.

Mitchell, J. W., and G. A. Livingston. 1968. Methods of studying plant hormones and growth-regulating substances. U.S. Dep. Agr. Agr. Handbk. 336:1–140.

Mitchell, R. 1963. Addition of fungal cell-wall components to soil for biological disease control. Phytopathology **53:**1068–1071.

Mitchell, R., and M. Alexander. 1962. Microbiological processes associated with the use of chitin for biological control. Proc. Soil Sci. Soc. Amer. **26:**556–558.

Mitchell, R. B., D. R. Hooten, and F. E. Clark. 1941. Soil bacteriological studies on the control of Phymatotrichum root rot of cotton. J. Agr. Res. **63:**535–547.

Mix, A. J. 1935. The life history of *Taphrina deformans*. Phytopathology **25:**41–66.

Moore, R. K. 1972. Aerated steam treatment of mushroom casing: I. Thermal sensitivity of selected fungi associated with *Agaricus bisporus* (Lange) Imbach; II. Soil colonization by *Verticillium malthousei* Ware. M. S. Thesis. Penn. State Univ., University Park, Penn. 60 pp.

Morgan, J. V., and H. B. Tukey, Jr. 1964. Characterization of leachate from plant foliage. Plant Physiol. **39**:590–593.

Morris, L. G., and K. W. Winspear. 1957. Some experiments on the steam sterilising of soil. I–II. J. Agr. Eng. **2**:262–270.

Mosse, B. 1973. Advances in the study of vesicular-arbuscular mycorrhiza. Annu. Rev. Phytopathol. **11**:171–196.

Mughogho, L. K. 1968. The fungus flora of fumigated soils. Trans. Brit. Mycol. Soc. **51**:441–459.

Müller-Kögler, E. 1938. Untersuchungen über die Schwarzbeinigkeit des Getreides und den Wirtspflanzenkreis ihres Erregers. (*Ophiobolus graminis* Sacc.). Arb. Biol. Reichsanst. Land- Forstwirt. Berlin-Dahlem **22**:271–319.

Munnecke, D. E. 1972. Factors affecting the efficacy of fungicides in soil. Annu. Rev. Phytopathol. **10**:375–398.

Munnecke, D. E., and P. A. Chandler. 1957. A leaf spot of *Philodendron* related to stomatal exudation and to temperature. Phytopathology **47**:299–303.

Nair, N. G., and P. C. Fahy. 1972. Bacteria antagonistic to *Pseudomonas tolaasii* and their control of brown blotch of the cultivated mushroom *Agaricus bisporus*. J. Appl. Bact. **35**:439–442.

Nash, S. M., and J. V. Alexander. 1965. Comparative survival of *Fusarium solani* f. *cucurbitae* and *F. solani* f. *phaseoli* in soil. Phytopathology **55**:963–966.

Nash, S. M., and W. C. Snyder. 1962. Quantitative estimations by plate counts of propagules of the bean root rot *Fusarium* in field soils. Phytopathology **52**:567–572.

Nash, S. M., and W. C. Snyder. 1965. Quantitative and qualitative comparisons of *Fusarium* populations in cultivated fields and noncultivated parent soils. Can. J. Bot. **43**:939–945.

Neal, D. C. 1935. Further studies on the effect of ammonia nitrogen on growth of the cotton-root-rot fungus, *Phymatotrichum omnivorum*, in field and laboratory experiments. Phytopathology **25**:967–968.

Neal, J. L., Jr. 1971. A simple method for enumeration of antibiotic producing microorganisms in the rhizosphere. Can. J. Microbiol. **17**:1143–1145.

Neal, J. L., Jr., T. G. Atkinson, and R. I. Larson. 1970. Changes in the rhizosphere microflora of spring wheat induced by disomic substitution of a chromosome. Can. J. Microbiol. **16**:153–158.

Neal, J. L. Jr., R. I. Larson, and T. G. Atkinson. 1973. Changes in rhizosphere populations of selected physiological groups of bacteria related to substitution of specific pairs of chromosomes in spring wheat. Plant Soil **39**:209–212.

Nelson, R. 1950. Verticillium wilt of peppermint. Mich. Agr. Exp. Sta. Tech. Bull. 221:1–259.

New, P. B., and A. Kerr. 1972. Biological control of crown gall: field measurements and glasshouse experiments. J. Appl. Bact. **35**:279–287.

Newhook, F. J. 1951. Microbiological control of *Botrytis cinerea* Pers. I. The role of pH changes and bacterial antagonism. II. Antagonism by fungi and actinomycetes. Ann. Appl. Biol. **38**:169–184, 185–202.

Newhook, F. J. 1957. The relationship of saprophytic antagonism to control of *Botrytis cinerea* Pers. on tomatoes. N. Z. J. Sci. Technol. (A), **38**:473–481.

Newhook, F. J., and F. D. Podger. 1972. The role of *Phytophthora cinnamomi* in Australian and New Zealand forests. Annu. Rev. Phytopathol. **10**:299–326.

Nordin, V. J., Ed. 1972. Biological control of forest diseases. XV Congress Internat'l. Union For. Res. Organ., Gainesville, Fla., 1971, Subject Group 2. Can. Forest Serv., Ottawa. 106 pp.

Odum, E. P. 1971. Fundamentals of Ecology. 3rd ed. W. B. Saunders, Philadelphia. 574 pp.

Ohr, H. D., D. E. Munnecke, and J. L. Bricker. 1973. The interaction of *Armillaria mellea* and *Trichoderma* spp. as modified by methyl bromide. Phytopathology **63**: 965–973.

Okafor, N. 1970. Influence of chitin on mycoflora and length of roots of wheat seedlings. Trans. Brit. Mycol. Soc. **55**:483–514.

Old, K. M., 1969. Perforation of conidia of *Cochliobolus sativus* in natural soils. Trans. Brit. Mycol. Soc. **53**:207–216.

Old, K. M., and W. M. Robertson. 1969. Examination of conidia of *Cochliobolus sativus* recovered from natural soil using transmission and scanning electron microscopy. Trans. Brit. Mycol. Soc. **53**:217–221.

Old, K. M., and W. M. Robertson. 1970. Growth of bacteria within lysing fungal conidia in soil. Effects of lytic enzymes and natural soil on the fine structure of conidia of *Cochliobolus sativus*. Trans. Brit. Mycol. Soc. **54**:337–341, 343–350.

Olsen, C. M., and K. F. Baker. 1968. Selective heat treatment of soil, and its effect on the inhibition of *Rhizoctonia solani* by *Bacillus subtilis*. Phytopathology **58**:79–87.

Ordish, G. 1967. Biological Methods in Crop Pest Control. Constable, London. 242 pp.

Ordish, G., and D. Dufour. 1969. Economic bases for protection against plant diseases. Annu. Rev. Phytopathol. **7**:31–50.

Owens, L. D., R. G. Gilbert, G. E. Griebel, and J. D. Menzies. 1969. Identification of plant volatiles that stimulate microbial respiration and growth in soil. Phytopathology **59**:1468–1472.

Palleroni, N. J., and M. Doudoroff. 1972. Some properties and taxonomic subdivisions of the genus *Pseudomonas*. Annu. Rev. Phytopathol. **10**:73–100.

Papavizas, G. C. 1968. Survival of root-infecting fungi in soil. VI. Effect of amendments on bean root rot caused by *Thielaviopsis basicola* and on inoculum density of the causal organism. Phytopathology **58**:421–428.

Papavizas, G. C. and C. B. Davey. 1960. Rhizoctonia disease of bean as affected by decomposing green plant materials and associated microfloras. Phytopathology **50**:516–522.

Papavizas, G. C., and J. A. Lewis. 1971. Survival of endoconidia and chlamydospores of *Thielaviopsis basicola* as affected by soil environmental factors. Phytopathology **61**:108–113.

Papavizas, G. C., P. B. Adams, and J. A. Lewis. 1968. Survival of root-infecting fungi in soil. V. Saprophytic multiplication of *Fusarium solani* f. sp. *phaseoli* in soil. Phytopathology **58**:414–420.

Papendick, R. I., V. L. Cochran, and W. M. Woody. 1971. Soil water potential and water content profiles with wheat under low spring and summer rainfall. Agron. J. **63:**731–734.

Papendick, R. I., and R. J. Cook. 1974. Plant water stress and the development of Fusarium foot rot in wheat subjected to different cultural practices. Phytopathology **64** (in press).

Papendick, R. I., M. J. Lindstrom, and V. L. Cochran. 1973. Soil mulch effects on seedbed temperature and water during fallow in eastern Washington. Proc. Soil Sci. Soc. Amer. **37:**307–314.

Parkinson, D., and J. S. Waid. 1960. The Ecology of Soil Fungi. Liverpool Univ. Press, Liverpool. 324 pp.

Parmeter, J. R., Jr., Ed. 1970. *Rhizoctonia solani:* Biology and Pathology. Univ. Calif. Press, Berkeley. 255 pp.

Peterson, G. M. 1937. Diminishing Returns and Planned Economy. Ronald Press, New York. 254 pp.

Phillips, D. J., and S. Wilhelm. 1971. Root distribution as a factor influencing symptom expression of Verticillium wilt of cotton. Phytopathology **61:**1312–1313.

Phillips, D. V., C. Leben, and C. C. Allison. 1967. A mechanism for the reduction of Fusarium wilt by a *Cephalosporium* species. Phytopathology **57:**916–919.

Pitt, D. 1964. Studies on sharp eyespot disease of cereals. I–II. Ann. Appl. Biol. **54:**77–89, 231–240.

Podger, F. D. 1972. *Phytophthora cinnamomi,* a cause of lethal disease in indigenous plant communities in Western Australia. Phytopathology **62:**972–981.

Pope, A. M. S. 1972. The decline phenomenon in take-all disease of wheat. Ph.D. Thesis. Univ. Surrey, Guildford, England. 161 pp.

Pope, A. M. S., and R. M. Jackson. 1973. Effects of wheatfield soil on inocula of *Gaeumannomyces graminis* (Sacc.) Arx and Oliver var. *tritici* J. Walker in relation to take-all decline. Soil Biol. Biochem. **5** (in press).

Potter, H. S., M. G. Norris, and C. E. Lyons. 1971. Potato scab control studies in Michigan using N-Serve nitrogen stabilizer for nitrification inhibition. Down to Earth **27**(3):23–24.

Powell, N. T. 1963. The role of plant-parasitic nematodes in fungus diseases. Phytopathology **53:**28–34.

Powell, N. T. 1971. Interactions between nematodes and fungi in disease complexes. Annu. Rev. Phytopathol. **9:**253–274.

Pratt, B. H., W. A. Heather, and C. J. Shepherd. 1972. Transcontinental occurrence of A1 and A2 strains of *Phytophthora cinnamomi* in Australia. Aust. J. Biol. Sci. **25:**1099–1100.

Pratt, B. H., W. A. Heather, and C. J. Shepherd. 1973. Recovery of *Phytophthora cinnamomi* from native vegetation in a remote area of New South Wales. Trans. Brit. Mycol. Soc. **60:**197–204.

Preece, T. F., and C. H. Dickinson. 1971. Ecology of Leaf Surface Microorganisms. Academic Press, New York. 640 pp.

Price, R. D., K. F. Baker, P. Broadbent, and E. H. Ridge. 1971. Effect on wheat plants of a soil or seed application of *Bacillus subtilis* either with or without the presence of *Rhizoctonia solani.* Proc. Aust. Conf. Soil Biol. **1971**:50–51.

Pugh, G. J. F., and J. H. van Emden. 1969. Cellulose-decomposing fungi in polder soils and their possible influence on pathogenic fungi. Neth. J. Plant Pathol. **75**:287–295.

Rabb, R. L., and F. E. Guthrie, Eds. 1970. Concepts of Pest Management. N. Car. State Univ., Raleigh. 242 pp.

Rahe, J. E., J. Kuć, Chien-Mei Chuang, and E. B. Williams. 1969. Induced resistance in *Phaseolus vulgaris* to bean anthracnose. Phytopathology **59**:1641–1645.

Ratliffe, G. T. 1929. A prolonged saprophytic stage of the cotton root-rot fungus. U.S. Dep. Agr. Circ. 67:1–8.

Rawlinson, C. J., D. Hornby, V. Pearson, and J. M. Carpenter. 1973. Virus-like particles in the take-all fungus, *Gaeumannomyces graminis.* Ann. Appl. Biol. **74**:197–209.

Reinking, O. A., and M. M. Manns. 1933. Parasitic and other fusaria counted in tropical soils. Z. Parasitenk. **6**:23–75.

Ricard, J. L. 1970. Biological control of *Fomes annosus* in Norway spruce (*Picea abies*) with immunizing commensals. Studia Forest. Suecica **84**:1–50.

Ricard, J. L., and W. B. Bollen. 1968. Inhibition of *Poria carbonica* by *Scytalidium* sp., an imperfect fungus isolated from Douglas-fir poles. Can. J. Bot. **46**:643–647.

Richardson, L. T. 1954. The persistance of thiram in soil and its relationship to the microbiological balance and damping-off control. Can. J. Bot. **32**:335–346.

Riggle, J. H., and E. J. Klos. 1972. Relationship of *Erwinia herbicola* to *Erwinia amylovora.* Can. J. Bot. **50**:1077–1083.

Rishbeth, J. 1957. *Fusarium* wilt of bananas in Jamaica. II. Some aspects of host-parasite relationships. Ann. Bot. **21**:215–245.

Rishbeth, J. 1963. Stump protection against *Fomes annosus.* III. Inoculation with *Peniophora gigantea.* Ann. Appl. Biol. **52**:63–77.

Roberts, W. 1874. Studies on biogenesis. Phil. Trans. Roy. Soc. London **164**:457–477.

Rohde, R. A., and W. R. Jenkins. 1958. Basis for resistance of *Asparagus officinalis* var. *altilis* L. to the stubby-root nematode *Trichodorus christiei* Allen 1957. Md. Agr. Exp. Sta. Bull. A-97:1–19.

Rombouts, J. E. 1953. The micro-organisms in the rhizosphere of banana plants in relation to susceptibility or resistance to Panama disease. Plant Soil **4**:276–288.

Ross, J. P. 1972. Influence of *Endogone* mycorrhiza on Phytophthora rot of soybean. Phytopathology **62**:896–897.

Rossetti, V., and A. A. Bitancourt. 1951. Estudos sôbre a "gomose de Phytophthora" dos citrus. II. Influência do estado de vegetação do hospedeiro nas lesões experimentais. Arqu. Inst. Biol. **20**(7):73–94.

Roth, L. M. 1961. A study of the odoriferous glands of *Scaptocoris divergens* (Hemiptera, Cydnidae). Ann. Entomol. Soc. Amer. **54**:900–911.

Rovira, A. D. 1969. Plant root exudates. Bot. Rev. **35**:35–57.

Rovira, A. D. 1972. Studies on the interactions between plant roots and micro-organisms. J. Aust. Inst. Agr. Sci. **38:**91–94.

Rovira, A. D., and R. Campbell. 1974. A scanning electron microscope study of the interactions between micro-organisms and *Gaeumannomyces graminis* (Syn. *Ophiobolus graminis*) on the roots of wheat. Physiol. Plant Pathol. **4:** (in press).

Ruinen, J. 1961–63. The phyllosphere. I. An ecologically neglected milieu. Plant Soil **15:**81–109. 1961. II. Yeasts from the phyllosphere of tropical foliage. Antonie van Leeuwenhoek **29:**425–438. 1963.

Saksena, S. B. 1960. Effect of carbon disulphide fumigation on *Trichoderma viride* and other soil fungi. Trans. Brit. Mycol. Soc. **43:**111–116.

Salas, J. A. 1970. Studies on the production of the perfect stage of *Mycena citricolor* (Berk. and Curt.) Sacc. Ph.D. Thesis. Univ. Calif., Berkeley. 117 pp.

Salas, J. A., and J. G. Hancock. 1972. Production of the perfect stage of *Mycena citricolor* (Berk. and Curt.) Sacc. Hilgardia **41:**213–234.

Sands, D. C., and A. D. Rovira. 1970. Isolation of fluorescent pseudomonads with a selective medium. Appl. Microbiol. **20:**513–514.

Sands, D. C., L. Hankin, and M. Zucker. 1972. A selective medium for pectolytic fluorescent pseudomonads. Phytopathology **62:**998–1000.

Sanford, G. B. 1941. Studies on *Rhizoctonia solani* Kühn. V. Virulence in steam sterilized and natural soil. Can. J. Res. (C) **19:**1–8.

Sanford, G. B. 1952. Persistence of *Rhizoctonia solani* Kühn in soil. Can. J. Bot. **30:**652–664.

Scherff, R. H. 1973. Control of bacterial blight of soybean by *Bdellovibrio bacteriovorus*. Phytopathology **63:**400–402.

Schildknecht, H., and K. Koob. 1971. Myrmicacin, the first insect herbicide. Angew. Chem. (Internat'l. Ed. Engl.) **10:**124–125.

Schippers, B., and A. K. F. Schermer. 1966. Effect of antifungal properties of soil on dissemination of the pathogen and seedling infection originating from Verticillium-infected achenes of *Senecio*. Phytopathology **56:**549–552.

Schroeder, W. T., and R. Provvidenti. 1969. Resistance to benomyl in powdery mildew of cucurbits. Plant Dis. Rep. **53:**271–275.

Schroth, M. N., and R. J. Cook. 1964. Seed exudation and its influence on pre-emergence damping-off of bean. Phytopathology **54:**670–673.

Schroth, M. N., and F. F. Hendrix, Jr. 1962. Influence of nonsusceptible plants on the survival of *Fusarium solani* f. *phaseoli* in soil. Phytopathology **52:**906–909.

Schroth, M. N., and W. C. Snyder. 1961. Effect of host exudates on chlamydospore germination of the bean root rot fungus, *Fusarium solani* f. *phaseoli*. Phytopathology **51:**389–393.

Schroth, M. N., A. R. Weinhold, A. H. McCain, D. C. Hildebrand, and N. Ross. 1971. Biology and control of *Agrobacterium tumefaciens*. Hilgardia **40:**537–552.

Schüepp, H., and E. Frei. 1969. Soil fungistasis with respect to pH and profile. Can. J. Microbiol. **15:**1273–1279.

Schüepp, H., and R. J. Green. 1968. Indirect assay methods to investigate soil fungistasis with special consideration of soil pH. Phytopathol. Z. **61**:1–28.

Schuster, R. M. 1972. Continental movements, "Wallace's line" and Indomalayan-Australasian dispersal of land plants: some eclectic concepts. Bot. Rev. **38**:3–86.

Scofield, C. S. 1919. Cotton rootrot spots. J. Agr. Res. **18**:305–310.

Sequeira, L. 1958. Bacterial wilt of bananas: dissemination of the pathogen and control of the disease. Phytopathology **48**:64–69.

Sewell, G. W. F. 1959. Direct observation of *Verticillium albo-atrum* in soil. Trans. Brit. Mycol. Soc. **42**:312–321.

Shaw, D. E., E. G. Cartledge, and D. J. Stamps. 1972. First records of *Phytophthora cinnamomi* in Papua New Guinea. Papua New Guinea Agr. J. **23**:46–48.

Shaw, E. J., Ed. 1965. Western Fertilizer Handbook. 4th ed. Soil Improvement Committee, California Fertilizer Association, Sacramento, Calif. 200 pp.

Shen, C. I. 1940. Soil conditions and the *Fusarium culmorum* seedling blight of wheat. Ann. Appl. Biol. **27**:323–329.

Shepherd, R. J., and F. J. Hills. 1970. Dispersal of beet yellows and beet mosaic viruses in the inland valleys of California. Phytopathology **60**:798–804.

Shigo, A. L. 1967. Successions of organisms in discoloration and decay of wood. Internat'l. Rev. Forest Res. **2**:237–299.

Shigo, A. L., and W. E. Hillis. 1973. Heartwood, discolored wood, and microorganisms in living trees. Annu. Rev. Phytopathol. **11**:197–222.

Shipton, P. J. 1969. Take-all decline. Ph.D. Thesis. Univ. Reading, England. 545 pp.

Shipton, P. J. 1972. Take-all in spring-sown cereals under continuous cultivation: disease progress and decline in relation to crop succession and nitrogen. Ann. Appl. Biol. **71**:33–46.

Shipton, P. J., R. J. Cook, and J. W. Sitton. 1973. Occurrence and transfer of a biological factor in soil that suppresses take-all of wheat in eastern Washington. Phytopathology **63**:511–517.

Siegle, H. 1961. Über Mischinfektionen mit *Ophiobolus graminis* und *Didymella exitialis*. Phytopathol. Z. **42**:305–348.

Simmonds, J. H. 1959. Mild strain protection as a means of reducing losses from the Queensland woodiness virus in the passion vine. Queensland J. Agr. Sci. **16**:371–380.

Simmonds, J. H. 1963. Studies in the latent phase of *Colletotrichum* species causing ripe rots of tropical fruits. Queensland J. Agr. Sci. **20**:373–424.

Simmonds, P. M. 1947. The influence of antibiosis in the pathogenicity of *Helminthosporium sativum*. Sci. Agr. **27**:625–632.

Sinden, J. W. 1971. Ecological control of pathogens and weed-molds in mushroom culture. Annu. Rev. Phytopathol. **9**:411–432.

Slatyer, R. O. 1967. Plant-Water Relationships. Academic Press, New York. 366 pp.

Slope, D. B., and R. Bardner. 1965. Cephalosporium stripe of wheat and root damage by insects. Plant Pathol. **14**:184–187.

Smiley, R. W. 1972. Relationship between rhizosphere pH changes induced by root absorption of ammonium- versus nitrate-nitrogen and root diseases, with particular reference to take-all of wheat. Ph.D. Thesis. Wash. State Univ., Pullman. 68 pp.

Smiley, R. W., and R. J. Cook. 1973. Relationship between take-all of wheat and rhizosphere pH in soils fertilized with ammonium vs. nitrate-nitrogen. Phytopathology **63**:882–890.

Smiley, R. W., R. J. Cook, and R. I. Papendick. 1970. Anhydrous ammonia as a soil fungicide against *Fusarium* and fungicidal activity in the ammonia retention zone. Phytopathology **60**:1227–1232.

Smiley, R. W., R. J. Cook, and R. I. Papendick. 1972. Fusarium foot rot of wheat and peas as influenced by soil applications of anhydrous ammonia and ammonia-potassium azide solutions. Phytopathology **62**:86–91.

Smith, A. L., and J. B. Dick. 1960. Inheritance of resistance to Fusarium wilt in Upland and Sea Island cottons as complicated by nematodes under field conditions. Phytopathology **50**:44–48.

Smith, A. M. 1972. Drying and wetting sclerotia promotes biological control of *Sclerotium rolfsii* Sacc. Nutrient leakage promotes biological control of dried sclerotia of *Sclerotium rolfsii* Sacc. Biological control of fungal sclerotia in soil. Soil Biol. Biochem. **4**:119–123, 125–129, 131–134.

Smith, A. M. 1973. Ethylene: a cause of fungistasis. Nature **246**:311.

Smith, A. M., and D. M. Griffin. 1971. Oxygen and the ecology of *Armillariella elegans* Heim. Aust. J. Biol. Sci. **24**:231–262.

Smith, S. N., and W. C. Snyder. 1971. Relationship of inoculum density and soil types to severity of Fusarium wilt of sweet potato. Phytopathology **61**: 1049–1051.

Smith, S. N., and W. C. Snyder. 1972. Germination of *Fusarium oxysporum* chlamydospores in soils favorable and unfavorable to wilt establishment. Phytopathology **62**:273–277.

Sneh, B., J. Katan, and Y. Henis. 1971. Mode of inhibition of *Rhizoctonia solani* in chitin-amended soil. Phytopathology **61**:1113–1117.

Snyder, W. C. 1963. Root diseases biologically controlled. Science **141**:835–837.

Snyder, W. C. 1969. Survival of *Fusarium* in soil. Ann. Phytopathol. **1**:209–212.

Snyder, W. C., M. N. Schroth, and T. Christou. 1959. Effect of plant residues on root rot of bean. Phytopathology **49**:755–756.

Stahmann, M. A. 1967. Influence of host-parasite interactions on proteins, enzymes, and resistance. pp. 357–369. In C. J. Mirocha and I. Uritani, Eds. The Dynamic Role of Molecular Constituents in Plant-Parasite Interaction. American Phytopathological Society, St. Paul, Minn. 372 pp.

Stanghellini, M. E., and J. G. Hancock. 1971. Radial extent of the bean spermosphere and its relation to the behavior of *Pythium ultimum*. Phytopathology **61**:165–168.

Stănková-Opočenská, E., and J. Dekker. 1970. Indirect effect of 6-azauracil on *Pythium debaryanum* in cucumber. Neth. J. Plant Pathol. **76**:152–158.

Staten, G., and J. F. Cole, Jr. 1948. The effect of pre-planting irrigation on pathogenecity of *Rhizoctonia solani* in seedling cotton. Phytopathology **38**:661–664.

Steinberg, R. A. 1952. Frenching symptoms produced in *Nicotiana tabacum* and *Nicotiana rustica* with optical isomers of isoleucine and leucine and with *Bacillus cereus* toxin. Plant Physiol. **27**:302–308.

Steiner, G. W., and J. L. Lockwood. 1969. Soil fungistasis: Sensitivity of spores in relation to germination time and size. Phytopathology **59**:1084–1092.

Steinhaus, E. A., Ed. 1963. Insect Pathology. An Advanced Treatise. 2 vols. Academic Press, New York. 1350 pp.

Stevenson, I. L., and S. A. W. E. Becker. 1972. The fine structure and development of chlamydospores of *Fusarium oxysporum*. Can. J. Microbiol. **18**:997–1002.

Stewart, R. B., and W. Robertson. 1968. Fungus spores from prehistoric potsherds. Mycologia **60**:701–704.

Stillwell, M. A., R. E. Wall, and G. M. Strunz. 1973. Production, isolation, and antifungal activity of scytalidin, a metabolite of *Scytilidium* species. Can. J. Microbiol. **19**:597–602.

Stolp, H. 1973. The bdellovibrios: bacterial parasites of bacteria. Annu. Rev. Phytopathol. **11**:53–76.

Stolzy, L. H., J. Letey, L. J. Klotz, and C. K. Labanauskas. 1965. Water and aeration as factors in root decay of *Citrus sinensis*. Phytopathology **55**:270–275.

Stotzky, G. 1965. Replica plating technique for studying microbial interactions in soil. Can. J. Microbiol. **11**:629–636.

Stotzky, G., and L. T. Rem. 1966–67. Influence of clay minerals on microorganisms. I–IV. Can. J. Microbiol. **12**:547–563, 831–848, 1235–1246; **13**:1535–1550.

Stover, R. H. 1962. Fusarial wilt (Panama disease) of bananas and other *Musa* species. Commonwealth Mycol. Inst. Phytopathol. Paper 4:1–117.

Street, H. E. 1966. The physiology of root growth. Annu. Rev. Plant Physiol. **17**:315–344.

Streets, R. B. 1969. Diseases of the Cultivated Plants of the Southwest. Univ. Ariz. Press, Tucson. 390 pp.

Swanback, T. R., and P. J. Anderson. 1947. Fertilizing Connecticut tobacco. Conn. (New Haven) Agr. Exp. Sta. Bull. 503:1–51.

Swift, M. J. 1968. Inhibition of rhizomorph development by *Armillaria mellea* in Rhodesian forest soils. Trans. Brit. Mycol. Soc. **51**:241–247.

Swinburne, T. R. 1973. Microflora of apple leaf scars in relation to infection by *Nectria galligena*. Trans. Brit. Mycol. Soc. **60**:389–403.

Tharp, W. H., and C. H. Wadleigh. 1939. The effects of nitrogen source, nitrogen level, and relative acidity on Fusarium wilt of cotton. Phytopathology **29**:756.

Theodorou, C. 1971. Introduction of mycorrhizal fungi into soil by spore inoculation of seed. Aust. Forest **35**:23–26.

Thiegs, B. J. 1955. Effect of soil fumigation on nitrification. Down to Earth **11**(1):14–15.

Thomason, I. J., D. C. Erwin, and M. J. Garber. 1959. The relationship of the root-knot nematode, *Meloidogyne javanica,* to Fusarium wilt in cowpea. Phytopathology **49**:602–606.

Thorne, G. 1942. Distribution of the root-knot nematode in high ridge plantings of potatoes and tomatoes. Phytopathology **32**:650.

Thorne, G. 1961. Principles of Nematology. McGraw-Hill, New York. 553 pp.

Timonin, M. 1961. The interaction of plant, pathogen, and *Scaptocoris talpa* Champ. Can. J. Bot. **39**:695–703.

Toussoun, T. A., R. V. Bega, and P. E. Nelson, Eds. 1970. Root Diseases and Soilborne Pathogens. Univ. Calif. Press, Berkeley. 252 pp.

Toussoun, T. A., W. Menzinger, and R. S. Smith, Jr. 1969. Role of conifer litter in ecology of *Fusarium:* stimulation of germination in soil. Phytopathology **59**:1396–1399.

Towers, B., and W. J. Stambaugh. 1968. The influence of induced soil moisture stress upon *Fomes annosus* root rot of loblolly pine. Phytopathology **58**:269–272.

Triantaphyllou, A. C. 1973. Environmental sex differentiation of nematodes in relation to pest management. Annu. Rev. Phytopathol. **11**:441–462.

Trinci, A. P. J. 1969. A kinetic study of the growth of *Aspergillus nidulans* and other fungi. J. Gen. Microbiol. **57**:11–24.

Tsao, P. H. 1964. Effect of certain fungal isolation agar media on *Thielaviopsis basicola* and on its recovery in soil dilution plates. Phytopathology **54**:548–555.

Tsao, P. H. 1970. Selection media for isolation of pathogenic fungi. Annu. Rev. Phytopathol. **8**:157–186.

Tveit, M., and M. B. Moore. 1954. Isolates of *Chaetomium* that protect oats from *Helminthosporium victoriae.* Phytopathology **44**:686–689.

Uhlenbroek, J. H., and J. D. Bijloo. 1960. Investigations on nematicides. III. Polythienyls and related compounds. Rec. Trav. Chim. Pays-bas **79**:1181–1196.

Vaartaja, O., J. Wilner, W. H. Cram, P. J. Salisbury, A. W. Crookshanks, and G. A. Morgan. 1964. Fungicide trials to control damping-off of conifers. Plant Dis. Rep. **48**:12–15.

Van den Heuvel, J. 1970. Antagonistic effects of epiphytic microorganisms on infection of dwarf bean leaves by *Alternaria zinniae.* Phytopathol. Lab. "Willie Commelin Scholten" Med. 84:1–84.

Van der Plank, J. E. 1968. Disease Resistance in Plants. Academic Press, New York. 206 pp.

Verona, O., and A. A. Lepidi. 1971. Introduzione allo studio dei micromiceti predatori di nematodi nel terreno agrario. L'Agr. Ital. **71** (26 n.s.):205–265.

Vojinović, Ž. D. 1972. Antagonists from soil and rhizosphere to phytopathogens. Final technical report. Inst. Soil Sci., Beograd, Yugoslavia. 130 pp.

Vojinović, Ž. D. 1973. The influence of micro-organisms following *Ophiobolus graminis* Sacc. on its further pathogenicity. Org. Eur. Med. Prot. Plantes Bull. **9**:91–101.

Vruggink, H. 1970. The effect of chitin amendment on actinomycetes in soil and on the infection of potato tubers by *Streptomyces scabies.* Neth. J. Plant Pathol. **76**:293–295.

Waksman, S. A. 1967. The Actinomycetes. A Summary of Current Knowledge. Ronald Press, New York. 280 pp.

Walker, J. C. 1969. Plant Pathology. 3rd ed. McGraw-Hill, New York. 819 pp.

Walker, J. C. 1971. Fusarium wilt of tomato. Amer. Phytopathol. Soc. Monogr. 6:1–56.

Walker, J. C., and W. C. Snyder. 1933. Pea wilt and root rots. Wisc. Agr. Exp. Sta. Bull. 424:1–16.

Wallace, J. M. 1956. Tristeza disease of citrus, with special reference to its situation in the United States. FAO Plant Prot. Bull. **4**:77–87.

Wallace, W. R. 1969. Progress Report on Jarrah Dieback Research in Western Australia. Mimeograph. Forests Department, Western Australia.

Wallwork, J. A. 1970. Ecology of Soil Animals. McGraw-Hill, Maidenhead, England. 283 pp.

Warcup, J. H., and K. F. Baker. 1963. Occurrence of dormant ascospores in soil. Nature **197:**1317-1318.

Ward, E. W. B. 1971. Leakage of metabolites from mycelium of a low-temperature basidiomycete at a supraoptimal temperature. Can. J. Bot. **49:**1049–1051.

Wargo, P. M. 1972. Defoliation-induced chemical changes in sugar maple roots stimulate growth of *Armillaria mellea.* Phytopathology **62:**1278-1283.

Warren, R. C. 1972. The effect of pollen on the fungal leaf microflora of *Beta vulgaris* L. and on infection of leaves by *Phoma betae.* Neth. J. Plant Pathol. **78:**89–98.

Watanabe, T., R. S. Smith, Jr., and W. C. Snyder. 1970. Populations of *Macrophomina phaseoli* in soil as affected by fumigation and cropping. Phytopathology **60:**1717–1719.

Waterston, J. M. 1941. Observations on the parasitism of *Rosellinia pepo* Pat. Trop. Agr. Trinidad **18:**174–184.

Watson, A. G., and E. J. Ford. 1972. Soil fungistasis—-a reappraisal. Annu. Rev. Phytopathol. **10:**327–348.

Watson, I. A. 1970. Changes in virulence and population shifts in plant pathogens. Annu. Rev. Phytopathol. **8:**209–230.

Weaver, J. E. 1926. Root Development of Field Crops. McGraw-Hill, New York. 291 pp.

Weber, D. J., and M. A. Stahmann. 1966. Induced immunity to Ceratocystis infection in sweetpotato root tissue. Phytopathology **56:**1066–1070.

Weber, N. A. 1972. The Attines: The fungus-culturing ants. Amer. Scientist **60:**448–456.

Weber, N. A. 1972. Gardening Ants. The Attines. American Philosophical Society, Philadelphia. 146 pp.

Weindling, R. 1932. *Trichoderma lignorum* as a parasite of other soil fungi. Phytopathology **22:**837–845.

Weindling, R. 1934. Studies on a lethal principle effective in the parasitic action of *Trichoderma lignorum* on *Rhizoctonia solani* and other soil fungi. Phytopathology **24:**1153–1179.

Weinhold, A. R., and T. Bowman. 1968. Selective inhibition of the potato scab pathogen by antagonistic bacteria and substrate influence on antibiotic production. Plant Soil **28:**12–24.

Weinhold, A. R., and T. Bowman. 1971. Virulence repression in *Rhizoctonia solani* by 3-0-methyl glucose. Phytopathology **61:**916.

Weinhold, A. R., J. W. Oswald, T. Bowman, J. Bishop, and D. Wright. 1964. Influence of green manures and crop rotation on common scab of potato. Amer. Potato J. **41:**265–273.

Weinke, K. E. 1962. The Influence of nitrogen on the root disease of bean caused by *Fusarium solani* f. *phaseoli.* Phytopathology **52:**757.

Wells, H. D., D. K. Bell, and C. A. Jaworski. 1972. Efficacy of *Trichoderma harzianum* as a biocontrol for *Sclerotium rolfsii.* Phytopathology **62:**442–447.

Wensley, R. N. 1971. The microflora of peach bark and its possible relation to perennial canker (*Leucostoma cincta* (Fr.) v. Hohnel (*Valsa cincta*)). Can. J. Microbiol. **17:**333–337.

Wensley, R. N., and C. D. McKeen. 1963. Populations of *Fusarium oxysporum* f. *melonis* and their relation to the wilt potential of two soils. Can. J. Microbiol. **9:**237–249.

Werner, H. O., T. A. Kiesselbach, and R. W. Goss. 1944. Dry-land crop rotation experiments with potatoes in northwestern Nebraska. Nebr. Agr. Exp. Sta. Bull. 363:1–43.

Weste, G. 1972. The process of root infection by *Ophiobolus graminis.* Trans. Brit. Mycol. Soc. **59:**133–147.

White, N. H. 1954. Decoy crops. The use of decoy crops in the eradication of certain soil-borne plant diseases. Aust. J. Sci. **17:**18–19.

Whitney, R. S., and R. Gardner. 1943. The effect of carbon dioxide on soil reaction. Soil Sci. **55:**127–141.

Whittington, W. J., Ed. 1969. Root Growth. Butterworths, London. 450 pp.

Wicker, E. F. 1968. Toxic effects of cycloheximide and phytoactin on *Tuberculina maxima.* Phytoprotection **49:**91–98.

Wicker, E. F., and J. Y. Woo. 1973. Histology of blister rust cankers parasitized by *Tuberculina maxima.* Phytopathol. Z. **76:**356–366.

Wiebe, H. H., Ed. 1971. Measurement of plant and soil water status. Utah Agr. Exp. Sta. Bull. 484:1–71.

Wilhelm, S. 1950. Verticillium wilt in acid soils. Phytopathology **40:**776–777.

Wilhelm, S. 1951. Effect of various soil amendments on the inoculum potential of the Verticillium wilt fungus. Phytopathology **41:**684:690.

Wilhelm, S. 1955. Longevity of the Verticillium wilt fungus in the laboratory and field. Phytopathology **45:**180–181.

Wilhelm, S., and J. B. Taylor. 1965. Control of Verticillium wilt of olive through natural recovery and resistance. Phytopathology **55:**310–316.

Williams, L. E., and D. D. Kaufman. 1962. Influence of continuous cropping on soil fungi antagonistic to *Fusarium roseum.* Phytopathology **52:**778–781.

Wilson, C. L. 1969. Use of plant pathogens in weed control. Annu. Rev. Phytopathol. **7:**411–434.

Winoto Suatmadji, R. 1969. Studies on the effect of *Tagetes* species on plant parasitic nematodes. Wageningen, Fonds Landbouw. Export Bur. Publ. **47**:1–132. H. Veenman and Zonen N. V., Wageningen.

Woltz, S. S., and R. H. Littrell. 1968. Production of yellow strapleaf of chrysanthemum and similar diseases with an antimetabolite produced by *Aspergillus wentii*. Phytopathology **58**:1476–1480.

Wong, TzeWeng. 1972. Effect of soil water on bacterial movement and Streptomycete-fungal antagonism. Ph.D. Thesis. Univ. Sydney, New South Wales. 167 pp.

Wood, F. A., and D. W. French. 1960. Bacteria in the perithecia of *Hypoxylon pruinatum* and their effect on ascospore germination and colony development. Phytopathology **50**:659.

Wood, R. K. S. 1967. Physiological Plant Pathology. Blackwell, Oxford. 570 pp.

Wood, R. K. S., and M. Tveit. 1955. Control of plant diseases by use of antagonistic organisms. Bot. Rev. **21**:441–492.

Wuest, P. J., and R. K. Moore. 1972. Additional data on the thermal sensitivity of selected fungi associated with *Agaricus bisporus*. Phytopathology **62**:1470–1472.

Wuest, P. J., K. F. Baker, and W. S. Conway. 1970. Sensitivity of selected mushroom pathogens to aerated steam. Phytopathology **60**:1274–1275.

Yarwood, C. E., A. P. Hall, and M. M. Nelson. 1953. Nutritive value of rust-infected leaves. Science **117**:326–327.

Zehr, E. I., Ed. 1972. Fungicide and Nematocide Tests. Results of 1972. Vol. 28: 1–219. American Phytopathological Society, St. Paul, Minn.

Zentmyer, G. A. 1963. Biological control of Phytophthora root rot of avocado with alfalfa meal. Phytopathology **53**:1383–1387.

Zentmyer, G. A. 1965. Bacterial stimulation of sporangium production in *Phytophthora cinnamomi*. Science **150**:1178–1179.

Zogg, H. 1959. Studien über die biologische Bodenentseuchung. II. Beeinflussung der Pathogenität von *Ophiobolus graminis* Sacc. durch die Mikrofloren verschiedener Böden mit verschiedenen Fruchtfolgen. Phytopathol. Z. **34**:432–444.

Zogg, H. 1969. Crop rotation and biological soil disinfection. Qual. Plant. Mater. Veg. **18**:256–273.

INDEX

Principal references are given in **boldface**